BACTERIAL
PROTEIN
TOXINS

BACTERIAL
PROTEIN
TOXINS

Edited by

Drusilla L. Burns

Joseph T. Barbieri

Barbara H. Iglewski

Rino Rappuoli

ASM
PRESS

Washington, D.C.

Copyright © 2003 ASM Press
American Society for Microbiology
1752 N Street, N.W.
Washington, DC 20036-2904

Library of Congress Cataloging-in-Publication Data

Bacterial protein toxins / edited by Drusilla L. Burns ... [et al.].
    p. ; cm.
Includes bibliographical references and index.
  ISBN 1-55581-245-7
  1. Bacterial toxins.
  [DNLM: 1. Bacterial Toxins. 2. Bacterial Proteins. QW 630.5.B2 B1313 2003] I. Burns, Drusilla L.

QP632.B3 B282 2003
579.3'165—dc21

                                                                            2002152890

Address editorial correspondence to: ASM Press, 1752 N St., N.W., Washington, DC 20036-2904, U.S.A.

Send orders to: ASM Press, P.O. Box 605, Herndon, VA 20172, U.S.A.
Phone: 800-546-2416; 703-661-1593
Fax: 703-661-1501
Email: books@asmusa.org
Online: www.asmpress.org

CONTENTS

CONTRIBUTORS

Klaus Aktories • Institut für Experimentelle und Klinische Pharmakologie und Toxikologie der Albert-Ludwigs-Universität Freiburg, Otto-Krayer-Haus, Albertstr. 25, D-79104 Freiburg, Germany

Joseph T. Barbieri • Department of Microbiology, Medical College of Wisconsin, 8701 Watertown Plank Rd., Milwaukee, WI 53226-0509

James B. Bliska • Department of Molecular Genetics and Microbiology, Center for Infectious Diseases, State University of New York at Stony Brook, Stony Brook, NY 11794-5222

Drusilla L. Burns • U.S. Food and Drug Administration, CBER FDA, Bldg. 29, Rm. 418, 8800 Rockville Pike, Bethesda, MD 20892

Bonny L. Dickinson • GI Cell Biology, Combined Program in Pediatric Gastroenterology and Nutrition, Children's Hospital; Harvard Digestive Diseases Center; and Department of Pediatrics, Harvard Medical School, Boston, MA 02115

Lawrence A. Dreyfus • Division of Cell Biology and Biophysics, School of Biological Sciences, University of Missouri-Kansas City, 5007 Rockhill Rd., Kansas City, MO 64110

Jeyanthy Eswaran • Department of Pathology, Cambridge University, Tennis Court Road, Cambridge CB2 1QP, United Kingdom

David J. FitzGerald • Laboratory of Molecular Biology, Center for Cancer Research, National Cancer Institute, Bldg. 37, Rm. 5124, Bethesda, MD 20892

Dara W. Frank • Department of Microbiology and Molecular Genetics, Medical College of Wisconsin, 8701 Watertown Plank Rd., Milwaukee, WI 53226

Mary C. Gray • Department of Medicine, Rm. 6828, Old Medical School, University of Virginia School of Medicine, Charlottesville, VA 22908

Jörg Hacker • Institute for Molecular Biology of Infectious Diseases, University of Würzburg, Röntgenring 11, D-97070 Würzburg, Germany

Erik L. Hewlett • Department of Medicine and Department of Pharmacology, Rm. 6832, Old Medical School, Box 800419, University of Virginia School of Medicine, Charlottesville, VA 22908

Wim Hol • Department of Biological Structure, University of Washington, SM-20, Seattle, WA 98185

Colin Hughes • Department of Pathology, Cambridge University, Tennis Court Road, Cambridge CB2 1QP, United Kingdom

Meredith L. Hunt • Department of Microbiology and Molecular Genetics, Medical College of Wisconsin, 8701 Watertown Plank Rd., Milwaukee, WI 53226

Barbara H. Iglewski • Department of Microbiology and Immunology, Box 672, Strong Memorial Hospital, 601 Elmwood Ave., University of Rochester, Rochester, NY 14642

Vassilis Koronakis • Department of Pathology, Cambridge University, Tennis Court Road, Cambridge CB2 1QP, United Kingdom

Robert J. Kreitman • Laboratory of Molecular Biology, Center for Cancer Research, National Cancer Institute, Bldg. 37, Rm. 5124, Bethesda, MD 20892

Wayne I. Lencer • GI Cell Biology, Combined Program in Pediatric Gastroenterology and Nutrition, Children's Hospital; Harvard Digestive Diseases Center; and Department of Pediatrics, Harvard Medical School, Boston, MA 02115

Stephen H. Leppla • Oral Infection and Immunity Branch, National Institute of Dental and Craniofacial Research, National Institutes of Health, Bldg. 30, Rm. 316, Bethesda, MD 20892

Vega Masignani • IRIS, Chiron SpA, Via Fiorentina 1, 53100 Siena, Italy

Angela Melton-Celsa • Department of Microbiology and Immunology, Uniformed Services University of the Health Sciences, F. Edward Hebert School of Medicine, 4301 Jones Bridge Rd., Bethesda, MD 20814-4799

Cesare Montecucco • Dipartimento di Scienze Biomediche, Università di Padova, Viale G. Colombo, 3, 35121 Padova, Italy

Matthew L. Nilles • Department of Microbiology and Immunology, University of North Dakota School of Medicine and Health Sciences, Grand Forks, ND 58202

Alison D. O'Brien • Department of Microbiology and Immunology, Uniformed Services University of the Health Sciences, F. Edward Hebert School of Medicine, 4301 Jones Bridge Rd., Bethesda, MD 20814-4799

Camilla Oxhamre • Microbiology and Tumor Biology Center, Karolinska Institutet, S-171 77 Stockholm, Sweden

Ira Pastan • Laboratory of Molecular Biology, Center for Cancer Research, National Cancer Institute, Bldg. 37, Rm. 5124, Bethesda, MD 20892

Shelley M. Payne • Section of Molecular Genetics and Microbiology, The University of Texas, Austin, TX 78712-1095

Everett C. Pesci • Department of Microbiology and Immunology, Brody School of Medicine, East Carolina University, Greenville, NC 27858

Mariagrazia Pizza • IRIS, Chiron SpA, Via Fiorentina 1, 53100 Siena, Italy

Gregory V. Plano • Department of Microbiology and Immunology, University of Miami School of Medicine, RMSB 3097A (R-138), 1600 NW 10th Ave., Miami, FL 33136

Rino Rappuoli • IRIS, Chiron SpA, Via Fiorentina 1, 53100 Siena, Italy

Agneta Richter-Dahlfors • Microbiology and Tumor Biology Center, Karolinska Institutet, S-171 77 Stockholm, Sweden

Ornella Rossetto • Dipartimento di Scienze Biomediche, Università di Padova, Viale G. Colombo, 3, 35121 Padova, Italy

Catharine B. Saelinger • Department of Molecular Genetics, Biochemistry and Microbiology, University of Cincinnati College of Medicine, 231 Albert Sabin Way, Cincinnati, OH 45267-0524

Maria Sandkvist • Department of Biochemistry, American Red Cross, Jerome H. Holland Laboratory, 15601 Crabbs Branch Way, Rockville, MD 20855

Kirsten Sandvig ● Department of Biochemistry, Institute for Cancer Research, The Norwegian Radium Hospital, Montebello, 0310 Oslo, Norway

Kurt Schesser ● Department of Microbiology and Immunology, University of Miami School of Medicine, 1600 NW 10th Ave., Miami, FL 33136

Patrick M. Schlievert ● Department of Microbiology, University of Minnesota Medical School, 420 Delaware St. SE, MMC 196, Minneapolis, MN 55455

Maria E. Scott ● Department of Biochemistry, American Red Cross, Jerome H. Holland Laboratory, 15601 Crabbs Branch Way, Rockville, MD 20855

Scott Stibitz ● U.S. Food and Drug Administration, 8800 Rockville Pike, Bethesda, MD 20892

Fiorella Tonello ● Dipartimento di Scienze Biomediche, Università di Padova, Viale G. Colombo, 3, 35121 Padova, Italy

F. Gisou van der Goot ● Department of Genetics and Microbiology, Centre Médicale Universitaire, 1, rue Michel-Servet, CH-1211 Geneva 4, Switzerland

Gloria I. Viboud ● Department of Molecular Genetics and Microbiology, Center for Infectious Diseases, State University of New York at Stony Brook, Stony Brook, NY 11794-5222

Franca R. Zaretzky ● Department of Medicine, Rm. 6828, Old Medical School, University of Virginia School of Medicine, Charlottesville, VA 22908

FOREWORD

As these words are being entered on the page, bacterial toxins are not far from the headlines. In the fall of 2001 the catastrophic attacks with fuel-laden airliners on the World Trade Center towers and the Pentagon were followed by assaults with *Bacillus anthracis* spores enclosed in letters sent through the United States postal system. The perpetrator of the anthrax assaults still remains unidentified, but 11 confirmed cases of inhalational anthrax and five deaths resulted. Now, facing the threat of bioterrorism realistically, the nation is struggling to identify countermeasures to protect the population in the event of a subsequent attack with anthrax or other agents—an attack of potentially much greater magnitude than that of late 2001. As a part of this effort, attention is being focused on the toxin produced by the anthrax bacillus that makes it so deadly, and on other toxins, such as botulinum toxin, that are directly relevant to bioterrorism.

In some sense these events take us back to the era in which bacterial toxins were first discovered. Evidence that certain bacteria produce diffusible toxic substances that are responsible for disease symptoms came late in the 19th century, soon after methods for isolating and growing bacteria in pure culture were developed. Injection of culture filtrates of, for example, the diphtheria bacillus into experimental animals produced death with symptoms and lesions in internal organs characteristic of the disease. Shortly thereafter it was found that sublethal amounts of diphtheria and tetanus toxins elicited the formation of substances in the bloodstream that could specifically neutralize these toxins. Thus toxins were pivotal in the discovery of the humoral immune system of mammals. This realization soon led to toxin-based vaccine strategies and eventually to the vaccines still used today to immunize against diphtheria and tetanus.

The century-long trek from the discovery of toxins to our current understanding of their chemistry and modes of action has been integrally linked with advances in fundamental knowledge of macromolecular chemistry, cell biology, and other relevant areas. In the 1930s diphtheria toxin was purified to a sufficient extent to ascertain that it was a protein. But concepts of the nature of proteins were primitive then, and not until the 1950s did understanding of their primary, secondary, and tertiary structures become firmly grounded. Similarly, significant data on the modes of action of toxins at the cellular and molecular levels have become available only as fundamental knowledge has advanced in areas such as how proteins are synthesized, how vesicle fusion occurs in neuronal cells, etc. Clearly it would have been impossible to determine that diphtheria toxin directly blocked protein synthesis before systems for studying this process had advanced to a certain level. As understanding of biological science has exploded in the second half of the 20th century, knowledge of toxins and their role in bacterial diseases has kept pace.

Perhaps progress in toxin research would have occurred more rapidly had it not been for certain misconceptions. In the post-World War II era, bacterial pathogenesis was not the "hot" area of research it is today. The discovery and application of antibiotics had created the widespread misconception that bacterial diseases had been conquered once and for all. Furthermore, toxins were considered odd curiosities of nature, far from the mainstream of modern biology, which was then focused on monumental questions—the genetic code, how proteins are encoded and synthesized, and the structure of proteins in three dimensions. Finally, the potential danger of working with toxins has probably always been consciously or unconsciously exaggerated in the minds of many able researchers who would otherwise have contributed to our understanding of this intriguing class of molecules.

These obstacles notwithstanding, the field of toxin research held its own through the post-World War II era, in considerable measure because of the interest of stalwart individuals such as A. M. Pappenheimer, Jr. ("Pap" to those of us who worked with him), W. E. ("Kits") van Heyningen, and Harry Smith. How a purified, potentially lethal toxin acted seemed to provide the most straightforward route to answer the question of how a bacterium could kill a human being.

As fundamental knowledge advanced in the 1960s and 1970s, seminal discoveries were made regarding the biochemical modes of action of several protein toxins—notably diphtheria, cholera, *Pseudomonas* exotoxin A, and the ricin family of plant toxins. Interest in these molecules grew as it became evident that a sizable fraction of bacterial and plant toxins were extraordinary enzymes having the capacity to enter mammalian cells and modify substrates within the cytosolic compartment. Hence, these proteins had the rare property of being able to cross membranes at some level. This fascinating property, combined with their interesting catalytic activities (e.g., ADP-ribosylation), brought toxins to the attention of mainstream biological researchers and made these proteins useful as tools to probe important metabolic pathways and processes.

In 1972 the first Gordon Conference on Bacterial Toxins was held (chaired by Sam Ajl), and the Conference has been a biennial event ever since. The European Workshop on Bacterial Protein Toxins came into existence later and has been held in various countries in alternate years. Fostering the exchange of ideas and advances in the area, these meetings have greatly elevated the stature of the field. As better methods to study the diverse array of virulence factors used by bacteria have evolved in recent years, interest in other aspects of bacterial pathogenesis has greatly increased, and the subject matter addressed in these conferences has broadened correspondingly. For decades toxins were about the only determinants of pathogenicity accessible to investigation. Now, all that has been changed by modern tools of biology.

The advent of recombinant DNA technology had a profound effect on toxin research, as it did on almost every other aspect of biology. Clearly the possibility of cloning and expressing a toxin in a heterologous organism required caution and regulation. Indeed, cloning of the most potent toxins (e.g., the clostridial neurotoxins) is still prohibited. Nonetheless, as structure-function relationships of various toxins were elucidated, it became evident that they were complex proteins and that certain parts were benign in the absence of the complementary parts. This allowed selected domains of the highly potent toxins to be cloned under minimal containment conditions and, in turn, permitted approaches such as directed mutagenesis and creation of chimeric molecules to be implemented. Among the results are "recombinant toxoids" and new types of targeted toxins.

The notion that a toxin might be used to generate a "magic bullet" that could target a specific tissue or subset of cells in the body is an old one that emerged in a new form in the 1980s. The concept of AB toxins, containing an enzymatic A moiety linked by a disulfide bridge to a receptor-binding B moiety, a motif found in many toxins (e.g., diphtheria and ricin toxins), immediately suggested a way to direct the action of these proteins to specific cells. Early attempts involved generating disulfide-linked chimeras containing, for example, the A chain of diphtheria or ricin toxin linked to a monoclonal antibody directed towards a tumor-specific antigen. Such chimeras were generally less potent and specific than hoped, but the early attempts in this direction generated widespread interest. Subsequently there have been isolated successes, including a recombinant chimeric toxin against T cells that has recently been licensed by the FDA for treatment of certain types of tumors.

The past two decades have witnessed the discovery of many new toxins and toxin-like virulence factors (e.g., the effectors introduced into cells by type III secretion systems), and their numbers will undoubtedly grow as the genomic sequences of bacterial pathogens are determined. The diverse strategies adopted by bacteria to subvert the cellular biochemistry and physiology of the host have provided tools for use in cellular and molecular biological investigations, new targets for drug development, and a better understanding of the selective advantages accruing to bacterial pathogens from toxin production. Over the same period, our understanding of the classical toxins has extended to finer levels of detail. The crystallographic structures of many of these molecules have been solved, providing a framework to understand how they recognize receptors, penetrate membranes, and recognize and modify substrates. This said, there is still no toxin for which we can claim a complete understanding, and for most of them, major gaps remain in our knowledge.

Now, at the beginning of the 21st century, we have in a sense come full circle. Despite the advances made over the past century, multiple threats—bioterrorism, the spread of antibiotic resistance, and the emergence of new bacterial pathogens—keep knowledge of bacterial virulence a subject as crucial to the health of mankind as it was at the beginning of the 20th century. This volume presents an excellent summary of current knowledge of bacterial toxins and will serve as a major resource for current practitioners and new students entering the field in the coming years. I wish them the thrill of discovery and the camaraderie with outstanding students and colleagues that I have experienced throughout my career.

R. John Collier
Harvard Medical School

PREFACE

For the past three decades, our understanding of the molecular aspects of bacterial toxins has grown exponentially. The seminal studies that identified diphtheria toxin as a protein with the intrinsic capacity to enter eukaryotic cells and inhibit protein synthesis through the posttranslational modification of a single eukaryotic protein provided a platform for studies on other medically important bacterial toxins. Initially, experts in protein structure, enzymology, and electrophysiology were drawn into our field to study toxin action, while more recent additions to our field have included cell biologists and investigators who study eukaryotic signal transduction. The study of bacterial toxins demands an integration of distinct fields of investigation, providing a synergy for our understanding of toxin action.

The initial concept of this book was spurred by our desire to provide a succinct reference source for students interested in bacterial toxins. The goal was to integrate historical experiences and contemporary concepts of toxin biosynthesis, structure, and function. One important concept that the authors share with students beginning their studies on bacterial toxins is the importance of pursuing quantitative analyses of toxins. This approach has allowed our field to advance to its current level of sophistication. The chapters include both written and pictorial representation of major concepts, provide references for additional reading, and are written by investigators who have direct expertise in the area. The book is divided into five sections which address (i) the genetics and regulation of toxin gene expression; mechanisms for toxin translocation across (ii) bacterial and (iii) eukaryotic membrane barriers; (iv) descriptions of toxins that covalently modify host target proteins and recently recognized toxins that modulate host protein function by noncovalent mechanisms; and (v) the current status of both the beneficial and harmful uses of bacterial toxins.

The study of bacterial toxins has led to novel strategies that use these toxins for medical purposes. However, in October of 2001, we witnessed the deplorable use of a bacterial organism as a reagent for terrorism, by an individual(s) who disregarded the ethical standards of our society. It is our obligation to work towards the beneficial use of bacterial toxins as vaccines, antidotes, and therapies.

This is the challenge we present to our junior microbiologists, and we provide this text as a reference for your studies.

Drusilla L. Burns
Joseph T. Barbieri
Barbara H. Iglewski
Rino Rappuoli

SECTION I
GENETICS AND REGULATION

The "life span" of a bacterial toxin extends from the time of its synthesis until it is destroyed within the eukaryotic cell after it has inflicted its damage. Major milestones in the existence of a toxin include synthesis and secretion from the bacterial cell, targeting of eukaryotic cells and subsequent entry into these cells, and manifestation of its toxic activity that results in alteration of a critical process within the eukaryotic cell. These topics will be considered in subsequent sections of this volume. A complete understanding of bacterial toxins also requires knowledge of the DNA that encodes the toxin and *cis*- and *trans*-regulatory elements that control the expression of toxin genes.

The genes encoding bacterial toxins may be located either on the bacterial chromosome or on extrachromosomal elements. Genes encoding diphtheria toxin, Shiga toxin, and cholera toxin are located on phages that have integrated into the pathogen's genome. Genes encoding some effector proteins that are delivered by type III and type IV secretion systems are contained within pathogenicity islands on the chromosome. Other genes encoding bacterial toxins are located on plasmids carried by the bacteria, such as the genes encoding anthrax lethal and edema toxins that are found on a large plasmid carried by *Bacillus anthracis* but not other common bacilli. The fact that certain toxin genes are located on elements that can be mobilized or may have been mobilized at some time in the past suggests possible mechanisms for horizontal transfer of toxin genes between bacteria.

Following their synthesis, protein toxins must be transported across bacterial membrane(s). Synthesis and transport of toxins are expensive undertakings for bacteria since a considerable amount of energy must be expended to accomplish these tasks. Therefore, bacteria have developed ways to tightly regulate the transcription of toxin genes so that toxins are produced only when they are needed. To effectively regulate toxin production, bacteria sense their environment and turn off transcription of toxin genes when toxin utilization would not be beneficial, since production of unneeded toxins puts the bacteria at a selective disadvantage for survival. In contrast, if in an environment in which toxins are beneficial, for example in the host environment, it is essential that the bacteria make toxins to help combat the host immune system or to escape hostile host environments.

1

A number of regulatory pathways have evolved that will be discussed in this section. These regulatory systems include two-component regulatory systems that have evolved to sense a number of environmental conditions, systems regulated by iron, and the AraC family of regulators. A particularly novel regulatory system is the quorum sensing system used by a variety of bacteria. In this system, the environment per se is not sampled, but rather, the system senses the presence of other bacteria. As we begin our study of bacterial toxins, let us first take a look at the genetics of toxins and the regulation of their synthesis.

Bacterial Protein Toxins
D. Burns et al., Editors
©2003 ASM Press, Washington, DC 20036

Chapter 1

Two-Component Systems

Scott Stibitz

INTRODUCTION

Two-component systems represent an elegant and successful solution to a problem posed to bacteria in the process of evolution by natural selection. Optimization of survival strategies requires the ability to adapt rapidly to different environments that a given bacterial species contacts in the particular niche(s) it inhabits. These environments are, by their nature, defined by conditions external to a bacterial cell, yet the changes in gene expression that allow adaptation are, by their nature, internal. How, therefore, can a bacterium respond to external changes, by corresponding changes in its interior, in the expression of genes that maximize survival under these "new" conditions?

The two-component solution to this problem rests on the transmembrane nature of one of its components, the sensor kinase protein (Fig. 1). By virtue of this membrane topology, environmental conditions outside the cytoplasmic compartment that impinge on the extracytoplasmic domain can effect signal transduction through the sensor protein itself and across the cytoplasmic membrane to affect the activity of its cytoplasmic component. This cytoplasmic component is then able to communicate these changes, via transfer of a phosphate group, to the second component, the response regulator protein, which in the vast majority of described systems directly affects gene expression. Phosphorylation of a response regulator protein typically results in increased binding affinity for regulated promoters. In this way multiple genes under the control of a given response regulator can be turned up or down in response to the environmental signals that affect its cognate sensor protein.

Transduction of signals across membranes does not, in and of itself, require two components. The transmembrane sensor protein accomplishes this act by itself. Proof that two separate components are not required to effect environmentally responsive gene regulation is evidenced by the ToxR protein of *Vibrio cholerae*. ToxR is a transmembrane protein that is also a DNA-binding protein and is thus like a "one-component" system. However, two-component systems

---

Scott Stibitz • Food and Drug Administration, 8800 Rockville Pike, Bethesda MD 20892.

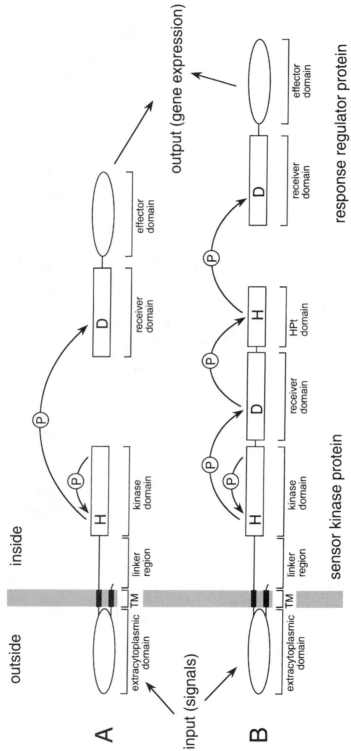

Figure 1 Two commonly encountered organizational schemes of two-component signaling modules. (A) The "paradigmatic" two-component system that undergoes phosphotransfer from the sensor kinase protein to the response regulator protein is pictured. (B) A "hybrid" or "unorthodox" sensor kinase, which participates in a phosphorelay mechanism, involves two additional modules but ultimately leads to the same result, phosphorylation of the receiver domain of the response regulator. Symbols: TM, transmembrane region; H, catalytic histidine; D, catalytic aspartic acid; P, phosphate. The path of phosphate transfer is shown by curved arrows. Rectangles represent conserved domains, and ellipses represent divergent domains. The extents of different domains that are discussed in the text are delineated below each example. The vertical gray bar represents the inner membrane of gram-negative organisms or the cytoplasmic membrane of gram-positive organisms. Nomenclature for the proteins and modules of two-component systems continues to evolve. The nomenclature presented here is used throughout this chapter.

are much more common. Why? Some advantages of the two-component organization are evident. One is the opportunity for amplification of an environmental signal. A single sensor molecule, when activated, can catalyze the phosphorylation of multiple response regulator proteins, thus amplifying the signal. Another advantage is the soluble nature of the second component, the response regulator protein, which gives it the capability of accessing the three-dimensional space of the cytoplasm rather than just the two-dimensional surface of the cytoplasmic membrane. Since not all of the genomic DNA of a bacterium can be physically in contact with the inner surface of the cytoplasmic membrane simultaneously, this provides a response regulator protein with equal access to the entire genome. Together, these properties make two-component systems highly suitable for coordinately regulating multiple genes of allied function that may be scattered throughout a bacterial genome (a regulon). It is not surprising, then, that they are often found fulfilling this role.

Two-Component Systems Are Involved in Regulation of Toxins and Other Virulence Factors

As with other bacterial processes, the expression and deployment of virulence strategies must be coordinated optimally to maximize survival and reproductive potential. It has thus been recognized, since the beginning of the molecular era of bacterial pathogenesis research, that regulation of virulence determinants is a critical part of the pathogenic personalities of medically relevant bacteria. Failure to express a particular virulence trait at a critical juncture or the inappropriate expression of a trait at the wrong time or place within a mammalian host can have consequences that are disastrous from the bacterium's point of view. It is not surprising, therefore, that, among the many regulatory schemes governing virulence potential, two-component systems have been found time and time again to play key roles. As the focus of this volume is on one particular type of virulence factor, bacterial toxins, a listing of some of the known two-component systems governing toxin production is presented in Table 1. One of these, the BvgAS system of *Bordetella pertussis*, will be examined in more detail as we proceed.

Sequence Conservation and the Modular Nature of Two-Component Systems

The evolutionary success of the two-component systems is attested to by their widespread distribution. This ubiquity manifests itself in two ways: the variety of gram-negative and gram-positive bacteria (encompassing most medically relevant bacteria) that utilize them, and the fact that most bacteria contain multiple two-component systems regulating different constellations of genes in response to different environmental cues. Recent determinations of the complete genomic sequence of a large number of bacterial species have contributed to this view. The *Pseudomonas aeruginosa* genome, as an extreme example, apparently contains genes for 64 response regulators and 63 histidine kinases. The large number of putative two-component systems uncovered by genomic sequence analysis brings us to the question: how are two-component systems recognized at the level of primary structure?

Table 1   Two-component systems regulating toxin production

Species	Response regulator	Sensor kinase	Toxins regulated
Bordetella pertussis	BvgA	BvgS	Pertussis toxin Adenylate cyclase toxin Dermonecrotic toxin
Clostridium perfringens	VirR	VirS	Perfringolysin O Collagenase Phospholipase C Protease, sialidase
Staphylococcus aureus	AgrC[a]	AgrA[a]	Hemolysins Exfoliative toxins Enterotoxins Proteases Toxic shock syndrome toxin
Streptococcus pyogenes	CovR	CovS	Streptokinase Streptolysin S Mitogenic factor
Pseudomonas marginalis	GacA	GacS	Pectate lyase
Pseudomonas syringae	CorR	CorS	Coronatine

*These affect target genes by controlling levels of a small RNA called RNAIII.

Membership in the class of two-component systems is based on protein sequence similarity of two conserved modules: the kinase domain of the sensor kinase proteins and the receiver domain of the response regulator proteins (Fig. 1). These two domains embody the core signaling mechanism between the two proteins. The other domains, such as the extracytoplasmic domain of the sensor protein and the DNA-binding effector domain of the response regulator protein, have diverged widely to meet the requirements of different regulatory pathways (Fig. 1). Figure 2 presents the distilled knowledge derived from sequence alignments of a large number of kinase and receiver domains. From this type of analysis we have learned that two catalytic amino acid residues are absolutely conserved (starred in Fig. 2). One of these is the histidine residue at which point the kinase domain autophosphorylates. The other is the aspartate residue in the receiver domain to which the phosphate group is transferred from its cognate sensor protein. Other residues that are highly conserved play vital roles in catalysis or other aspects of signal transduction, as will be described below.

Typically the degree of conservation in these two modules is in the range of 25 to 30% amino acid sequence identity. Given the high degree of functional conservation, this relatively high amount of divergence may seem surprising. It is less so, however, when one considers that the common condition of multiple two-component systems in a given species requires that some specificity be built into the system. In most cases one would reason that a given response regulator should be activated only by its cognate sensor protein and thus only in response to the environmental signals appropriate to it. Clearly, specificity of interaction between the two components requires structural uniqueness and therefore sequence divergence relative to other co-resident systems.

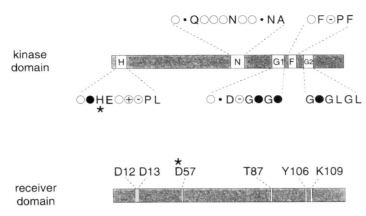

Figure 2 Highly conserved residues among the conserved domains of two-component systems. For the histidine kinase, these residues are represented by the H, N, G1, F, and G2 boxes indicated, with consensus sequences shown. For the receiver domain, important residues are shown, numbered as for the CheY protein of *E. coli*. The actual residue numbers of different receiver domains vary slightly, although the overall linear relationship is highly conserved. For positions where at least 70% of the sequences show the same residue, letters indicate that residue. For positions where 50% of aligned sequences have an amino acid of the same chemical family, symbols indicate that family. Positions with less than 50% conservation are indicated by black dots. Symbols: open circles, nonpolar (I, L, M, V); filled circles, polar (A, G, P, S, T); circles with plus signs, basic (H, K, R); circles with minus signs, acidic or amidic (D, E, N, Q). Residues directly participating in phosphotransfer are starred. Adapted from reference 6 with permission.

The modular nature of two-component systems is reflected in the variety of ways these two modules have been incorporated into the organization of two-component system proteins (6). However, two organizational plans account for the majority of these. These two plans are shown in Fig. 1. Figure 1A shows the paradigmatic minimal two-component system as it is most often encountered in nature. In this case, the flow of phosphate after autophosphorylation of the kinase histidine residue involves only one additional transfer, to the response regulator protein aspartate. The second plan, shown in Fig. 1B, depicts what was once termed an "unorthodox" sensor kinase protein organization. In this design, phosphate is transferred from the kinase histidine to the aspartate of an auxiliary response regulator domain, and from there to the histidine of a histidine phosphotransfer (HPt) domain, before being ultimately passed to the response regulator domain attached to the effector domain. This multistep process is often referred to as a "phosphorelay" mechanism and is hypothesized to provide the evolutionary advantage of multiple checkpoints for additional regulatory input.

Our understanding of the details of how these modules function and interact has undergone an explosion in recent years, largely from the results of X-ray crystallographic and nuclear magnetic resonance (NMR) structural analyses, which have provided the atomic details of the three-dimensional structures of multiple examples of kinase, receiver, and HPt domains. The combination of these studies with the detailed genetic analyses that have been performed over the years has provided great insight into the secrets of these amazing molecules, which allow bacteria to feel and respond to their environment. In an attempt to organize this mass of knowledge, the structure-function relationships of the various domains of two-component systems will be examined in turn, following the path of information from outside to inside a bacterial cell, as a signal is transduced and converted to a meaningful response. For each domain the general state of knowledge will be presented as it relates to two-component systems in general, and then illustrated, when appropriate, with examples from the BvgAS system governing expression of toxins and other virulence factors in *B. pertussis*.

STRUCTURE-FUNCTION RELATIONSHIPS IN TWO-COMPONENT SYSTEM DOMAINS

The Extracytoplasmic Sensor Domain

The two-component paradigm dictates that signals that will ultimately affect patterns of gene expression first impinge on a given two-component system at the location occupied by the extracytoplasmic sensor domain of the sensor kinase protein. In specific cases this assumption is often based on the expectation that a given signal is extracytoplasmic and on the extracytoplasmic nature of this domain of the sensor kinase protein. However, evidence supporting an actual physical interaction of the sensor domain with the signals being sensed is often lacking. Mutations that affect the sensing abilities of the sensor kinase protein often do not affect the extracytoplasmic domain. In addition, for several sensor kinases, for example, VirA of *Agrobacterium tumefaciens* and EnvZ of *Escherichia coli*, removal or substitution of the entire periplasmic domain can lead to a sensor kinase that still responds correctly to its cognate environmental signals (1). The idea that some signals sensed by sensor kinases are sensed at an intracellular location has received support from the recent recognition of p-aminosalicylic acid (PAS) domains present in a large number of sensor kinase proteins (9). PAS domains are domains involved in sensing such conditions as light, oxygen, small ligands, and redox potential, often through the presence of cofactors such as chromophores and heme. In all cases, these domains have been found at positions in the proteins predicted to have intracellular locations, suggesting that the signals they sense are also intracellular. Clearly, in many cases, signals perceived by two-component sensor proteins are sensed by the extracytoplasmic domain. However, the student of two-component systems should examine this assumption critically on a case-by-case basis.

The PhoQ sensor protein of *Salmonella enterica* serovar Typhimurium is one of the best examples of a case where there exists clear evidence of the interaction between an environmental cue, the presence of Mg^{2+} ion, and the periplasmic sensor domain. This comes from biochemical studies that have shown that this domain binds Mg^{2+} and that conformational changes take place upon binding. In addition, genetic studies have identified specific acidic (negatively charged) residues, substitution of which results in decreased binding of Mg^{2+} and an altered regulatory response in vivo. Evidence for the direct involvement of the periplasmic domain of the BvgS protein is less direct, but not insubstantial. In this case, mutations isolated in the laboratory that dramatically affect the ability of this protein to sense the signals of magnesium sulfate, nicotinic acid, and temperature have mapped to an intracellular location, the linker region (see below). However, a difference in sensitivity to these signals between two closely related species, *B. pertussis* and *Bordetella bronchiseptica,* was found through "domain swapping" experiments to be attributable to sequence differences in their periplasmic domains. It should be noted that BvgS is one of the sensors predicted to contain an intracellular PAS, the implications of which have not yet been explored experimentally.

The Transmembrane Region

The extracytoplasmic and the cytoplasmic domains of the sensor kinase proteins are connected by the transmembrane region, which traverses the inner membrane of gram-negative bacteria and the cytoplasmic membrane of gram-positive bacteria. The actual number of transmembrane segments can vary, depending on membrane topology, and has generally been identified because of their length (20 to 30 amino acids), their hydrophobic character, the presence of positively charged residues on the predicted cytoplasmic flanking sequence, and their location between domains of the protein identified as extracytoplasmic or cytoplasmic by other methods, often employing gene fusions to alkaline phosphatase. These sequences thus play an important role in dictating the membrane topology of sensor kinases and in their membrane localization. Beyond this, they may play an important role in transducing signals from the periplasm to the cytoplasm. In some cases, such as the EnvZ protein of *E. coli,* mutations in the transmembrane region can affect signaling. Additionally, structural and genetic studies of the periplasmic region of the aspartate chemotaxis receptor have indicated that movement of transmembrane helices relative to each other is involved in signal transduction across the membrane. This finding may also be relevant to sensor kinase proteins because the two appear to have a common mechanism of signal transduction. This is evidenced by creation of functional protein fusions of the periplasmic aspartate-binding domains of chemotaxis receptors with the cytoplasmic histidine kinase domains of EnvZ. These fusions couple EnvZ-mediated regulation of porin to the presence of aspartate.

The BvgS protein has two transmembrane segments that flank the extracytoplasmic region. The first of these segments (TM1) presumably acts as a signal peptide, which, during protein synthesis, directs sequences following it (the sensor domain) to the periplasm, and the second (TM2) acts as a "stop-transfer" sequence to allow the remainder, the C-terminal part of the protein, to reside in the cytoplasm. No evidence of a further role for these segments is available in this system.

The Linker Region

Cytoplasmic to the transmembrane region, and N-terminal to the conserved kinase domain, lies a region that has been termed the linker region of sensor kinase proteins. As these regions also serve to physically connect the sensory and kinase domains, they are also believed to play an important part in propagation of signals transduced across the inner membrane. A pivotal role is supported by the many examples of mutations isolated in this region of a number of sensor kinase proteins that have the effect of biasing, sometimes to a great extent, signaling by the kinase domain, toward either the nonstimulated state or the stimulated state. Although sequence conservation within this region is poor, predicted structural characteristics have been noted in the linker regions of a large number of sensor kinases. These have been termed type P linkers and have two sequences, AS-I and AS-II, approximately 20 amino acids in length, which are predicted to form amphipathic α-helices and which are separated by an approximately 10-amino-acid segment. This has led to a proposal whereby motions of the spatially proximal TM2 transmembrane segment relative to the membrane dictate the association of AS-I with either the membrane or AS-II (12). In the nonstimulated (no external signal encountered) state, AS-I is envisioned to interact with the inner face of the membrane, and under conditions of stimulation, movement of the transmembrane segment would cause AS-I to switch its interaction to AS-II. AS-II would thus be available to interact with the histidine kinase domain or not, depending on the state of stimulation. In this way signals could be passed through the linker from the external sensor domain to the kinase domain to affect its activity. Whether this will eventually be shown to be part of the mechanism of signal transduction by sensory kinase proteins, it seems likely from genetic studies that the linker region plays a key role in this process.

BvgS does not appear to contain a type P linker. However, a role for the linker region in BvgS function is suggested from mutational studies. Mutations that result in a BvgS constitutive phenotype, in other words, that result in a permanently active form of the histidine kinase, and constitutive expression of the virulence determinants regulated by BvgAS have mapped to the linker region. This region, which is 161 amino acids in length, possesses a highly positively charged region directly adjacent to the TM2 sequence. Interestingly, a number of the constitutive mutations affect this charged region. The ground state of BvgS appears to be the kinase active mode, and modulating environmental signals thus are predicted to result in a reduction of phosphotransfer to BvgA. Thus, the effect of these mutations can be seen as uncoupling the histidine kinase from negative input by the sensor domain. However, the sensor domain apparently fulfills a role beyond simply inhibiting BvgS activity, as deletions or substitutions within it result in a Bvg⁻ phenotype. This requirement for an intact periplasmic domain can be overcome, however, by the presence of the constitutive mutations in the linker region described above.

The Histidine Kinase Domain

The histidine kinase domains of a large number of two-component sensor kinases have been shown to have autocatalytic activity in vitro. Using ATP as a substrate, this domain phosphorylates itself at a conserved histidine residue and it is from this histidine that the phosphate is transferred to an aspartate residue in the receiver domain. Recently, it has become clear that the kinase domain itself is made up of two domains of separable function. One domain is the core domain, which contains the "H" motif recognized as a conserved sequence within the sensor kinase proteins and within which resides the catalytic histidine. The other domain is the enzymatic kinase domain, which actually catalyzes the transfer of a high-energy phosphate bond from ATP to this histidine. The kinase domain contains the N, G1, F, and G2 conserved motifs. Evidence for functional autonomy comes from the observation that the two domains of EnvZ, and other histidine kinases, can be separated genetically, synthesized separately, and then mixed to reconstitute kinase activity in vitro. The phosphorylated core domain so produced is also capable of donating its phosphate group to its cognate response regulator domain, OmpR in this case.

It has been recognized for some time that sensor kinase proteins are active as dimers. This realization has come largely from genetic studies in which two otherwise inactive mutant forms of a sensor kinase protein, affected in different domains, would confer activity when expressed together in the same bacterial strain. Such "intracistronic complementation" experiments led to the further conclusion that the histidine kinase domain of one monomer in a dimer normally phosphorylates the core domain of its partner, and vice versa. Recently, the structure of the isolated EnvZ core domain at an atomic level was determined (10). From this it can be seen that the core domain is also responsible for dimerization of EnvZ. Hence, this domain is also referred to as the dimerization domain. Each monomer of this dimeric structure is composed of two antiparallel α-helices connected by a loop. The two monomers dimerize by forming a four-helix bundle on which the two catalytic histidines are displayed on the outside (Fig. 3) (see Color Plates following p. 256). This domain thus performs the dual functions of mediating dimerization and acting as a catalytic partner, or substrate, of the kinase domain.

The structure of the isolated catalytic subdomain of the EnvZ histidine kinase (and those of several other histidine kinases) has also been solved recently

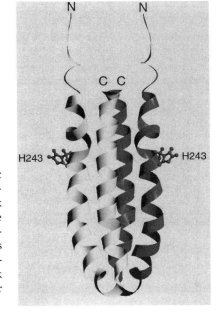

Figure 3 NMR solution structure of the dimeric core domain of EnvZ. The protein structure is depicted in ribbon format showing the four-helix bundle formed from two identical monomers. The catalytic histidine at position 243 is shown in ball-and-stick format. Based on structural coordinates deposited in the Research Collaboratory for Structural Bioinformatics (RCSB) Protein Data Bank (PDB), as described in reference 10. (See Color Plates following p. 256.)

(8). From these it can be seen that what had previously been defined as the conserved N, G1, F, and G2 sequence motifs make up the nucleotide-binding site for the ATP substrate (Fig. 4) (see Color Plates following p. 256). Although not revealed in these studies of isolated domains, the catalytic domain is attached by its N terminus to one end of the core dimerization domain by a linkage that is presumably flexible. Thus, it is situated such that it can associate with the core domain to catalyze phosphorylation of the catalytic histidine. However, it must also be free to move out of the way to allow subsequent access of the response regulator domain to effect phosphotransfer.

Besides autophosphorylation and phosphorylation of their cognate response regulator proteins, many histidine kinases are known to catalyze an additional reaction, the dephosphorylation of their cognate response regulator. The balance struck between these opposing activities determines the level of intracellular phosphorylated response regulator protein. While the phosphatase activity apparently resides within the greater histidine kinase domain, a coherent picture of the structural basis for this activity has not emerged from the work on disparate two-component systems. It appears possible, however, that the relative spatial relationship of the core dimerization domain and the catalytic domain may, at least in some cases, dictate the balance between these opposing activities. Since a functional pair of core and catalytic subdomains resides on different monomers, their relative spatial relationships are subject to the effects of movements of their respective linker regions and/or transmembrane segments, either relative to each other or to the membrane. In this way, changes in the external milieu, represented by conformational changes in the extracytoplasmic sensor domain, linked to movements of the transmembrane region and/or linker region,

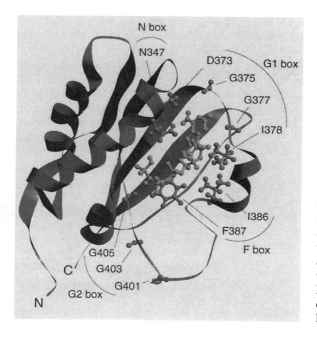

Figure 4 NMR solution structure of the catalytic kinase domain of EnvZ. The protein structure is depicted in ribbon format. The segments corresponding to the conserved N, G1, F, and G2 boxes are colored blue, and key residues from these are shown in bll-and-stick format. This structure was determined in the presence of the nonhydrolyzable ATP analog AMP-PNP, which is shown in green ball-and-stick format. Based on structural coordinates deposited in the RCSB PDB, as described in reference 8. (See Color Plates following p. 256.)

may result in changes in the kinase-phosphatase balance and thus affect levels of phosphorylated response regulator protein.

The BvgS kinase domain sequence appears unremarkable in terms of its possession of, and linear organization of, the conserved H, N, G1, F, and G2 sequences, but structural data in this system are not available. As expected, mutations in the key histidine residue in the H motif abolish activity and result in a Bvg⁻ phenotype. Other mutations in the kinase domain appear to activate the kinase, rendering it less sensitive to down-modulation by environmental signals.

The Receiver Domain

To date, the structures of eight receiver domains have been determined. These show remarkable conservation of overall structure. CheY of *E. coli* is presented in Fig. 5 (see Color Plates following p. 256) as representative of this class. The structure consists of a five-stranded parallel β-sheet in which the β-strands alternate with five α-helices, which surround the central β-sheet. At one pole, the loops connecting the α-helices and β-strands form an acidic pocket. This pocket, which houses the invariant catalytic aspartic acid (number 57 in CheY), also contains the highly conserved aspartate residues at positions 12 and 13 and lysine 109. Two other nearby residues, threonine 87 and tyrosine 109, also play key roles in activation (see below). Although once thought of as a passive partner in the phosphotransfer reaction with a histidine kinase, it is now recognized that the receiver domain is the major catalytic force behind this reaction. This has come largely from the experimental demonstration that many response regulators can catalyze their own phosphorylation when presented with small high-energy phosphate donors such as acetyl phosphate or carbamoyl phosphate.

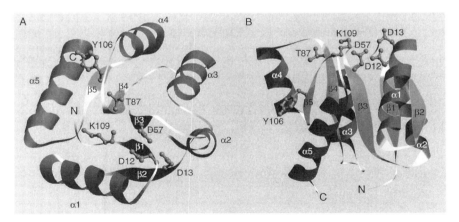

Figure 5 CheY structure determined by X-ray crystallography. The structure of CheY is presented as representative of the common structure of receiver domains from many sources. The protein structure is shown in ribbon format, and the α-helical and β-strand segments are numbered. Key residues introduced in Fig. 2 are shown in blue ball-and-stick format. Two views are shown for clarity. Based on structural coordinates deposited in the RCSB PDB, as described in reference 7. (See Color Plates following p. 256.)

Thus, the kinase core subdomain (or HPt domain [see below]) can be seen as a specific phosphodonor substrate for the action of the receiver domain of its cognate response regulator.

A research area of great interest and activity concerns the mechanism of activation of the response regulator protein. How does phosphorylation at an aspartate residue in one domain of a protein result in activation of a linked domain? Once again, refined structural studies have pointed the way to an understanding. For four response regulators (CheY of *E. coli*, FixJ of *Sinorhizobium meliloti*, NtrC of *S. enterica* serovar Typhimurium, and Spo0A of *Bacillus stearothermophilus*) the structure of the activated form of the receiver domain has been determined and can be compared with the nonactivated structure. Details revealed in these structures suggest a common mechanism for activation. In all cases, upon phosphorylation, the conserved threonine residue homologous to threonine 87 in CheY reorients itself, apparently to hydrogen bond with an oxygen of the phosphate group. The tyrosine residue corresponding to tyrosine 106 in CheY is then able to rotate inward from its protruding position in the unactivated form. In this manner, structural changes induced by phosphorylation in the phosphate-accepting pocket can be propagated to the outside surface of the receiver domain (4). In FixJ and NtrC, approximately half of the structure of this domain experiences alteration (α3 to β5 and connecting loops). In the other two, the structural changes are more modest, affecting the β4-α4 loop, α4 and part of α5 in CheY, and the β3-α3 loop, the β4-α4 loop, and α4 in Spo0A. The strategies by which these structural changes in the response regulator domain are harnessed to effect the activation of the C-terminal effector domain are likely to be varied. However, two possibilities, especially relevant to response regulator proteins that affect gene expression, are suggested. In FixJ, changes in the receiver domain result in the strong dimerization of this domain. As a result, the C-terminal DNA-binding domain effectively dimerizes as well. As a number of transcriptional regulatory proteins are more active as dimers, this suggests a straightforward mechanism for activation of this type of effector domain. Another possibility is evident from the structure of the entire NarL response regulator protein of *E. coli*. In this case, the C-terminal DNA-binding domain is sterically blocked by the presence of the N-terminal receiver domain. Activation is thought to involve a change in the conformation such that the N-terminal domain swings away to allow access of the C-terminal effector domain to DNA-binding sites. As the two domains in the inactivated form are held together by interactions at the domain interface, phosphorylation and the accompanying alterations in this interface could relieve the inhibition of DNA binding and result in activation. The two activation mechanisms of multimerization and relief of steric inhibition are not mutually exclusive and may operate simultaneously.

HPt Domains

Once thought to be relatively unique and termed "unorthodox" sensor proteins, some, more complicated, two-component sensor kinase proteins participate in a mechanism whereby phosphate is relayed through two additional steps compared to the simplest two-component system whose signaling mechanism is composed of only the domains described so far (Fig. 1A). This is sometimes referred to as a His-Asp-His-Asp phosphorelay to indicate the additional steps. The additional Asp and His residues involved are contained within an additional receiver domain and an HPt domain, respectively. In a number of sensor kinase proteins, including ArcB, BarA, and EvgA of *E. coli,* as well as BvgS of *B. pertussis,* the two additional domains are part of a large sensor kinase protein, as shown schematically in Fig. 1B. However, these domains will function normally if liberated from the sensor protein and expressed as isolated domains in vivo or examined in vitro. Indeed, phosphorelay systems in which these four domains are on four different proteins are known to occur naturally (e.g., the KinA, Spo0F, Spo0B, and Spo0A system of *Bacillus subtilis*).

The auxiliary response regulator domains are not expected to differ markedly in structure from other response regulator domains. Their presence was, in fact, recognized initially by virtue of their sequence similarity and the presence of key conserved residues, including, of course, the active site aspartic acid. The role of the HPt domains, however, was not originally suspected as they lack significant sequence similarity with histidine kinase domains. Recently the structures of three bacterial HPt domains have been solved. Two of these are monomeric HPt domains and most representative of the HPt domains of complex sensor kinases such as BvgS (although the sensor kinases themselves are active as dimers, the HPt domains are monomeric). These are the C-terminal HPt of the ArcB sensor protein and the P1 domain of CheA, both from *E. coli.* Interestingly, the structure of these monomeric HPts, represented in Fig. 6 (see Color Plates following p. 256) by the ArcB HPt, is reminiscent of the structure of the core dimerization and phosphotransfer domain of histidine kinases, such as EnvZ described above. In both cases the structure comprises a four-helix bundle. The dimeric core domains, being formed from two identical subunits, contain two catalytic histidine residues capable of accepting and donating phosphate, whereas the monomeric domains contain only a single active histidine. Another structural difference is that the helices are somewhat bent in the monomeric

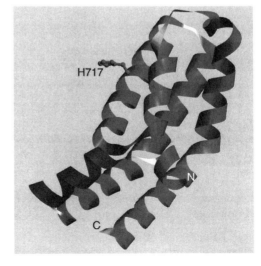

Figure 6 Structure of the monomeric HPt domain from the ArcB sensor kinase of *E. coli* determined by X-ray crystallography. Protein structure is shown in ribbon format, with the catalytic histidine at position 717 shown in blue ball-and-stick format. Based on structural coordinates deposited in the RCSB PDB, as described in reference 3. (See Color Plates following p. 256.)

domains but quite straight in the dimeric core domains. The monomeric HPt domains can thus be viewed as structurally and functionally analogous to the histidine kinase core domains, a difference being that phosphate is received from a response regulator domain rather than a kinase catalytic domain prior to passage to another response regulator. The C-terminal HPt domain of BvgS is analogous in function to the ArcB HPt domain, and it is expected to be structurally similar as well, although sequence similarity between the two is restricted to the region immediately surrounding the active site histidine.

Interactions of Response Regulator and HPt Domains

As discussed above, given the large number of two-component signaling systems that can be operative simultaneously in some species of bacteria, specificity of interaction between the sensor kinase protein and the response regulator protein is a critical issue. While "cross talk" between systems has sometimes been demonstrated, often in the presence of perturbing mutations, specificity of interaction appears to be the rule rather than the exception. Genetic studies have implicated residues in the core histidine-containing dimerization subdomains of the sensor kinase and residues surrounding the acidic pocket of the response regulator in this recognition. The best structural approximation of the response regulator-kinase interaction has come from the X-ray crystal structure of the receiver domain from Spo0F of *B. subtilis,* complexed with a cognate HPt domain (13). In this case, the HPt domain is that of the Spo0B protein. Unlike the HPts of CheA and ArcB, Spo0B is dimeric and thus may be a reasonable substitute for the core subdomain of histidine kinases, having a four-helix bundle with two catalytic histidine residues (Fig. 7A) (see Color Plates following p. 256). Attached to this dimerized core are two identical domains, which are similar in structure to kinase catalytic domains. However, these domains lack kinase activity because they do not possess a functional ATP-binding site composed of the N, G1, F, and G2 motifs. In overall structure, however, Spo0B serves as a reasonable surrogate for an intact histidine kinase domain. As shown in Fig. 7B, Spo0B and the Spo0F receiver domain dock with each other in such a way that the critical histidine and aspartate residues are aligned so that phosphotransfer can easily occur. As expected, residues surrounding the acidic pocket of Spo0F and those surrounding the histidine of Spo0B make specific contacts with each other. It can also be seen that the α1-helix of each response regulator makes extensive contacts with the α1-helix of its respective Spo0B monomer. When the protein surface by which Spo0F interacts with Spo0B is examined in terms of the degree of conservation of individual residues, it can be seen that it is a mosaic of conserved and variant residues. The variant residues are expected to dictate specificity of interaction of the proteins with their cognate HPt or kinase domains, and the conserved residues are expected to preserve the critical alignment necessary for the phosphotransfer reaction to occur.

The Effector Domain

At the end of this long line of causality is the final step whereby environmental cues are at last converted to a corresponding action. In the vast majority of sys-

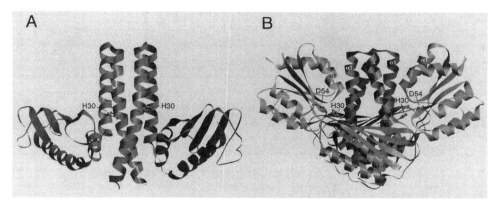

Figure 7 Structure of Spo0B (A) and a complex between Spo0F and Spo0B (B) determined by X-ray crystallography. (A) The structure of a Spo0B dimer is presented as representative of the overall structure of histidine kinase domains. Protein structure is shown in ribbon format while catalytic histidines are shown in blue ball-and-stick format. Based on structural coordinates deposited in the RCSB PDB, as described in reference 11. (B) The structure of a complex between a Spo0B dimer and two cognate receiver domains of Spo0F. Based on structural coordinates deposited in the RCSB PDB, as described in reference 13. (See Color Plates following p. 256.)

tems that have been characterized, this corresponds to activation of gene expression by stimulation of the transcription of specific genes. Three major classes of effector domains of this type have been recognized, based on sequence and, where known, functional similarities. Named for representative members, they are the NarL family, the OmpR family, and the NtrC family. Of these three families, NtrC has the most unique mechanism of transcriptional activation, using the energy of ATP hydrolysis to stimulate the productive interaction of a nonvegetative form of RNA polymerase (containing σ^{54}) at regulated promoters. Compared to the other two, the site of NtrC binding with respect to the regulated promoter is less critical, and in fact, in vitro, NtrC can stimulate transcription even when not bound to its DNA target. It is thus often referred to as a transcriptional enhancer, analogous to such proteins in eukaryotes. Members of the OmpR and NarL families have sequences consistent with the known structures of the effector domains of these two proteins. In the case of NarL, this is a fairly typical helix-turn-helix motif, whereas OmpR displays a variation known as a "winged helix-turn-helix" motif. Members of the OmpR and NarL families activate transcription by the more common mechanism of binding to specific sequences directly upstream of regulated promoter elements and stimulating the binding of, and/or initiation of transcription by, RNA polymerase by making specific protein-protein contacts with parts of the multimeric polymerase. Within this general schema, specific details of the nature and location of DNA-binding sites relative to the transcriptional initiation site can vary a great deal. However, at least one theme has emerged. For many of these regulators, and certainly for the eponymous members OmpR and NarL, regulation involves multiple binding sites in promoter regions. In many cases these sites are of differing affinities, such that the degree of occupation of the different sites by the response regulator depends on its degree of phosphorylation. This has been extensively studied in the case of OmpR regulation of the alternate porin proteins OmpC and OmpF in *E. coli*. In this case, multiple OmpR-binding sites in each promoter region result in a situation whereby relatively low levels of phospho-OmpR (occurring under conditions of low osmolarity of the growth medium) result in OmpF expression but lack of OmpC expression. At high levels of phospho-OmpR (high osmolarity), OmpF expression is repressed and OmpC is activated. Many regulators of the NarL subfamily also use a configuration of binding sites at regulated promoters with high-affinity sites upstream and lower-affinity binding sites more proximal to the promoter. Occupation of the downstream sites occurs by cooperative binding, assisted by protein-protein interactions with the response regulator bound to the high-affinity upstream sites. It has been suggested that the dependence of transcriptional activation on occupation of low-affinity sites is an adaptation that allows the response regulator to dissociate frequently and diffuse to the membrane, thus sampling the state of activation of its cognate sensor kinase and allowing adjustment of its phosphorylation status to be in keeping with prevailing environmental conditions.

BvgA belongs to the NarL family of effector domains. From studies of a number of BvgA-regulated promoters, a common theme has emerged. At the *ptx*, *cya*, and *fha* promoters, phospho-BvgA binds to high-affinity primary binding sites situated upstream of the -35 and -10 regions, at a distance that varies between promoters. Activation occurs when cooperative binding of phospho-BvgA dimers extends from this primary site, through a region of secondary binding lacking any apparent recognition sequences for BvgA. When the most promoter-proximal sequences are filled by phospho-BvgA, transcription activation can occur, involving, at least in the case of the *fha* promoter, contacts between BvgA and the C-terminal domain of the α-subunit of RNA polymerase. Other interactions with RNA polymerase may also be involved. The responsiveness of different promoters to the level of phospho-BvgA varies and is apparently determined by the affinity of the primary binding site (less responsive promoters have lower-affinity primary binding sites) and by the length of DNA that must be occupied by cooperatively bound BvgA (longer in less responsive promoters). It has been suggested that this may effect a temporal program of gene expression whereby, upon introduction to a host environment, activation of the BvgAS system leads to rising levels of phospho-BvgA, as BvgAS is autoregulated. This results in sequential activation of virulence genes in an order that optimizes success and thus reproductive fitness (see below). Whether or not the differential sensitivities of promoters to phospho-BvgA translate into a temporal program, additional support for the existence and importance of varying phospho-BvgA concentrations in the natural history of *B. pertussis* has come from the discovery of a new class of genes. These genes, the intermediate-phase genes, are expressed maximally when the BvgAS system is partially activated but not when it is fully or only weakly activated. Analysis of the promoter of one of these genes, the *bipA* gene, suggests that occupation of intermediate-affinity BvgA-binding sites upstream of the core promoter results in activation at intermediate levels of phospho-BvgA (2). Lower levels are insufficient to fully induce, and at higher levels, binding to lower-affinity sites further downstream represses transcription. Thus, *bipA* regulation is analogous to that of the *ompF* gene of *E. coli* by the response regulator OmpR as described above.

ROLE OF THE BvgAS TWO-COMPONENT SYSTEM IN TOXIN REGULATION

The BvgAS Regulon

Whooping cough, or pertussis, is often used as a prime example of a toxin-mediated disease. The *B. pertussis* bacterium synthesizes a number of sophisticated toxins, which are believed to be responsible for the major symptoms of this severe respiratory ailment. Chief among these is pertussis toxin (Ptx), a complex AB-type toxin comprising a B pentamer composed of four different subunits and an A toxic subunit, which, upon internalization into a host cell, catalyzes the ADP-ribosylation of specific G-protein α-subunits of the Gi family, involved in regulation of host adenylate cyclase (see chapters elsewhere in this volume). *B. pertussis* also elaborates an invasive adenylate cyclase toxin (Cya), a member of the RTX family of toxins. Cya shares its mechanisms of secretion and host cell internalization with other members of the RTX family. The toxic portion of the Cya molecule is a calmodulin-stimulated adenylate cyclase enzyme, which confers elevated cAMP levels to host cells upon internalization. Dermonecrotic toxin (Dnt), related to the cytotoxic necrotizing factor of *E. coli,* is another addition to the toxin repertoire of *B. pertussis.* The biochemical action of Dnt has recently been shown to involve activation of the host Rho GTPase by transglutamination. Ptx and Cya have profound inhibitory effects on host immune cells such as lymphocytes, neutrophils, macrophages, and monocytes and, as such, constitute an evasion of the host immune defenses. A role for Dnt in the pathogenesis of pertussis has not been established. In addition to toxins, *B. pertussis* elaborates a number of adhesins, which contribute to its ability to adhere to host tissues during infection. These include filamentous hemagglutinin (Fha), pertactin, and fimbriae. This list of virulence factors is not comprehensive but will serve to illustrate the principles discussed here.

All of these factors, as well as several others not described here, are positively regulated by the *bvgAS* locus. The *bvgAS* locus, originally termed the *vir* locus, was discovered as a result of a transposon-induced mutation, which simultaneously abolished the expression of many different virulence genes. Virtually all protein-based virulence factors have since been shown to be Bvg-regulated, and a number of candidate virulence factors have been identified based on their activation by the *bvgAS* locus (the *vag*'s, for *vir*-*a*ctivated *g*enes). Together these constitute a regulon, a group of genes that are regulated by a similar mechanism and governed by the same environmental signals, in keeping with their demonstrated or inferred congruity of function. Virulence regulons are extremely common, being found in practically all virulent bacterial species. These regulons can be relatively simple, as in the case of BvgAS, where a single two-component system appears to govern the synthesis of all virulence genes, or complex, as in the case of *S. enterica* serovar Typhimurium, where multiple two-component systems and other regulators are integrated into a sophisticated network capable of sensing multiple environments within the mammalian host and responding to each through expression of the appropriate sets of genes.

Regulatory Networks/BvgR and the BvgAS-Repressed Genes

Two-component systems are found regulating not only the structural genes for toxins and other virulence genes but also other regulators, including other two-component systems. The regulatory networks that result can be extremely complex. The advantage to the bacterium for such networks is apparently the ability to improve the sophistication of their responses and provide for a larger number of environmental contingencies. For example, a given gene may be expressed only when two or three different conditions are fulfilled or only in the absence of a condition. A complete discussion of complex regulatory networks is beyond the scope of this chapter. However, a relatively simple network commonly encountered in virulent bacteria is evidenced by the presence of reciprocally regulated genes. Reciprocal, in this case, is relative to more standard virulence factors such as toxins, adhesins, and mechanisms for evading the host response. In the case of *B. pertussis*, the reciprocally regulated genes are called *vrg*'s (for *vir*-repressed genes) and were originally identified through randomly introduced gene fusions, which were activated when the *vag*'s were repressed through additives such as magnesium sulfate or nicotinic acid, which down-modulate the BvgAS regulon. These genes are regulated via the BvgR repressor, whose gene is itself transcriptionally activated by the BvgAS system. In *B. pertussis*, the function of the *vrg*'s is unknown. In the closely related veterinary pathogen *B. bronchiseptica*, the genes reciprocally regulated include those required for synthesis and function of flagella. *B. bronchiseptica*, apparently being capable of survival in the natural environment, uses the BvgAS regulatory system, at least in part, to choose between a free-living lifestyle, or survival in the environment (motility turned on, virulence genes off), and existence in an animal host (motility turned off, virulence genes on). The role of the *vag*'s in the natural history of *B. pertussis*, being an obligate human parasite, is less clear. It has been suggested that these genes are turned on and aid in intracellular survival when *B. pertussis* is taken up by host cells.

Is Regulation of Virulence Determinants Itself a Virulence Determinant?

The notion that regulation of virulence factors plays a critical role in the fitness of pathogenic bacterial species should, in theory, be testable by experiment. In fact, there are a number of examples of mutations in key regulatory genes, including two-component systems, either those that knock out those systems or those that result in constitutive expression of their regulon, that have a negative impact on virulence in experimental animal models. However, the existence of reciprocally regulated genes means that either type of mutation results in the inability of an organism to express part of its relevant genetic repertoire. The discovery and characterization of the BvgR repressor made it possible to construct a *B. pertussis* strain that constitutively expressed both the *vag*'s and the *vrg*'s and is thus phenotypically regulation$^-$. This strain had a constitutive mutation affecting the BvgS linker region (see above) and a lesion knocking out the

BvgR repressor (5). When tested in an animal model, this strain was, in fact, less virulent. In related experiments, this reduction in virulence was found to be attributable to the BvgR knockout. Thus, it appeared that the inability to repress the *vrg*'s was a liability in vivo. Similar results had previously been obtained with *B. bronchiseptica*. Thus, the BvgAS system makes at least two contributions to virulence, being involved in expression of virulence factors such as toxins and also being involved in repression of genes that are detrimental in vivo.

It has also been suggested that an additional role of BvgAS is to act as an orchestrator of a temporal program of gene expression. This comes from the observation following a shift of *B. pertussis* from growth conditions under which the BvgAS system is not active to those under which it is. Following such a shift, phospho-BvgA concentrations in the cell are rising. The genes for synthesis of the adhesin Fha are activated quickly, within minutes, whereas those for toxins such as Ptx and Cya are not activated for several hours. We now understand that the time of induction is related to promoter structure and relative responsiveness to phospho-BvgA (see above). It is supposed that *B. pertussis*, upon introduction into a new host, would need to express Fha first, to adhere and gain a foothold, and only later to express toxins, to evade the host response. Orchestrating this sequential expression would be the BvgAS system. Attempts have also been made to test this hypothesis experimentally. By swapping promoter regions of the *fha* and *ptx* genes, *ptx* expression was made to occur early. Similarly, in another strain, *fha* expression was delayed. In both cases, a measurable reduction in virulence in a mouse model of infection was observed, again supporting the important role of regulation in the fitness of bacterial pathogens.

TWO-COMPONENT SYSTEMS: ACHILLES' HEEL OF BACTERIAL PATHOGENS?

As more of previously dependable antibiotic therapies are becoming less effective against bacterial infections, other modalities for the control of infectious disease are being investigated. Two approaches to this problem are the development of new vaccines and the discovery or creation of new anti-infective compounds. Two-component systems are under consideration in both efforts.

Vaccines

One approach to vaccine development involves the attenuation of organisms to create a live vaccine strain that can colonize and engender a strong and effective immune response without causing a harmful infection. Historically, this has been accomplished simply by growth of the bacterium in vitro for many generations to allow it to lose virulence characteristics. We now know that this occurs by mutation or loss of key virulence genes. The molecular era in microbiology affords many opportunities to create attenuated strains by a more directed approach. One way to do this in a global way is to affect the regulatory systems governing virulence factors. For example, either null or constitutive mutations in the PhoPQ virulence regulatory system of *S. enterica* serovar Typhimurium lead to much decreased virulence in a mouse model, yet engender a protective immune response. Salmonella vaccines based on this approach are currently being tested in human clinical trials. Another (non-two-component) class of regulatory mutations, affecting the host DNA methylation system, has also shown great promise in early studies.

Anti-Infective Compounds

Most antibiotics in use today owe their existence to bacteria, which created them or their chemical predecessors over the ages through the process of evolution. Essentially all of these are aimed at vital processes of the bacterial cell, including the synthesis of DNA (replication), RNA (transcription), and protein (translation). As the need for new anti-infective compounds is approached with the knowledge of the molecular bases of virulence, another approach presents itself: compounds that interfere with the virulence mechanisms of bacterial pathogens. Several features of two-component systems suggest them as potential targets for hopeful drug designers. They have a shared mechanism of signaling; thus, compounds that disrupt this basic mechanism should be active against many different two-component systems. Their widespread distribution in the bacterial world also suggests broad applicability. Finally, they are apparently not part of the human biochemical makeup, not being detected within the recently completed human genome sequence. Either by inhibiting activation of key virulence genes or by inducing the inappropriate expression of deleterious genes (such as the *vrg*'s of *Bordetella* species), it is hoped that new compounds could help tip the balance in a bacterial infection toward the interests of the host. A number of compounds that inhibit two-component systems have been reported, and a number of them display antibacterial activity. Much work will be required to assess the utility of this approach.

CONCLUSION

Despite the tremendous growth of knowledge about the functioning of two-component systems in the regulation of bacterial toxins and other virulence factors, a number of major questions remain largely unanswered. It has been relatively straightforward, given the tools molecular biology has provided, to demonstrate the presence of two-component systems and their critical role in regulation of bacterial phenotypes. However, much progress in determining the chemical or physical nature of the relevant environmental cues to which these systems react and the nature of their interaction with sensor kinase proteins has yet to be made. Such an understanding is a crucial aspect of understanding the evolution of bacterial virulence mechanisms and how, possibly, to subvert them. In addition, although beautiful molecular structures have created mental images of the operation of these remarkable systems, the central phenomenon of signal transduction across the cytoplasmic membrane is still poorly understood. Future research should lead to answers to these challenging questions.

REFERENCES

1. **Chang, C.-H., and S. C. Winans.** 1992. Functional roles assigned to the periplasmic, linker, and receiver domains of the *Agrobacterium tumefaciens* VirA protein. *J. Bacteriol.* **174:**7033–7039.

2. **Deora, R., H. J. Bootsma, J. F. Miller, and P. A. Cotter.** 2001. Diversity in the *Bordetella* virulence regulon: transcriptional control of a Bvg-intermediate phase gene. *Mol. Microbiol.* **40:**669–683.

3. **Kato, M., T. Mizuno, T. Shimizu, and T. Hakoshima.** 1997. Insights into multistep phosphorelay from the crystal structure of the C-terminal HPt domain of ArcB. *Cell* **88:**717–723.

4. **Kern, D., B. F. Volkman, P. Luginbühl, M. J. Nohalle, S. Kustu, and D. E. Wemmer.** 1999. Structure of a transiently phosphorylated switch in bacterial signal transduction. *Nature* **402:**894–898.

5. **Merkel, T. J., S. Stibitz, J. M. Keith, M. Leef, and R. Shahin.** 1998. Contribution of regulation by the *bvg* locus to respiratory infection of mice by *Bordetella pertussis*. *Infect. Immun.* **66:**4367–4373.

6. **Parkinson, J. S., and E. C. Kofoid.** 1992. Communication modules in bacterial signaling proteins. *Annu. Rev. Genet.* **26:**71–112.

7. **Stock, A. M., E. Martinez-Hackert, B. F. Rasmussen, A. H. West, J. B. Stock, D. Ringe, and G. A. Petsk.** 1993. Structure of the Mg^{2+}-bound form of CheY and mechanism of phosphoryl transfer in bacterial chemotaxis. *Biochemistry* **32:**13375–13380.

8. **Tanaka, T., S. K. Saha, C. Tomomori, R. Ishima, D. Liu, K. I. Tong, H. Park, R. Dutta, L. Qin, M. B. Swindells, T. Yamazaki, A. M. Ono, M. Kainosho, M. Inouye, and M. Ikura.** 1998. NMR structure of the histidine kinase domain of the *E. coli* osmosensor EnvZ. *Nature* **396:**88–92.

9. **Taylor, B. L., and I. B. Zhulin.** 1999. PAS domains: internal sensors of oxygen, redox potential, and light. *Microbiol. Mol. Biol. Rev.* **63:**479–506.

10. **Tomomori, C., T. Tanaka, R. Dutta, H. Park, S. K. Saha, Y. Zhu, R. Ishima, D. Liu, K. I. Tong, H. Kurokawa, H. Qian, M. Inouye, and M. Ikura.** 1999. Solution structure of the homodimeric core domain of *Escherichia coli* histidine kinase EnvZ. *Nature Struct. Biol.* **6:**729–734.

11. **Varughese, K. I., Madhusudan, X. Z. Zhou, J. M. Whiteley, and J. A. Hock.** 1998. Formation of a novel four-helix bundle and molecular recognition sites by dimerization of a response regulator phosphotransferase. *Mol. Cell* **2:**485–493.

12. **Williams, S. B., and V. Stewart.** 1999. Functional similarities among two-component sensors and methyl-accepting chemotaxis proteins suggest a role for linker region amphipathic helices in transmembrane signal transduction. *Mol. Microbiol.* **33:**1093–1102.

13. **Zapf, J. W., U. Sen, Madhusudan, J. A. Hoch, and K. I. Varughese.** 2000. A transient interaction between two phosphorelay proteins trapped in a crystal lattice reveals the mechanism of molecular recognition and phosphotransfer in signal transduction. *Struct. Fold. Des.* **8:**851–862.

Bacterial Protein Toxins
D. Burns et al., Editors
©2003 ASM Press, Washington, DC 20036

Chapter 2

Regulation of Bacterial Toxin Synthesis by Iron

Shelley M. Payne

Pathogenic bacteria synthesize toxins and other virulence factors when they are within their host, but the genes encoding these virulence factors are usually repressed when the bacteria are growing in typical laboratory conditions. This relationship between virulence gene expression and host environmental factors has been known since the 1930s, when attempts to produce toxins in sufficient quantities for characterization revealed that bacteria grown in the laboratory typically synthesized low levels of toxin. This result led to experiments designed to identify the host factors that regulate toxin gene expression.

There are a number of differences between the conditions within the mammalian host environment and conditions outside the host that could influence the expression of virulence genes. These variables include temperature, osmolarity, nutrient availability, and ion concentrations. One of these conditions that has been clearly demonstrated to regulate bacterial toxin synthesis is the availability of iron. In most cases, the effect of iron is repression of toxin synthesis (Table 1).

Bacteria are likely to encounter a low-iron environment when colonizing the human. Estimates of the concentrations of free iron at sites within the human body are on the order of 10^{-18} to 10^{-9} M, and there have been a number of studies showing that the iron-poor environment of the host limits bacterial growth. Although iron limitation has an inhibitory effect, it also serves as one of the signals to the bacterium that it is within a host.

Bacteria respond to low iron concentration by increasing the expression of virulence factors that may promote their growth and survival and ultimately cause damage to the host. These include genes for iron acquisition systems that allow microorganisms to obtain enough iron for growth and metabolism in an iron-restricted environment. Pathogens may synthesize surface receptors for host iron sources such as hemoglobin, heme, transferrin, and lactoferrin that allow binding and ultimately removal of iron from these host sources. Many microorganisms also synthesize siderophores, low-molecular-weight, high-affinity iron chelators that are secreted into the environment. Siderophores are able to solu-

---

Shelley M. Payne • Section of Molecular Genetics and Microbiology, The University of Texas, Austin, TX 78712-1095.

Table 1 Toxins and other secreted virulence proteins whose
synthesis is negatively regulated by iron

Toxin	Produced by
Diphtheria toxin	*Corynebacterium diphtheriae*
Shiga toxin	*Shigella dysenteriae*
Shiga-like toxin 1	*Escherichia coli*
Hemolysin	*E. coli*
Hemolysin	*Vibrio cholerae*
Exotoxin A	*Pseudomonas aeruginosa*
Tetanus toxin	*Clostridium tetani*
VacA cytotoxin	*Helicobacter pylori*
CHO cell elongation factor	*Plesiomonas shigelloides*

bilize iron or remove it from other chelates, including those of the host, and
transport the iron back into the cell via specific receptors and transport proteins.

Whereas iron is essential for growth of most pathogens, free iron is toxic.
Excess iron can lead to Fenton reactions and the generation of toxic oxygen rad-
icals. Thus, the expression of iron uptake systems must be tightly regulated to
control the amount of iron within the bacterial cell. In addition, iron acquisition
genes are coordinately regulated with genes involved in the oxidative stress response. Iron availability is one of the regulators of catalase and superoxide dis-
mutase gene expression, and the oxidative stress-response regulators OxyR and
SoxRS influence expression of the iron regulator gene *fur* in *Escherichia coli*. Iron
availability also regulates expression of some metabolic pathway genes, including those encoding aconitase and fumarase. Thus, in many bacteria, iron regulation of toxin is part of a global regulatory network coordinating the cellular
response to alterations in available iron levels.

There are several possible advantages to bacteria in using iron concentration
to regulate toxin synthesis. Iron-regulated promoters are common in bacteria and
allow expression of the downstream gene in the low-iron environment of the
host. Coupling virulence factor expression to this in vivo signal allows the virulence determinant to be expressed in the host but not in the external environment where it would not be needed. Additionally, exotoxin synthesis may cause
release of intracellular iron by damaging host cells, thereby increasing the availability of iron to the bacterium. Thus, secretion of toxins may represent an additional mechanism for bacterial iron acquisition in the host.

Although some details of iron-mediated regulation remain to be determined,
many of the regulatory pathways have been elucidated at the molecular level,
and a basic understanding of the mechanisms by which iron regulates toxin
production has been gained for several human pathogens. The following discussion of three iron-regulated toxins, diphtheria toxin, Shiga toxin, and *Pseudomonas*
exotoxin A, illustrates the similarities and differences in the control mechanisms.
Diphtheria toxin exhibits a relatively simple model of regulation in which a bacterial protein represses synthesis in the presence of iron. Shiga toxin has a similar
mechanism of regulation but may be further controlled by phage regulators.
Pseudomonas exotoxin A has the most complex regulation of the three; regulation

by an iron-binding repressor is one step in a cascade of positive and negative regulators that act in concert to control toxin expression.

DIPHTHERIA TOXIN REGULATION

One of the best-understood examples of environmental regulation of toxin production is synthesis of diphtheria toxin by *Corynebacterium diphtheriae*. This toxin, which ADP-ribosylates EF-2, inhibiting host protein synthesis, was one of the first toxins for which a relationship between iron availability and toxin production was shown. By adding iron to broth that had relatively low amounts of endogenous iron, Pappenheimer showed that toxin production initially increased with increasing iron concentration, reaching a maximum at approximately 2.5 μM of Fe. This increased toxin production was due to iron stimulating the growth of the toxin-producing bacteria. Larger amounts of iron, however, drastically reduced toxin synthesis; no toxin was detected when the concentration of added iron was increased approximately fourfold. Maximal toxin production occurred at an iron concentration that is slightly less than optimal for growth of the bacteria. This pattern of expression (Fig. 1) is typical for iron-regulated bacterial toxin production.

Mapping of genes required for diphtheria toxin production revealed that the structural gene for the toxin, *tox*, is found in the genome of corynebacteriophages such as β, and toxin is only produced by *C. diphtheriae* strains that are carrying a Tox⁺ prophage. The *tox* gene is adjacent to the bacteriophage *att* site, mapping at the junction between the phage and host bacterial DNA. Although toxin is synthesized by lysogens, toxin levels are low unless the bacteria are grown under iron-limiting conditions.

In vitro transcription-translation was used to show that the phage *tox* locus was actually the structural gene for the toxin and to show that iron regulation

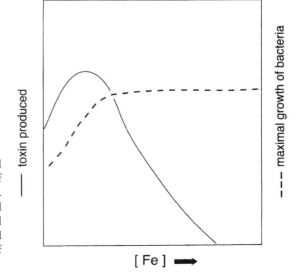

Figure 1 Bacterial growth and toxin production as a function of iron concentration of the medium. The solid line indicates the total amount of toxin in the culture, and the dashed line indicates the final cell density as the concentration of iron increases.

was mediated by a *C. diphtheriae* protein. Phage β DNA was added to *E. coli* cell-free extracts, and it was demonstrated that toxin was synthesized from the phage DNA. Adding iron to the *E. coli* reaction did not inhibit toxin synthesis in vitro, indicating that an additional factor was required for regulation by iron. The addition of *C. diphtheriae* cell extract to the *E. coli* system inhibited toxin synthesis, even when the extract was prepared from nonlysogenic strains. These data showed that the synthesis of toxin is inhibited by a *trans*-acting factor encoded on the *C. diphtheriae* chromosome. Isolation and mapping of mutants that were constitutive for toxin synthesis led to the model of toxin regulation in which a chromosomal gene encodes a repressor that acts at the operator sequence upstream of the phage toxin gene. Characterization of the bacterial gene, designated *dtxR*, confirmed that it encodes an iron-responsive repressor. The DtxR protein binds to the *tox* operator under conditions of high iron concentration and inhibits transcription of the *tox* gene (Fig. 2).

DtxR is a member of the IdeR (iron-dependent regulator) family of proteins found in gram-positive and acid-fast bacteria. DtxR is found predominantly in organisms with a high GC content. Members of the IdeR family regulate expression of a number of genes, including those encoding toxin and iron acquisition systems. In *C. diphtheriae*, DtxR regulates genes involved in the synthesis and transport of siderophores, including corynebactin, and genes encoding heme uptake and utilization proteins, as well as toxin. Additional genes regulated by DtxR have been identified based on the ability of their promoters to bind DtxR in vitro. The functions of some of these iron-regulated promoter (*irp*) genes have not been determined as yet.

DtxR is a 226-amino-acid polypeptide that forms homodimers. The protein dimerizes weakly in the absence of iron. In the presence of ferrous iron, the dimers are stabilized and bind to a specific operator sequence upstream of the

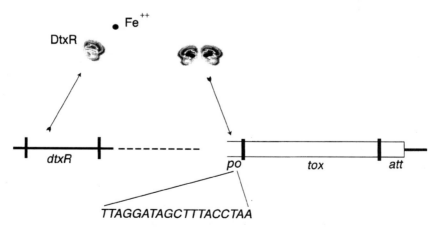

Figure 2 Regulation of diphtheria toxin gene (*tox*) expression. The line indicates bacterial DNA, and the open box indicates prophage sequences. The prophage is drawn in the opposite orientation so that the *tox* gene is shown 5′ to 3′. The sequence below the *tox* promoter and operator region is the DtxR-binding site.

coding sequence. The binding sequence in the *tox* operator (*toxO*) is a 9-bp palindrome interrupted by a single base. This sequence is conserved among the operators of the DtxR-regulated genes, and a comparison of a large number of these sequences led to the identification of a consensus DtxR binding site (Fig. 3). The DtxR-binding site overlaps the *tox* promoter sequence. Thus, binding of DtxR prevents RNA polymerase binding and transcription of the downstream gene.

The structure of DtxR has been determined by X-ray crystallography. Comparison of the structures of the protein in the presence and absence of the metal corepressor provided information about how the protein dimerizes. The protein has also been cocrystallized with target DNA to characterize the protein-DNA interaction. These data, along with results of mutational analyses, have revealed that DtxR has three domains: the DNA-binding domain (domain 1), the dimerization domain (domain 2), and a Src homology domain 3 (SH3)-like region (domain 3) that contacts both of the other two domains. The DNA-binding domain contains a helix-turn-helix region characteristic of DNA-binding proteins. The SH3-like domain has the metal-binding sites and binds two molecules of iron per monomer. The iron is chelated through Glu, His, and Cys residues at each metal-binding site. There is an anion-binding site associated with metal-binding site I, indicating that an anion, predicted to be phosphate, may function intracellularly as a corepressor. Binding of the anion appears to be essential for DtxR function.

In the presence of ferrous iron, the DNA-binding helices of domain 1 of DtxR are shifted 3 to 5 Å compared to the apo-repressor. This is likely a result of an alteration in the conformation of the flexible domain 3. The hinge-like motion that shifts the DNA-binding helices allows the DNA-binding site of DtxR to fit into the major groove of the DNA duplex at the operator site. Single mutations have been isolated in the SH3-like domain that change the conformation of DtxR such that iron is not required for DNA binding by the mutant DtxR proteins. The expression of DtxR-regulated genes in these mutants is repressed even in low-iron conditions.

Figure 3 Comparison of DtxR-binding sites of selected *C. diphtheriae* promoters. The sequences for DtxR binding to *tox, hmuO* (heme oxygenase), and *irp1* (an iron-regulated promoter) are shown for comparison. The consensus sequence derived from these and other iron-regulated promoters is shown below, with arrows indicating the inverted repeat. Bold type indicates the two bases shown to be critical for DtxR binding to the *tox* promoter.

tox	TT**A**GGATAGCTTTACC**T**AA
hmuO	TGAGGGGAACCTAACCTAA
irp1	TTAGGTTAGCCAAACCTTT
Consensus	TTAGGTTAGCCTAACCTAA

←—————— ——————→

Crystallographic analysis of DtxR bound to the *tox* operator, together with site-directed mutagenesis data, indicates that the operator sequence interacts with glutamine and arginine residues in domain 1 of DtxR. Two dimeric repressor proteins bind to opposite sides of the *tox* operator. DNA footprinting shows that DtxR protects a 27-bp sequence that includes the 9-bp palindrome (Fig. 3). Mutational analysis of the DtxR-binding site indicates that deletion of one of the 9-base inverted repeats eliminates binding. The T at the +7 and the A at the −7 position in the *toxO*-binding site (Fig. 3) are critical for DtxR binding, and changing these residues significantly reduces binding and regulation.

The model for *tox* regulation shown in Fig. 2 summarizes these data and illustrates the regulatory pathway.

SHIGA TOXINS

Shiga toxins are a family of related toxins that disrupt eukaryotic cell protein synthesis by deadenylation of the 28S ribosomal RNA at a specific site. These are AB-type toxins in which the B subunit binds to target cells and delivers the enzymatically active A1 fragment to the host cell cytoplasm. Shiga toxin was originally identified in *Shigella dysenteriae* type 1 and was shown in the 1940s to be subject to iron regulation. Poor yields of the toxin in vitro led to an analysis of environmental factors similar to that applied to diphtheria toxin synthesis. And, like for diphtheria toxin, iron was shown to be the component of media that reduced production of the *S. dysenteriae* Shiga toxin. Members of the Shiga toxin family, though related in their structure and function, differ in their genetics and regulation. They may be chromosomally or prophage encoded, and those that are on prophages are subject to regulation by phage promoters.

A wide range of serotypes of *E. coli*, including *E. coli* O157:H7, produce toxins of the Shiga toxin family. Stx1 (StxI), synthesized by a number of pathogenic *E. coli* strains, is essentially identical to *S. dysenteriae* Shiga toxin. Comparison of the deduced amino acid sequences indicates a single amino acid difference in the A subunits of the two proteins; the B subunits are identical. Stx2 (StxII) and its variants are less closely related to Shiga toxin but have the same mechanism of action. *E. coli* O157:H7 and other hemorrhagic *E. coli* strains produce one or both of the Shiga toxins, which are significant factors in the pathogenesis of disease caused by these organisms. A role for Shiga toxin in pathogenesis of shigellosis is less clear. Shiga toxin may contribute to the severity of disease caused by *S. dysenteriae* type 1, but the toxin is not required for *S. dysenteriae* pathogenesis and is not produced by other *Shigella* species.

The A and B subunits of both Stx1 and Stx2 are encoded by separate genes, *stxA* and *stxB*, with *stxB* immediately downstream of *stxA* (Fig. 4). Only 12 base pairs separate the A and B coding sequences, and there is no evidence for a separate promoter for the B subunit. However, the B subunit is synthesized in greater quantity than the A subunit. The 5:1 ratio of B/A subunits in the holotoxin may result from different efficiencies of ribosome binding or other forms of posttranscriptional control.

In Stx1- and Stx2-producing *E. coli*, lambdoid prophages encode the toxins (Fig. 4). The *stx* structural genes are located in the phage late gene region, downstream of the gene encoding the homolog of the phage lambda Q antiterminator.

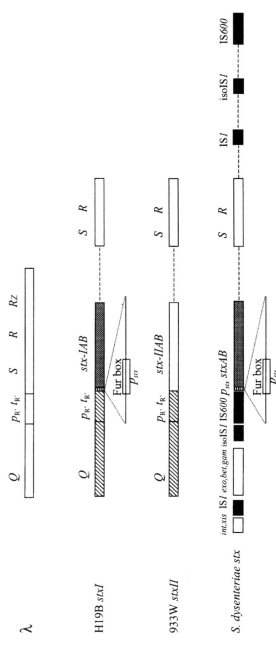

Figure 4 Organization of the Shiga toxin genes in *S. dysenteriae* and *E. coli*. Portions of the lambdoid phages encoding Stx1 (StxI; *stxI*) and Stx2 (StxII; *stxII*) are shown. The organization of the same region of λ is shown for comparison. The dashed lines indicate that the distances are not drawn to scale. Highly homologous genes are indicated by the same hatching. Solid boxes are insertion sequences.

Induction of the prophages leads to increased toxin synthesis. This increase is partially due to the increased number of copies of the phage genome as it replicates but is also due to enhanced transcription of the *stx* genes from the phage late promoter p_R'. In addition, *stx1* has its own promoter, and the toxin genes can be transcribed independently of phage induction. The gene arrangement for the *stx2* phages is similar to that of *stx1*, but *stx2* does not appear to have a promoter other than the phage late promoter. Downstream of the Stx1 and Stx2 toxin genes are prophage genes S and R encoding the holin and endolysin, respectively. It has been suggested that coupling of expression of the toxin and lysis genes may facilitate release of the toxin from the bacterial cell.

In *S. dysenteriae* type 1, Shiga toxin is a chromosomal gene and is not part of an intact prophage (Fig. 4). However, there are genes flanking the toxin genes that have homology to phage genes, suggesting that the toxin gene may have originally been acquired on a prophage with many of the phage genes having subsequently been lost. Upstream of *stxA* and *stxB* are genes with homology to lambda genes *exo, bet,* and *gam*, while the S and R genes downstream of the toxin genes are closely related to those found on the *stx* prophages. No sequences with homology to Q or p_R' are located in the *S. dysenteriae tox* region. The *S. dysenteriae tox* promoter is identical to the promoter immediately upstream of *E. coli stx1*. The *S. dysenteriae* toxin genes are flanked by copies of several insertion sequences, giving this region the appearance of a composite transposon. Thus, in both *E. coli* and *Shigella* species, the Shiga toxin genes are on mobile genetic elements. Movement of the *stx1* gene between strains of *E. coli* has been demonstrated both in vitro and in vivo, and tandem amplification of the *S. dysenteriae* region flanked by IS600 has been observed.

Although iron was known to repress Shiga toxin synthesis, this was not explored at the molecular level until studies on the regulation of iron transport systems in *E. coli* resulted in characterization of an iron-regulatory protein termed Fur. Like DtxR, Fur binds ferrous iron, and the iron-bound form binds DNA. Most commonly, Fur binding represses expression of genes, but positive regulatory effects have been reported, and posttranscriptional as well as transcriptional regulation has been attributed to Fur. The *E. coli* Fur protein is the prototype of a family of repressors found in a large number of gram-negative bacterial species and in some gram-positive bacteria as well. Fur regulates many genes related to iron acquisition, including genes for siderophore synthesis and for transport of siderophores and other iron-containing compounds. However, the Fur regulon encompasses a much wider range of genes than those directly related to iron uptake. Toxins, oxidative stress and acid shock response proteins, chemotaxis and swarming proteins, components of metabolic pathways, and Fur itself are regulated by iron and the Fur protein. Fur is more abundant in the cell than most regulatory proteins, which may reflect the large number of Fur-binding sites in the genome. Fur may also have other functions, such as binding excess iron in the cell to prevent iron toxicity. The multiple roles of Fur may help explain the complex regulation of *fur* expression. In addition to autoregulation, regulation of *fur* by the cAMP-binding protein CRP and by the oxidative stress response regulators has been shown.

Although Fur and DtxR are not homologous at the amino acid level, they have a number of features in common. The overall three-dimensional structures of the proteins are similar. Both form homodimers in the presence of iron and, upon binding of iron, undergo conformational changes that position the DNA-binding domain for contact with the DNA-binding site. Both Fur and DtxR bind to operators of iron-regulated genes to repress transcription of the downstream genes. However, the DNA-binding sequences that are recognized by Fur and DtxR are distinct, and the amino acid residues of the proteins that appear to be the contact points with the DNA are different. Neither protein can substitute for the other.

Fur is a 27-kDa protein that forms dimers through the carboxy-terminal domain. This domain also contains the metal-binding site, but dimerization does not require that iron be bound to the protein. Structural studies indicate that the protein binds one iron atom per Fur monomer, and the metal is coordinated with histidines and an aspartate or glutamate residue. Additionally, Fur has a zinc-binding site that appears to be required for Fur activity.

DNA binding by Fur occurs through the amino-terminal helix-turn-helix domain of the protein. The Fur consensus binding site, or Fur box, GATAATGA-TAATCATTATC (Fig. 5), has been determined by comparison of promoter sequences of Fur-regulated genes. Like DtxR, the binding site was originally proposed to be a region of dyad symmetry, a 9-bp inverted repeat separated by a single base pair. DNase I and hydroxyl radical footprinting of the sequences with purified Fur protein has confirmed that the region of DNA protected by the protein contains this repeat. However, the binding site may be more complex than a single inverted repeat. Additional analysis of natural and synthetic Fur-binding sites suggests that the binding site may actually consist of an array of smaller direct or inverted repeats based on a core NAT(A/T)AT repeat unit. In this model, the consensus Fur box is composed of three of these repeats (Fig. 5). The total number of repeats and their orientations can vary, but many of the natural promoters have two direct and one inverted repeat. The thymines in each

Figure 5 Comparison of the Fur boxes of iron-regulated promoters of *Shigella*, *E. coli*, and *P. aeruginosa*. The DNA sequence of the known or predicted Fur binding sites in the promoters for *stx1*, *S. dysenteriae* Shiga toxin gene, the aerobactin siderophore operon (*iuc*), *E. coli fur*, and the *P. aeruginosa pvdS* and *tonB* genes. The consensus binding site for Fur in gram-negative bacteria is shown. The longer arrows indicate the palindrome, and the shorter arrows and bold letters indicate the core repeat units.

stx1	GAATATGATTATGATTTTC
Shiga *tox*	GAATATGATTATGATTTTC
iuc	GATAATGAGAATCATTATT
fur	TATAATGATACGCATTATC
P.aeruginosa pvdS	GTAATTGACAATCATTATC
P.aeruginosa tonB	CTGAATGATAATAATTATC
Consensus	**GATAATGATAATCATTATC**

repeat are the primary sites of interaction with Fur, and the Fur protein appears to wrap helically around the DNA.

Putative Fur-binding sites have been identified in the promoters for the Shiga toxin gene in *S. dysenteriae* (Fig. 6) and for *stx1* in *E. coli*. The promoters for these genes are identical (Fig. 4 and 5), with a Fur box 108 to 129 bp upstream of the translation initiation codon. The Fur box sequence overlaps the proposed −10 sequence of the promoter. This Fur box sequence differs at several sites from the consensus sequence and lacks the core ATAAT units, suggesting that it may bind Fur with somewhat lower affinity than the canonical site.

Repression of *stx1* expression by Fur was demonstrated by measuring expression of a toxin-reporter gene fusion in wild-type and *fur* mutant strains of *E. coli*. The *fur* mutation resulted in loss of repression in high iron. These studies were done with the cloned toxin genes, and the construct lacked phage sequences that could influence regulation. Thus, the effect of iron and Fur, independent of potential prophage regulators, was determined in those studies. In other studies, in which toxin synthesis from Fur$^+$ and Fur$^-$ strains carrying intact prophages was measured, the results suggest more complex regulation. Toxin synthesis was not repressed by iron in the *fur* mutant, confirming Fur-mediated iron regulation, but the levels of toxin synthesis were lower under both high- and low-iron conditions in a *fur* mutant than in the parent strain. Fur may have indirect effects on toxin synthesis related to ability to influence expression of a large number of other genes within the cell.

In contrast to the *stx1* promoter, *stx2* does not have a Fur box sequence, and synthesis of this toxin is not sensitive to repression by iron. The DNA sequences upstream of the *stx1* and *stx2* open reading frames are different. The putative −10 and −35 sequences of *stx2* do not match the *stx1* promoter or the *E. coli* promoter consensus. If *stx2* has its own promoter, that promoter is not iron regulated. Stx2 synthesis, like Stx1 synthesis, is lower in a *fur* mutant than in the Fur$^+$ parent, suggesting an effect of Fur that does not involve direct binding of Fur on the *stx2* promoter.

Phage sequences upstream of the *E. coli stx* genes regulate production of both Stx1 and Stx2. Both *stx1* and *stx2* are located downstream of the phage

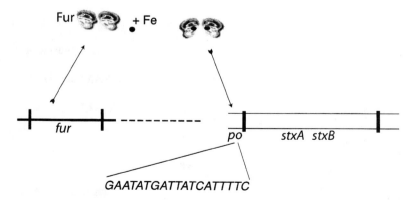

Figure 6 Regulation of Shiga toxin gene (*stxAB*) expression. The sequence below the *stx* promoter and operator region is the Fur-binding site or Fur box.

promoter p_R'. Q antiterminator acts at the DNA *qut* site that includes sequences in the p_R' promoter and allows expression of downstream genes. Production of significant amounts of Stx2 in an *E. coli* O157:H7 strain was found to be dependent upon transcription from p_R', and Stx1 synthesis is increased by expression from p_R'. Thus, toxin synthesis is linked to prophage induction and hence to expression of the downstream lysis genes. Prophage induction increases synthesis of the toxin, which is subsequently released from the bacterial cell upon lysis by the phage proteins. The *S. dysenteriae* Shiga toxin gene appears to have identical iron regulation as *stx1*. However, the *S. dysenteriae* chromosomal segment, while having some of the lambda phage characteristics, lacks the p_R' promoter, and thus lacks the transcription associated with phage induction.

PSEUDOMONAS EXOTOXIN A

Pseudomonas aeruginosa exotoxin A has the same activity as diphtheria toxin, inhibition of host protein synthesis by ADP-ribosylation of EF-2, and like diphtheria toxin, exotoxin A synthesis is repressed by iron. However, the toxin structures are different, and the mechanisms by which iron represses synthesis are also distinct.

Exotoxin A, like many *Pseudomonas* species virulence factors, is regulated by multiple environmental signals. In addition to iron, oxygen tension, temperature, and the concentration of certain nucleotides influence the amount of toxin synthesized. *P. aeruginosa* has a Fur protein with homology to *E. coli* Fur, and this was a likely candidate for the mediator of iron regulation. The *fur* gene was isolated by complementation of an *E. coli* mutant, indicating that there are functional similarities in the two proteins, and DNA sequence analysis indicated that the two proteins are approximately 50% identical. One area of significant difference in the structures of the *Pseudomonas* Fur protein and those of other species is the lack of the highly conserved cysteine motifs [Gly-X-Cys-(2-5)X-Cys and His-X-X-Cys-X-X-Cys] that may play a role in zinc binding by other Fur proteins.

The regulation of expression of *P. aeruginosa fur* also differs from *E. coli fur*. In *E. coli*, autoregulation has been observed; there is a Fur-binding site in the *fur* promoter, and Fur represses transcription of its own gene. The *P. aeruginosa fur* gene does not have a recognizable Fur-binding site, and the purified Fur protein does not bind to the *P. aeruginosa fur* promoter. The *P. aeruginosa fur* gene has two transcription start sites, P1 and P2. The proximal P2 promoter has a sigma-70 consensus sequence, whereas the distal P1 promoter does not have characteristics of known promoters. Synthesis of the T1 transcript from promoter P1 may require alternative sigma factors or other environmentally influenced regulators, but these have not been identified. The distal promoter appears to be important, however, because insertions between the two promoters significantly reduce *fur* expression.

Analysis of the effects of Fur on exotoxin synthesis was complicated by the inability to isolate *fur* null mutants. That these mutants cannot be isolated suggests that *fur* is an essential gene in *P. aeruginosa*. Point mutants that have altered Fur function can be isolated, however. Some of the *fur* point mutants of *P. aeruginosa* have the predicted Fur$^-$ phenotype of constitutive expression of siderophore synthesis and iron transport proteins. These *fur* mutants also produce high levels of exotoxin A in the presence of iron. This indicates that wild-type Fur is required for repression of toxin synthesis by iron. However, the exotoxin A (*toxA*) promoter has no homology with the consensus *Pseudomonas* Fur-binding site, and Fur does not bind to the *toxA* promoter in vitro. The effect of Fur on ToxA is indirect. *toxA* expression is controlled by a complex cascade of positive and negative regulators. A preliminary model that explains many of the features of *toxA* regulation has been developed (Fig. 7) and is discussed below.

The *toxA* promoter is directly regulated at the transcriptional level by the positive regulator RegA. RegA binds to the exotoxin A promoter and increases transcription. Little or no toxin synthesis is seen in the absence of RegA. High iron levels reduce the amount of RegA and therefore reduce toxin synthesis. The *regA* gene appears to be a part of an operon with a second, downstream open reading frame designated *regB*. However, there is no evidence that this encodes a protein required for toxin regulation.

During growth in low iron, accumulation of *toxA* and *regA* transcripts is biphasic. There is a peak of transcript accumulation early in the growth phase and a second peak during the late phase of growth. There are two *regA* promoters and two transcripts can be distinguished. The longer T1 transcript that initiates 164 bp upstream of the translation initiation codon is detected early and is weakly iron regulated. The shorter T2 transcript initiating at P2 is only detected in low-iron conditions and is made later in the growth phase. Although both P1

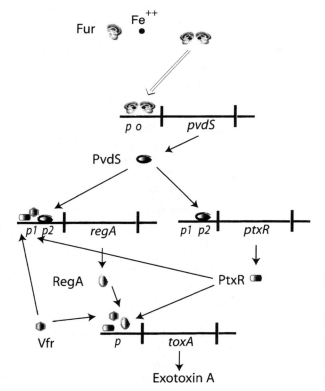

Figure 7 Model of iron regulation of *P. aeruginosa* exotoxin A synthesis. Genes and gene products that regulate expression of *toxA* are shown. Solid arrows indicate positive regulation by the indicated protein. Open arrow indicates negative regulation.

and P2 transcripts are regulated by iron and Fur, there are no apparent Fur-binding sites in the *regA* promoters to explain the regulation, suggesting that additional factors mediate the effect of Fur and iron on *regA* and *toxA* expression.

A second regulator of exotoxin A synthesis, PtxR, was also identified as having a role in iron regulation of toxin synthesis. PtxR is a member of the LysR family of transcriptional regulators and has the helix-turn-helix motif characteristic of this group of DNA-binding proteins. Increased amounts of PtxR lead to increased levels of exotoxin A production. The expression of *ptxR* mimics that of *regA*; there is an early phase of *ptxR* transcription from the constitutive P1 promoter and a late, iron-repressible phase of transcription from P2. Iron-regulated transcription from P2 is dependent on Fur in microaerobic but not aerobic conditions. Despite the presence of sequences with homology to the *P. aeruginosa* Fur-binding site in the *ptxR* promoter, binding of Fur to the region has not been detected.

Both *regA* and *ptxR* are positively regulated by PvdS. PvdS was first identified as a factor required for siderophore synthesis in *P. aeruginosa*. PvdS has homology to members of the ECF (extracytoplasmic factor) family of alternative sigma factors. This family includes proteins that regulate expression of siderophore genes in other pseudomonads and also *E. coli* FecI, which regulates expression of the ferric citrate receptor gene. The genes encoding members of the ECF family have Fur-binding sites in the promoters and are directly regulated by Fur.

PvdS not only has homology to the ECF family of iron-regulated sigma factors but also has functional and biochemical properties characteristic of sigma factors. It copurifies with RNA polymerase (RNAP) from *P. aeruginosa*, and isolated PvdS protein forms 1:1 complexes with core RNA polymerase. The PvdS-RNAP holoenzyme binds in vitro to PvdS-regulated promoters. The *pvdS* promoter contains a Fur-binding site, and expression of the gene is negatively regulated by Fur. In high iron, the Fur repressor binds upstream of *pvdS* and prevents expression of the sigma factor. In low iron, no repression occurs, and PvdS is synthesized and directly activates transcription of *regA* and *ptxR*. PvdS initiates transcription of the late, T2, transcripts of both *regA* and *ptxR*. The product of *ptxR* appears to further enhance transcription of *regA*, with RegA subsequently acting at *toxA* to induce synthesis of the toxin. Thus, it is through PvdS that Fur and iron regulate expression of *regA* and *ptxR* and ultimately *toxA* (Fig. 7).

This model does not explain all aspects of regulation of *toxA*. Complicating the picture of iron regulation of toxin synthesis is the observation that Fur mediates iron regulation of exotoxin A synthesis only under microaerobic conditions. In highly aerobic conditions, iron-regulated expression of toxin appears to be independent of Fur. It is difficult to show this unequivocally since *fur* null mutants of *P. aeruginosa* are not available. The *fur* point mutants may retain some activity under certain conditions. However, there may also be another unidentified regulator that is able to repress expression of *regA* and *ptxR* during aerobic growth in iron-rich media. Further, there are mediators of other environmental signals that may enhance, interfere with, or independently regulate exotoxin A synthesis. One of these, Vfr, has homology to the *E. coli* cAMP-binding protein CRP. Vfr enhances exotoxin A synthesis both directly and through *regA,* but the environmental signals to which Vfr responds are not yet known. Therefore, while much is understood of the molecular basis for regulation of toxin synthesis, the picture is incomplete. The regulation is complex and operates at multiple levels, allowing precise adjustments in expression in response to multiple signals.

Repression of toxin synthesis by iron is observed in a number of unrelated pathogens. In each case, an iron-binding repressor protein provides the link between elevated iron levels in the environment and repression of toxin synthesis. While the overall effect is the same, the details of each system are distinct. There is little homology among the iron-regulated promoters of these different groups of bacteria or among the protein repressors that mediate the regulation by iron. In the case of *P. aeruginosa*, the iron regulation is indirect, working through a cascade of positive regulators. Thus, a variety of mechanisms have evolved to use a common environmental signal, iron limitation, to control expression of virulence factors.

SUGGESTED READING

1. **Calderwood, S. B., and J. J. Mekalanos.** 1987. Iron regulation of Shiga-like toxin expression in *Escherichia coli* is mediated by the *fur* locus. *J. Bacteriol.* **169:**4759–4764.
2. **Escolar, L., J. Pérez-Martín, and V. de Lorenzo.** 1999. Opening the iron box: transcriptional metalloregulation by the Fur protein. *J. Bacteriol.* **181:**6223–6229.
3. **Frank, D. W., D. G. Storey, M. S. Hindahl, and B. H. Iglewski.** 1989. Differential regulation by iron of *regA* and *toxA* transcript accumulation in *Pseudomonas aeruginosa*. *J. Bacteriol.* **171:**5304–5313.
4. **Hantke, K.** 2001. Iron and metal regulation in bacteria. *Curr. Opin. Microbiol.* **4:**172–177.
5. **McDonough, M. A., and J. R. Butterton.** 1999. Spontaneous tandem amplification and deletion of the Shiga toxin operon in *Shigella dysenteriae* 1. *Mol. Microbiol.* **34:**1058–1069.
6. **Murphy, J. R., A. M. Pappenheimer, Jr., and S. Tayart de Borms.** 1974. Synthesis of diphtheria *tox*-gene products in *Escherichia coli* extracts. *Proc. Natl. Acad. Sci. USA* **71:**11–15.
7. **Pappenheimer, A. M., Jr., and S. J. Johnson.** 1936. Studies in diphtheria toxin production. I: the effect of iron and copper. *Br. J. Exp. Pathol.* **17:**335–341.
8. **Qiu, X., C. L. Verlinde, S. Zhang, M. P. Schmitt, R. K. Holmes, and W. G. Hol.** 1995. Three-dimensional structure of the diphtheria toxin repressor in complex with divalent cation co-repressors. *Structure* **3:**87–100.
9. **Sung, L. M., M. P. Jackson, A. D. O'Brien, and R. K. Holmes.** 1990. Transcription of the Shiga-like toxin type II variant operons of *Escherichia coli*. *J. Bacteriol.* **172:**6386–6395.
10. **Vasil, M. L., and U. A. Ochsner.** 1999. The response of *Pseudomonas aeruginosa* to iron: genetics, biochemistry and virulence. *Mol. Microbiol.* **34:**399–413.
11. **White, A., X. Ding, J. C. van der Spek, J. R. Murphy, and D. Ringe.** 1998. Structure of the metal-ion-activated diphtheria toxin repressor/tox operator complex. *Nature* **394:**502–506.

Bacterial Protein Toxins
D. Burns et al., Editors
©2003 ASM Press, Washington, DC 20036

Chapter 3

AraC Family Regulators and Transcriptional Control of Bacterial Virulence Determinants

Dara W. Frank and Meredith L. Hunt

The integration of environmental signals and activation or repression of the transcriptional machinery allow bacteria to respond appropriately for the niche in which they reside. Such adaptive responses can enhance growth and replication, optimize the ability to resist noxious substances or temperatures, and in some cases permit bacteria to evade the normal immune mechanisms of a host. Many of these regulatory events involve either transcriptional activation or repression of genes with related functions that are organized in chromosomal clusters called operons. As more information accumulates from nucleotide sequence analyses and annotation of bacterial genomes, regulatory molecules fall into related families such as the quorum sensing factors, two-component families, the LysR family, and the AraC/XylS family. This review will focus on selected members of the AraC/XylS family of regulatory proteins as a potential model system to study the induction of genes related to bacterial virulence in human infections.

DEFINITION OF AN AraC FAMILY MEMBER

The founding member of the AraC family was discovered as a positive regulator of the L-arabinose operon of *Escherichia coli* in 1971. Between 1971 and 1993 the family expanded to approximately 27 proteins, all containing a conserved stretch of 99 amino acids generally located in the carboxyl terminus of the protein. To define a profile of conserved sequences, Gallegos et al. (4) aligned the 27 known members and used this profile to search for new members in the PIR and SWISSPROT protein databases. Newer members were added to the profile and a consensus sequence was derived. The more refined profile was then used to search the translated products of predicted open reading frames within nucleic acid databases. The final profile can be accessed as PS01124 in the PROSITE database and defines over 100 proteins that are most often transcriptional acti-

Dara W. Frank and Meredith L. Hunt • Department of Microbiology and Molecular Genetics, Medical College of Wisconsin, 8701 Watertown Plank Rd., Milwaukee, WI 53226.

vators (4). Predicted and known functions of AraC family members include the control of diverse cellular functions such as stress response, sugar catabolism, and virulence factor production. As most of these functions represent unique targets for therapeutic intervention, information on the mechanisms of transcriptional activation may be useful in the design of drugs to prevent microbial growth in certain environments.

STRUCTURE AND FUNCTION ASPECTS OF THE FAMILY

On the basis of structure-function analyses, the members of the AraC family can generally be divided into two groups (9). One protein group controls sugar catabolism and associates as dimers. The dimerization domain and the ligand-binding domain are usually localized to an N-terminal domain. Ligand binding (such as arabinose) acts as a signal for the protein to change to an active conformation and induce transcription from operons that encode enzymes involved in substrate processing. The C-terminal domain contains two helix-turn-helix (HTH) motifs that comprise the highly homologous regions defining AraC membership. The primary DNA targets of this group are class II promoter regions in which the activator-binding site overlaps the −35 signal for RNA polymerase (RNAP) binding. The activation of transcription in these cases is often aided by the contribution of the cAMP receptor protein.

Members of the second subfamily are active as monomers, tend to be smaller proteins, and are generally involved in stress responses. They bind to asymmetric DNA sequences composed of two conserved motifs (A box and B box) separated by seven variable bases. MarA, encoded by *E. coli* and *Salmonella enterica* serovar Typhimurium, and Rob, encoded by *E. coli*, are examples of this subfamily whose structures have been solved. MarA is a relatively small protein (129 amino acids) possessing only a DNA-binding domain with two HTH motifs (14). Rob (289 amino acids), on the other hand, possesses an N-terminal domain with DNA-binding activity and a C-terminal domain with an unknown function. The C-terminal domain of Rob, however, is structurally analogous to GalT, an *E. coli*-encoded protein that converts galactose-1-phosphate and UDP-glucose to glucose-1-phosphate and UDP-galactose (8). Monomeric members of the AraC family can bind at sites located upstream of the −35 hexamer (class I promoters) and at sites that overlap the −35 region (class II promoters).

STRUCTURAL ANALYSIS OF THE AraC FAMILY

The C-Terminal Domain, DNA-Binding Activity

Although the AraC family of proteins has been characterized extensively by genetic approaches, information on tertiary structure is still hampered because of the low solubility of a majority of the members. Rhee et al. (14) reported the first AraC family structure by cocrystallizing MarA and its double-stranded 22-bp DNA target fragment (Fig. 1) (8) (see Color Plates following p. 256). In vivo MarA activates genes of the *mar* regulon, which mediates a multiple antibiotic resistance phenotype. The seven α-helices of MarA fold into two structurally similar N and C subdomains connected by helix-4. DNA recognition is mediated

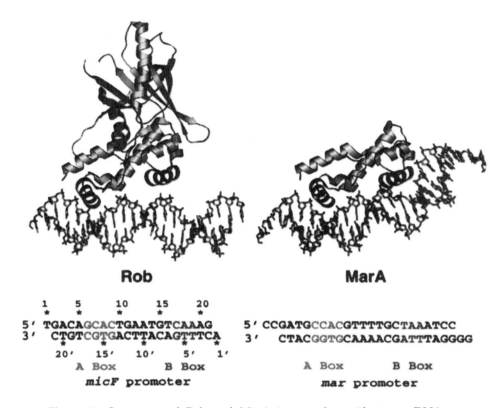

Figure 1 Structures of Rob and MarA in complex with target DNA sequences. The structurally similar N-terminal DNA binding domains are colored orange in the ribbon diagrams. The C-terminal domain unique to Rob is blue. MicF and MarA sequences used for cocrystallization are shown with A-box sequences in orange and B-box sequences in light blue type. Only the N-terminal HTH motif of Rob directly contacts bases of the binding sequence. MarA bends the DNA target so that both HTH motifs are located in adjacent major groove surfaces on one side of the DNA. From reference 8 with permission. (See Color Plates following p. 256.)

by a bipartite HTH motif consisting of residues 31 to 52 of helix-3 and residues 79 to 102 of helix-6, which protrude from the same face of the protein. The recognition helices insert into two adjacent segments of the major groove and cause an overall bend in the DNA of approximately 35° toward MarA (Fig. 1). Nucleotide bases that interact with the recognition helices of MarA, as determined by DNase I protection studies, match those in the cocrystal structure. MarA residues Trp-42 (N subdomain) and Gln-91 and Thr-95 (C subdomain) interact with bases C32, T7, and T8, respectively, via van der Waals contacts, which, with sequence-specific hydrogen bonds, are postulated to contribute to binding specificity.

The structural analysis of a second AraC family member, Rob, demonstrated some similarities to MarA but also some significant differences, suggesting that alternative modes of DNA binding may influence transcriptional activation relative to different promoter contexts (Fig. 1) (8). Rob is most closely related to SoxS, MarA, and TetD, proteins that mediate transcriptional activation of gene products protecting bacteria from oxidative damage, organic solvents, and antibiotics. Interestingly, Rob can activate many of the same genes as SoxS and MarA, such as the *micF* gene. Cocrystallization of *micF* DNA and Rob resulted in two independent complexes. In one complex Rob contacted the major groove of the DNA backbone over the consensus binding site. The other complex was approximately 5 to 6 bp away on the opposite face of the DNA and may be artifactual due to the high protein concentrations utilized for crystal growth. The DNA-binding domain of Rob is located at the N terminus, which contains two HTH motifs connected by a rigid central helix. Structural analyses suggest that only the N-terminal HTH contacts DNA in the major groove over the A-box sequence (Fig. 1) (8). Both van der Waals and hydrogen bond interactions appear to confer binding specificity. The C-terminal HTH of Rob engages the DNA backbone but lacks the major groove interactions with the B-box sequences seen with the C-terminal HTH of MarA (Fig. 1). Interaction with the DNA backbone may contribute to the binding affinity of Rob, which is similar to the affinity of MarA. The C-terminal domain of Rob is positioned on top of the N-terminal DNA-binding domain and is structurally homologous to GalT, suggesting that this region of the protein may bind some kind of small effector. A likely candidate could be an intermediate of galactose metabolism. Effector binding would presumably regulate the transcriptional activation activity of Rob. The interaction of potential ligands with a *micF*-Rob complex, however, has not been demonstrated (8).

The *micF*-Rob complex does not appear to bend DNA as was seen in the MarA structure. The absence of DNA bending may be due to the positioning of an acidic loop that connects β-strands 3 and 4 in the C-terminal domain of Rob. This loop is absent from MarA but is located close to Rob's DNA-binding surface. The loop may prevent DNA bending in the crystal structure. Alternatively, binding of an unknown ligand may swing the loop away from the DNA-binding surface and allow tighter DNA interactions.

How do the protein structures and DNA-binding activities relate to the biologic activity of AraC and other family members? AraC binds to two separate regions of a 17-bp consensus site that are analogous to the A- and B-box sites. In the p_{BAD} promoter, two binding sites, I_1 and I_2, are arranged in a direct repeat (see Fig. 2 for the relative positions of I_1, I_2, and p_{BAD}). The spacing and orientation of these sequences are important for transcriptional activation. On the basis of genetic analyses, the A box and B box of I_1 are engaged by the N- and C-terminal HTH motifs of AraC. The I_2 site, which is promoter proximal and lacks a good consensus B box, appears to be occupied by the C-terminal HTH motif. On the basis of structural data, Kwon et al. (8) proposed that one AraC protomer binds to the A box and B box of a bent *araI₁* DNA site, much like the binding observed between MarA and its cognate sequence. The second protomer specifically interacts with only the A box of the *araI₂* site, similar to the binding observed with Rob. This model would be consistent with chemical footprinting data showing little interaction of AraC with the B box of the I_2 site and fit the two types of recognition events observed in the structures solved to date. It may also imply

Figure 2 Regulatory region of the *araBAD* operon, the binding sites for AraC (I_1, I_2, O_1, and O_2), and promoters p_C and p_{BAD}. The domains of AraC consisting of the N-terminal arm, the dimerization domain, the linker domain, and the DNA-binding domain are marked. The light switch mechanism is illustrated. In the absence of arabinose the N-terminal arms interact with the DNA-binding domain and cause the protein to assume an extended conformation and preferential binding to distal I_1 and O_2 sites, forming a DNA loop and repressing transcription from p_{BAD} and p_C. Arabinose-binding changes the orientation of the arms and allows binding to adjacent I_1 and I_2 sites. AraC bound to the O_1 site in the presence of arabinose is shown in gray to illustrate partial occupancy. From reference 6 with permission.

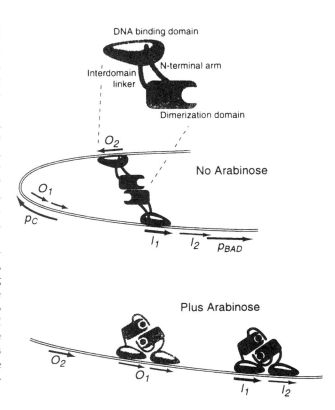

that differences in the binding of the second protomer could leave some protein surfaces open for interaction with RNAP subunits.

The N-Terminal Domain, Dimerization, and Ligand-Binding Activities

In the absence of arabinose, AraC binds widely separated half sites on the DNA (I_1 and O_2), causing an intervening 210-bp loop to form that represses transcription from the p_{BAD} and p_C promoters (Fig. 2). Arabinose binding causes a shift in the DNA-binding specificity of AraC to two adjacent half sites (I_1 and I_2), preventing the loop formation and enhancing transcription. AraC exists as a dimer in either the absence or presence of arabinose. The ability to form dimers is mediated by the N-terminal domain, which is composed of amino acids 1 to 170. High-resolution structures were obtained for the N-terminal domain with or without L-arabinose (15). In complex with arabinose, the dimerization domain of AraC is composed of an eight-stranded antiparallel β-barrel (β1 to β8) followed by a long linker containing two turns of 3_{10} helix and a ninth β-strand (β9) forming one sheet of the β-barrel. Two α-helices pack against the outer surface of the barrel. L-Arabinose binds within a pocket of the β-barrel lined with both polar and nonpolar side chains. One molecule of arabinose is enclosed by residues 7 to 18 of AraC, which loop around to close off the end of the β-barrel. Sugar binding is stabilized by hydrogen bonds from side chains and water molecules. In the crystal, two monomers associate by an antiparallel coiled coil formed between helix-2 of each monomer. Each end of the coiled coil possesses three leucine residues, Leu-150 and Leu-151 from one monomer and Leu-161 from the second monomer, that pack together in a "knobs into holes" motif. The central portion of the coiled-coil interface is stabilized by hydrogen bonds from both monomers.

Although the ligand-bound and unliganded AraC structures are superimposable (between residues 19 and 167), there are differences between molecules with and without arabinose (15). In the absence of L-arabinose AraC monomers have two potential dimerization interfaces. One of the interfaces is nearly the same coiled-coil interaction seen in liganded AraC crystals; however, the unliganded protein appears to have a distortion that reorients the leucine triad at the ends of the coiled coil. As a result, there is a slight rotation of one monomer relative to another. The second dimer of unliganded AraC buries a large surface area by bringing together two β-barrels in a face-to-face manner. In both cases the N-terminal arm consisting of residues 7 to 18 that closes the ligand-binding site in the presence of arabinose is disordered and unstructured. The side chain of Tyr-31 takes the place of arabinose and packs against the indole ring of Trp-95. This conformation is stabilized by hydrophobic interactions, hydrogen bond formation, and van der Waals interactions. When arabinose binds AraC, the β-barrel oligomerization surface is blocked due to the displacement of the tyrosine residue of the opposing monomer and the folding of the N-terminal arm leading to the antiparallel coiled-coil structure. It is notable that in the β-barrel dimer the DNA attachment points of the C-terminal binding DNA domain are 60 Å apart while in the coiled-coil dimer the distance is reduced to 37 Å. It is postulated that the reduced distance could make DNA looping unfavorable and increase affinity for the I_1 and I_2 sites. Although regulation of AraC activity by changes in the dimerization interface in the absence or presence of arabinose is an attractive model, genetic evidence and the construction of chimeric proteins or DNA targets strongly favor alternative models (the light switch model, see following sections).

LIGAND-INDEPENDENT EXPRESSION

Overproduction of some members of the AraC family of regulators (e.g., XylS) from strong promoters can lead to transcriptional activation of target promoters in the absence of the effector (4). These data suggest that an equilibrium may exist between active and inactive regulators and that overproduction may increase the total number of molecules and eventually overcome a threshold of induction. On the basis of structural data, changes in protein conformation upon ligand binding could contribute to the regulation of activity by modifying the number or quality of available dimerization surfaces. It could also be postulated that an increase in protein concentration in vivo may provide sufficient opportunities for the proteins to assume an active conformation. The activation of the AraC family regulators by their cognate ligand becomes an interesting variable when considering the number of AraC family proteins likely involved in the induction of virulence-related genes. What is the nature of the ligands involved in signaling the bacterium that it is within a specific host environmental niche? Are there tissue-specific ligands? Genetic and structural information as well as information regarding whether ligands exist is clearly needed to design potential inhibitory molecules that might be able to shut down particular pathways and prevent transcriptional activation of operons controlling toxins, adhesins, or other factors enhancing bacterial survival within host tissues.

LIGHT SWITCH MECHANISM FOR THE ACTION OF AraC

The modulation of DNA binding by small molecules specific for AraC family members is a key issue when considering whether disruption of the activation process could be a useful therapeutic strategy. In the absence of arabinose the AraC dimer prefers to bind to half sites that are separated by a DNA loop, resulting in transcriptional repression. In the presence of arabinose the protein's affinity for adjacent half sites increases approximately 50-fold, the loop is no longer formed, and transcription of the target promoter is enhanced. Structural data indicate that the binding of arabinose may trigger the use of a different dimerization surface that leads to the binding preference for adjacent sites. An important caveat to this hypothesis is that the structural data, derived from the crystallization of only the N-terminal domain, may not be an accurate depiction of the dynamics of the entire protein.

In subsequent studies, an N-terminal arm of 18 amino acid residues that extends from the dimerization domain was identified as being a key regulator of how AraC interacts with various DNA sequences (Fig. 2) (6). Recall that in the absence of arabinose this part of the molecule appears disordered, but in the presence of ligand, direct and water-mediated hydrogen bonds, stabilizes the sugar-binding pocket, and closes off a β-barrel. In the context of the entire molecule Harmer et al. proposed that the arms could interact with the DNA-binding domain to keep the molecule in a fairly rigid and extended configuration (6). Arabinose binding frees the arms from their interaction with the DNA-binding domain, allowing a preferential interaction with the dimerization domain. This model (Fig. 2) is referred to as the light switch mechanism because the arms switch expression of the system on and off. As predicted by the model, deletion of the N-terminal arms allows AraC to bind to adjacent half sites in the absence of arabinose (6). Mutations in the DNA-binding domain that weaken the interaction between the arm and the DNA-binding domain should result in a constitutive phenotype. Such mutants were isolated with a combination of alanine and glutamic acid scans of surface residues of the DNA-binding domain and a systematic examination of residues in the N-terminal arm that conferred constitutive activity. In addition, when the dimerization domains are eliminated and the DNA-binding domains are connected with a flexible linker of either 13 or 19 amino acids, AraC binds to adjacent half sites. Wild-type AraC also binds with equal affinity in the absence and presence of arabinose to half sites connected by a flexible single-stranded DNA linker. Combined, these data support the notion that AraC modulates its DNA-binding preference by changing the relative position of the DNA-binding domains. In the off position, the arms engage the

DNA-binding domains, which hold the protein in a state that prefers to bind to distal half sites, forming a DNA loop that represses transcription. The rigidity of the DNA also contributes to the binding preference. In the on position, the arms fold over the bound arabinose and the protein assumes a less rigid conformation, allowing the DNA-binding domains to interact with adjacent sites (6).

REGULATION OF VIRULENCE DETERMINANTS BY SELECTED MEMBERS OF THE AraC FAMILY

Although the studies on AraC remain the gold standard for understanding the mechanistic aspects of this protein family, information is beginning to accumulate on the diverse activities and complex regulatory mechanisms that AraC members and auxiliary proteins play in the control of bacterial replication in infected hosts. We have chosen to highlight three members of the AraC family, ExsA of *Pseudomonas aeruginosa*, UreR of the *Enterobacteriaceae*, and Rns of *E. coli*. ExsA is highly homologous to LcrF and VirF of *Yersinia* species. All three proteins are key factors in the transcriptional control of the type III secretion and translocation system where protein toxins are directly injected into eukaryotic cells. These systems appear not only to rely on AraC family transcriptional activators but also possess unique auxiliary factors that may provide important signaling information to the bacterium about whether it is in contact with host cells and whether the secretion system is functional. UreR is not only represented in the *E. coli* genome but has almost identical relatives encoded by *Proteus*, *Providencia*, and *Salmonella* species. UreR controls the expression of an operon involved in urease production and represents a virulence-related system where the ligand is known. The first steps of infection usually rely on adherence to host tissues. Rns of enterotoxigenic *E. coli* controls the expression of bacterial adherence organelles called pili or fimbriae and represents an example of an AraC family member in which activator-binding sites can be arranged in unusual configurations not commonly seen for prokaryotic activators.

ExsA of *P. aeruginosa*

ExsA was initially discovered as part of a regulatory operon involved in the production of an extracellular protein, exoenzyme S (3). Exoenzyme S possesses ADP-ribosyltransferase activity, and its expression is correlated with the ability of *P. aeruginosa* to cause a fatal sepsis in patients who suffer burn wound or acute lung infections. The C terminus of ExsA exhibits the conserved HTH motifs of the AraC family. This region of the molecule has 94% identity to the C-terminal 80 residues of VirF and LcrF of *Yersinia* species. VirF, LcrF, and ExsA enhance the transcription of several operons encoding components of type III secretion systems. Type III secretion refers to the ability of gram-negative bacteria to inject toxins directly into the cytosol of mammalian cells, bypassing the usual mechanisms of receptor-mediated endocytosis and subsequent toxin trafficking. Several toxins act on cellular cytoskeletal components to paralyze phagocytic activity or aid bacterial replication by being cytotoxic toward macrophage or other cells of the innate immune response. The regulatory networks not only control the expression of the components of the secretory apparatus but also control the expression of the toxins that will be injected by this specialized delivery mechanism. The operons controlled by ExsA are depicted in Fig. 3 (see Color Plates following p. 256). Toxins that are delivered by the type III apparatus (ExoS, ExoT, ExoY, and ExoU) are coordinately regulated by ExsA but are not linked to operons encoding the secretory and translocation proteins.

In the early stages of the study of ExsA-mediated transcriptional control, only the regulatory locus had been identified. In addition to ExsA, the sequenced chromosomal locus contained predicted open reading frames corresponding to ExsC, ExsB, and a partial open reading frame located just downstream of ExsA, ExsD (Fig. 3 and 4). Promoter fusion analysis with intergenic regions established that ExsCBA formed an operon and that ExsD began a second regulatory unit (17). Polar insertions in ExsC, -B, or -A disrupted transcriptional initiation from the two operons and the unlinked ExoS structural gene, indicating that ExsA regulated its own transcription as well as genes predicted to encode potential secretory functions (the locus beginning with ExsD) and the actual effector toxins (ExoS). As for most of the AraC family ExsA was notoriously difficult to purify in an active form, and the only protein that demonstrated specific binding activity was a maltose-binding protein fusion (7). DNase I protection analyses of known promoters, initially mapped by transcriptional fusion analysis, identified a consensus binding site for ExsA as TnAAAAnA. The arrangement of consensus elements showed three different patterns (Fig. 4). The p_C promoter (controlling the ExsCBA operon) possesses direct (p_{C1} and p_{C2}) repeats separated by 12 bp and an inverted arrangement separated by 17 bp (p_{C2} and p_{C3}). Convergent inverted repeat sites separated by 7 bp were identified in the p_D promoter region. The ExoS promoter region contained an inverted set of repeat consensus sequences that were 23 bp apart. The p_S promoter region was functionally dissected into two independent but coordinately regulated genes, each with a single ExsA consensus binding site (Fig. 4). Mutagenesis of the internal adenosine bases con-

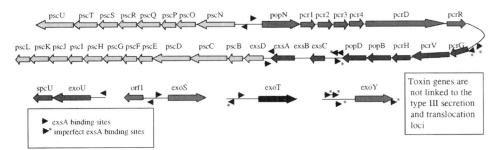

Figure 3 The operons of *Pseudomonas aeruginosa* type III secretion system and effector toxins ExoU, ExoS, ExoT, and ExoY are shown. Individual gene products are indicated as colored arrows (the sizes of protein products are approximate and not to scale). Different colors are used to identify operons based on promoter mapping, transcriptional fusion, or DNA-binding analyses. The toxins and known chaperone proteins are located in different chromosomal locations than the set of operons that encode the secretory, translocation, and regulatory proteins. ExsA consensus and imperfect binding sites are illustrated. (See Color Plates following p. 256.)

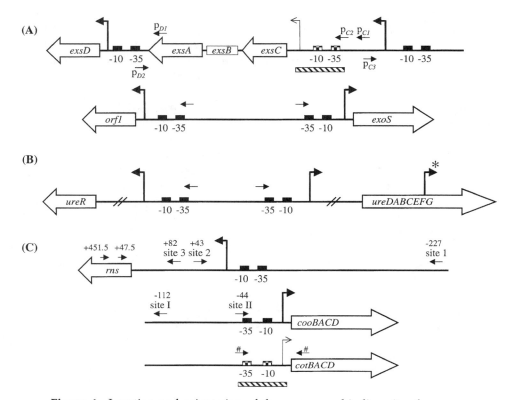

Figure 4 Location and orientation of the consensus binding sites for members of the AraC protein family: (A) ExsA, (B) UreR, and (C) Rns. Open reading frames are represented by open arrows. Bent arrows illustrate experimentally defined transcriptional start sites, and filled boxes indicate the sigma-70 promoter hexamers. A striped bar indicates a putative promoter region identified by sequence homology. An asterisk located within the *ureDABCEFG* operon is used to identify a UreR + urea-dependent start site present only in the plasmid-encoded locus. Positions of the Rns-binding sites 1, 2, and 3 and sites I and II are numbered relative to the transcriptional start sites. The two binding sites within *rns* are numbered relative to the translational start codon. Putative Rns-binding sites that overlap with the −35 hexamer or are within the *cotB* coding region are marked with a # sign. Information used to assemble this summary figure was derived from references 7, 10, 11, and 16. Figure is not to scale.

firmed the importance of this site for transcriptional initiation. Thus, genes that are coordinately regulated with the type III secretion and translocation system possess one or more ExsA consensus elements centered approximately 17 bp upstream of the −35 hexamer of the RNAP-binding site. This pattern holds true for most promoter regions known to encode type III-related proteins, except for the regulatory operon encoding ExsCBA (p_C). Reexamination of sequences encompassing p_C suggests that one promoter likely responds to ExsA-mediated induction but a second and uncharacterized promoter may mediate transcription in response to different environmental cues (Fig. 4).

Regulation of the type III secretory system in *Pseudomonas aeruginosa* incorporates an AraC member as only one component. Other important aspects include the potential ligands that might be interacting with the nonconserved N-terminal region of ExsA. Stimuli that induce the transcription of the *Pseudomonas* type III system include a calcium-poor environment and contact with serum components or eukaryotic cells. Likely these signals are not all working via the same mechanisms and may or may not be directly related to the activity of ExsA. ExsB, reported to be a potential open reading frame with homology to an expressed protein in *Yersinia* species, is transcribed as part of the regulatory operon but is not expressed (5). Mutations of potential start codons for ExsB have no effect on the expression of type III proteins in *P. aeruginosa*. On the other hand, deletion of *exsB* prevents expression of ExoS or other type III proteins. The RNA region of *exsB* appears to be required for *exsA* translation, and a postulated RNA structure involving *exsB* and perhaps part of *exsC* may regulate the amount of ExsA produced. ExsC has no homolog in other type III systems sequenced to date. Complementation studies indicate that ExsC may play a role in the translation of ExoS as stop mutations in *exsC* decrease the final yield of this toxin. Attempts to delete *exsC* resulted in DNA rearrangements of the regulatory locus, indicating that *exsB* and *exsA* might be a deleterious combination in *Pseudomonas aeruginosa*. Deletion of both *exsC* and *exsB*, however, is stable and allows the complementation of *exsA*-specific insertional mutants. This result indicates that the translational regulation of ExsA expression can be overcome.

In summary, the model for the regulation of the *Pseudomonas* type III pathway is incomplete but has some tantalizing hints as to how an AraC member, ExsA, may induce key operons that assemble an apparatus capable of injecting at least four proteins with toxic activity into eukaryotic cells. The input signals that may activate ExsA binding are unclear and difficult to interpret without further studies on the structural features of this protein. There is no information on whether ExsA dimerizes or whether the protein changes in activity or DNA recognition site in response to ligand binding. Other information suggests that auxiliary proteins (ExsC) and RNA regions (*exsB*) may influence the level of ExsA expression in vivo. As in *Yersinia* species, there is evidence that the absence of a functional secretory apparatus negatively affects the transcription of all the operons. A firm regulatory connection between the secretory apparatus and ExsA activity, however, has not been made.

UreR Regulation of Urease

The activation of particular members of the AraC family by certain effector ligands is thought to be mediated by interaction with the nonconserved domain (usually at the N terminus) of the activator. To date, the identity of very few ligands that interact with virulence gene activators of the AraC family is known. One example of a known ligand/virulence gene activator is the urea/UreR system.

Many pathogens produce the virulence factor urease, a multisubunit enzyme that catalyzes the breakdown of urea to carbon dioxide and ammonium. The

hydrolysis of urea to ammonium benefits bacteria by providing a readily ab-
sorbed nitrogen source and by increasing the pH of the local environment. The
ability to thrive even in tissue where an acute inflammatory response is ongoing
has definite advantages. Infections with bacteria producing urease often result in
cystitis and acute pyelonephritis and can progress to bacteremia. The rise in pH
mediated by high urease expression also causes the formation of urinary stones.

The urease locus consists of seven genes (*ureDABCEFG*) that encode structural and accessory proteins, required to form the active urease enzyme, and the
divergently transcribed transcriptional activator of this gene cluster, UreR. The
locus is present as a single copy on the chromosome of *Proteus mirabilis* and is
also carried on large plasmids in *E. coli*, *Salmonella* species, and *Providencia stuar-
tii*. The plasmid-encoded loci carried by these pathogens are almost identical and
as such have been termed the *Enterobacteriaceae* plasmid-encoded urease locus.
Additionally, UreR from *Proteus mirabilis* and the plasmid-encoded urease loci
are functionally interchangeable (2). Urea is the ligand that directly activates
UreR. Two UreR + urea-dependent promoters, p_{ureR} and p_{ureD}, were identified in
Proteus mirabilis. In addition to these two promoters, a third UreR + urea-
dependent promoter, p_{ureG}, was identified in the plasmid-encoded urease locus
(1) (Fig. 4).

UreR domain structure
On the basis of protein homology, UreR consists of a nonconserved N-terminal
domain, a conserved C-terminal DNA-binding domain, and a short linker connecting these functional regions. Construction of chimeric proteins demonstrated
that the C-terminal (164 to 293) UreR domain activated transcription from the
p_{ureD} and bound DNA fragments by gel mobility shift analysis (13). Chimeric
proteins encoding N-terminal UreR amino acids 1 to 182 function as dimerization
domains for the repressor LexA. Although UreR contains three leucine residues
(147, 148, and 158) that model to the AraC leucines (150, 151, and 161), mutational analysis of the individual residues did not affect transcriptional activity.
As in AraC, mutagenesis of multiple residues (147 and 148 or all three), however,
resulted in a significant reduction in activity. The reduction of activity was not
associated with the loss of dimer formation as both wild-type and triple leucine-
to-alanine substitutions eluted as dimeric proteins during gel filtration chromatography (13).

The *ureR, ureD* intergenic region
UreR is divergently transcribed from the *ureDABCEFG* operon. The intergenic
region between *ureR* and *ureD* in *Proteus mirabilis* and the *Enterobacteriaceae*
plasmid-encoded loci is composed of either 492 or 414 bp, respectively (Fig. 4).
Two UreR-binding sites were identified in this region, one overlapping the consensus −35 region of the *ureR* promoter (p_{ureR}) and one overlapping the poorly
conserved −35 region of p_{ureD}. Comparison of bases protected in DNase I footprinting experiments with a recombinant histidine-tagged UreR protein preparation (rUreR) resulted in the derivation of the consensus sequence, T(A/G)(T/
C)(A/T)(T/G)(C/T)T(A/T)(T/A)ATTG. rUreR binds to each site in p_{ureR} and p_{ureD}
in the absence or presence of urea as measured in vitro by electrophoretic mobility shift assays (16). Thomas and Collins (16) proposed that in the absence of

urea, RNAP binds to the -35 site of p_{ureR} and transcribes *ureR*, providing a low constitutive intracellular concentration of the activator. However, RNAP cannot make the same productive contact with the p_{ureD} promoter without rUreR + urea due to the lack of a consensus -35 element upstream of *ureD*. Measurements of the strength of rUreR interactions at both promoters clearly indicate an increase in affinity for rUreR to its binding sites in the presence of urea, and, importantly, the strongest affinity in the presence of urea was found to p_{ureD}. Thus, when urea is present, rUreR binding is stabilized and it effectively recruits RNAP to p_{ureD} to initiate transcription of the urease operon. UreR activates transcription from p_{ureR} by approximately 20-fold in the presence of urea (16).

The binding of rUreR to p_{ureG} in the plasmid-encoded locus could not be demonstrated using the experimental conditions conducive for rUreR binding to other sites (Fig. 4) (16). The authors noted that the upstream region of *ureG* contained two sites that only partially matched an intact UreR consensus site. Further investigation is required to determine if these sites play a role in UreR-mediated activation of *ureG*.

Ligand binding

Urea does not directly bind to the C-terminal DNA-binding domain or contact the DNA target site, suggesting that the mechanism of transcriptional activation involves urea binding to the N-terminal domain of UreR. This step is postulated to result in a conformational change in the protein such that a tighter interaction between the HTH domains and the DNA-binding site occurs. Although direct evidence of urea binding to UreR is still lacking, histidine mutations in the dimerization domain (His-102) or in the linker region between the two domains (His-175) result in constitutive UreR activation independent of urea (13). The authors speculate that His-102 likely contributes to urea binding whereas His-175 could affect the overall conformation of UreR, resulting in specific and high-affinity DNA binding. Thus, it is not a change in binding site specificity but a change in binding site affinity that is mediated by the ligand urea.

Rns Regulation of Adhesion Factors

For many bacterial pathogens the expression of adherence factors is a critical initial step to permit colonization of the host to establish an infection. Enterotoxigenic *E. coli* expresses CS1 and CS2 pili, encoded by the operons *cooBACD* and *cotBACD*, respectively, that are required for colonization of the gastrointestinal tract, leading to severe diarrheal disease. The plasmid-encoded Rns activates itself and the expression of these virulence factors. It is not known if an effector ligand is involved in Rns activation. Work to date on this AraC family member demonstrates that Rns utilizes an unusual arrangement of activator binding sites to mediate transcription from its own promoter but has a more conventional arrangement of sites for the activation of pilin promoters (Fig. 4).

The similarity between the 31-kDa Rns and other AraC activators is confined to the C-terminal conserved DNA-binding domain containing two HTH motifs. Both HTH motifs are predicted to contact DNA within two consecutive major grooves on the same face of DNA. The nonconserved N-terminal region of Rns has not been extensively studied. Thus, the contribution of this region to ligand binding or dimerization is unknown.

Rns-binding sites; p_{coo} and p_{cot}

Due to the inherent insoluble nature of many AraC proteins upon purification, an N-terminal maltose-binding protein fusion to Rns was used to provide a stable, soluble, and active form of Rns to identify DNA target sequences (11). With this experimental approach, two Rns-binding sites were identified, separated by 31 bp in the p_{coo} promoter. One site was centered at −112 bp relative to the transcriptional start site (site I) and the other site centered at −44 bp, which overlaps the −35 hexamer (site II) (Fig. 4). Three specific thymines in both p_{coo} sites I and II were identified as critical for binding by uracil interference assays (11). In a uracil interference assay, uracil, which lacks a C5-methyl group, is substituted for thymine. The C5-methyl groups are required for hydrophobic interactions with Rns. Comparison of the spatial arrangement of these three thymines at both sites demonstrated the pattern 3'-TAT-six nucleotides-T-5' where the third thymine is found on the opposite strand. This arrangement indicated Rns, like other AraC regulators, bound to the major groove of DNA with two thymines in one region and the third in an adjacent region of the major groove. Mutational analysis of sites I and II indicates that although both binding sites are required for Rns-mediated activation of the CS1 pilin locus, site II is most important. Most likely this result is due to a need for interaction between Rns and RNAP at this site, as has been documented for other AraC activators that bind close to or overlapping the −35 hexamer. Aligning the thymine triads of p_{coo} sites I and II identified an asymmetric consensus sequence for Rns binding. Munson and Scott (11) used this sequence to search for similar sites upstream of other Rns-activated virulence genes. A single Rns-binding site with the same orientation and distance from the open reading frame as site II of p_{coo} was identified overlapping the putative −35 hexamer of p_{cot} (Fig. 4). This analysis indicated that activation of Rns-regulated genes involves binding proximal to the −35 hexamer (11). However, Rns does not bind to this region in its own promoter.

Autoregulation of Rns expression

Rns regulates its own expression. DNase I footprinting identified three Rns-binding sites upstream of the *rns* coding region organized in a configuration rarely used by transcriptional activators (12). Site 1 was located upstream from the transcription start point centered at bp −227 and distal to the *rns* promoter region (Fig. 4). Sites 2 and 3 were found downstream of the transcriptional start point centered at bp +43 and +82, respectively (Fig. 4). Uracil interference assays of the p_{rns} site 3 identified three conserved thymines. Comparisons of p_{rns} sites 1, 2, and 3 and p_{coo} sites I and II indicated that all three thymines were conserved and enabled a refinement of the Rns-binding site to a 16-bp consensus, (T/A)(C/T/G)**A**(T/A)(T/A)(A/T)(A/T)(A/C/T)(A/T)**TAT**(C/T)(T/A/G)N(T/A/C) (10). By using this consensus sequence to search for potential binding sites in the *rns* vicinity, two additional areas were identified within the *rns* coding region centered at +47.5 and +451.5 relative to the start codon (Fig. 4). Rns binding has only been demonstrated to the +47.5 site; binding at the +451.5 has not been studied (10).

From mutational analysis of these Rns-binding sites and DNase I footprinting, sites 1 and 3 were found to be absolutely required for Rns-mediated autoregulation (12). The role of site 2 is unclear. Although not an absolute requirement for activation of *rns*, some evidence suggests Rns binding to site 2 may cooperatively increase the binding of Rns to site 3. The Rns-binding sites identified within the coding region of the protein do not appear to be involved in autoregulation. Rns, despite binding at sites distal to the p_{rns} −35 region, functions to promote the formation of a single open complex by RNAP. In prokaryotes, binding sites located downstream of the promoter are commonly used by repressors. Thus, the identification of Rns-binding sites in this "repressor zone" and the fact that Rns must bind to site 3 in this region for complete activation represent a new arrangement not recognized previously for transcriptional activators (12).

Identification of other Rns-like regulators with similar binding site and recognition sequences

Using promoter-*lacZ* reporter fusions, Scott's group found that Rns and a number of other AraC virulence gene activators were functionally interchangeable. VirF, an activator of invasion genes in *Shigella flexneri*, and two activators of pilin synthesis, AggR (enteroaggregative *E. coli*) and CfaR (enterotoxigenic *E. coli*), were found to mediate transcription from p_{rns}. Transcriptional activation required the downstream site 3. This observation suggested that AggR and CfaR could recognize Rns-like binding sites. Indeed, the HTH motifs of these regulators are significantly conserved, indicating that the DNA recognition sequences may be similar. A search for Rns-binding sites surrounding these activators and their cognate operons identified many potential binding sites (10). Various arrangements of Rns-binding sites were seen upstream and within coding regions of the activators and the operons they regulate. An arrangement of sites nearly identical to the Rns-binding sites at p_{rns} was found surrounding genes encoding CfaR and CsvR, AraC activators of CS4 pilin production in enterotoxigenic *E. coli*. These results indicate that the unusual arrangement and use of binding sites by Rns for transcriptional activation from p_{rns} are not unique. The authors propose that Rns is the prototype of a group of activators within the AraC family that are not limited to binding sites upstream of the transcriptional start site in order to induce transcription (10).

SUMMARY

The AraC proteins represent a diverse family that controls mRNA expression from a number of bacterial genes. Although initially grouped together because of the conservation of HTH DNA-binding motifs, there seems to be a wide variety of mechanisms for enhancing transcriptional initiation and DNA recognition. Structural analysis has been slowed because of the general aggregation properties of these proteins, but new information is accumulating through the recognition of different functional domains, the construction of chimeric proteins, and the use of genetic approaches. The AraC family includes many transcriptional activators that enhance the production of bacterial virulence determinants. Determining whether these activators respond to environmental stimuli or ligands present in host tissues may lead to the development of a new generation of therapeutics aimed at diminishing the survival advantage of pathogenic bacteria in vivo.

REFERENCES

1. **D'Orazio, S. E. F., and C. M. Collins.** 1995. UreR activates transcription at multiple promoters within the plasmid-encoded urease locus of the *Enterobacteriaceae. Mol. Microbiol.* **16:**145–155.

2. **D'Orazio, S. E. F., V. Thomas, and C. M. Collins.** 1996. Activation of transcription at divergent urea-dependent promoters by the urease gene regulator UreR. *Mol. Microbiol.* **21:**643–655.

3. **Frank, D. W.** 1997. The exoenzyme S regulon of *Pseudomonas aeruginosa. Mol. Microbiol.* **26:**621–629.

4. **Gallegos, M.-T., R. Schleif, A. Bairoch, K. Hofmann, and J. L. Ramos.** 1997. AraC/XylS family of transcriptional regulators. *Microbiol. Mol. Biol. Rev.* **61:**393–410.

5. **Goranson, J., A. K. Hovey, and D. W. Frank.** 1997. Functional analysis of *exsC* and *exsB* in the regulation of exoenzyme S production from *Pseudomonas aeruginosa. J. Bacteriol.* **179:**1646–1654.

6. **Harmer, T., M. Wu, and R. Schleif.** 2001. The role of rigidity in DNA looping-unlooping by AraC. *Proc. Natl. Acad. Sci. USA* **98:**427–431.

7. **Hovey, A. K., and D. W. Frank.** 1995. Analyses of the DNA-binding and transcriptional activation properties of ExsA, the transcriptional activator of the *Pseudomonas aeruginosa* exoenzyme S regulon. *J. Bacteriol.* **177:**4427–4436.

8. **Kwon, H. J., M. H. J. Bennik, B. Demple, and T. Ellenberger.** 2000. Crystal structure of the *Escherichia coli* Rob transcription factor in complex with DNA. *Nat. Struct. Biol.* **7:**424–430.

9. **Martin, R. G., and J. L. Rosner.** 2001. The AraC transcriptional activators. *Curr. Opin. Microbiol.* **4:**132–137.

10. **Munson, G. P., L. G. Holcomb, and J. R. Scott.** 2001. Novel group of virulence activators within the AraC family that are not restricted to upstream binding sites. *Infect. Immun.* **69:**186–193.

11. **Munson, G. P., and J. R. Scott.** 1999. Binding site recognition by Rns, a virulence regulator in the AraC family. *J. Bacteriol.* **181:**2110–2117.

12. **Munson, G. P., and J. R. Scott.** 2000. Rns, a virulence regulator within the AraC family, requires binding sites upstream and downstream of its own promoter to function as an activator. *Mol. Microbiol.* **36:**1391–1402.

13. **Poore, C. A., C. Coker, J. D. Dattelbaum, and H. L. T. Mobley.** 2001. Identification of the domains of UreR, an AraC-like transcriptional regulator of the urease cluster in *Proteus mirabilis. J. Bacteriol.* **183:**4526–4535.

14. **Rhee, S., R. G. Martin, J. L. Rosner, and D. R. Davies.** 1998. A novel DNA-binding motif in MarA: the first structure for an AraC family transcriptional activator. *Proc. Natl. Acad. Sci. USA* **95:**10413–10418.

15. **Soisson, S. M., B. MacDougall-Shackleton, R. Schleif, and C. Wolberger.** 1997. Structural basis for ligand-regulated oligomerization of AraC. *Science* **276:**421–425.

16. **Thomas, V. J., and C. M. Collins.** 1999. Identification of UreR binding sites in the *Enterobacteriaceae* plasmid-encoded and *Proteus mirabilis* urease gene operons. *Mol. Microbiol.* **31:**1417–1428.

17. **Yahr, T., and D. W. Frank.** 1994. Transcriptional organization of the *trans*-regulatory locus which controls exoenzyme S synthesis in *Pseudomonas aeruginosa. J. Bacteriol.* **176:**3832–3838.

Bacterial Protein Toxins
D. Burns et al., Editors
©2003 ASM Press, Washington, DC 20036

Chapter 4

Quorum Sensing

Everett C. Pesci and Barbara H. Iglewski

Although bacteria are unicellular organisms, many species have developed a mechanism to communicate with each other. Quorum sensing, or cell-to-cell signaling, allows a population of bacteria to act as a group rather than as separate individuals. This method of genetic regulation includes multiple clever systems that provide bacteria with the ability to regulate specific genes in a cell density-dependent manner. In gram-positive bacteria, the most well-studied cell-to-cell signal systems involve peptide pheromone signals that control a variety of different cellular functions, including the production of toxins associated with virulence. Alternatively, many gram-negative organisms have been found to contain an acyl-homoserine lactone signaling system that allows temporal control of many genes. This chapter will focus on quorum sensing in gram-negative bacteria with a special emphasis on the well-studied intercellular communication network found in *Pseudomonas aeruginosa*.

ACYL-HOMOSERINE LACTONE-BASED QUORUM SENSING

Many gram-negative bacteria have the means to communicate intercellularly. This ability was first discovered in the marine symbiont *Vibrio fischeri* (see references 5 and 6 for complete reviews of *V. fischeri* quorum sensing). This bacterium lives either freely in the marine environment or as part of a homogeneous population within the light organ of some marine fishes. As a free-living organism *V. fischeri* remains at a very dilute cell density and does not luminesce. However, as an endosymbiont in the light organ of a fish, this bacterium can achieve a very high cell density. When this happens, the *V. fischeri* population luminesces en masse to produce a visible light. The cause of this curious phenomenon began to appear when it was reported that an extracellular factor, or an "autoinducer," was responsible for regulating the population-dependent light production by *V. fischeri*. Several years later, the autoinducer produced by *V. fischeri* was identified

Everett C. Pesci • Department of Microbiology and Immunology, Brody School of Medicine at East Carolina University, Greenville, NC 27858. *Barbara H. Iglewski* • Department of Microbiology and Immunology, University of Rochester School of Medicine and Dentistry, Rochester, NY 14642.

as *N*-3-(oxohexanoyl)homoserine lactone (3-oxo-C$_6$-HSL). Shortly after this, the luminescence *lux* operon of *V. fischeri* was cloned, which led to the identification of the transcriptional activator protein LuxR and the autoinducer synthase LuxI. The LuxR and LuxI proteins were unique at the time of their discovery and are considered the prototypic members of their respective families. LuxR is a 250-amino-acid protein that becomes activated in the presence of 3-oxo-C$_6$-HSL. Upon activation by 3-oxo-C$_6$-HSL, LuxR induces the transcription of the genes required for luminescence. Members of the LuxR family have two highly conserved regions. The amino two-thirds of LuxR form an autoinducer-binding domain, and the carboxyl third of the protein contains a helix-turn-helix motif that serves as a DNA-binding site. The DNA-binding site recognizes a 20-base-pair DNA sequence with dyad symmetry centered just upstream from the -35 region of the promoter. This element is referred to as a *lux* box and has been found near the promoters of many different quorum sensing-controlled genes. The autoinducer synthase LuxI is a 193-amino-acid protein that is sufficient for 3-oxo-C$_6$-HSL production when expressed in *Escherichia coli*. It has been shown that the enzymatic action of LuxI catalyzes the formation of the autoinducer from *S*-adenosylmethionine and an acyl chain (from an acyl carrier protein). The 3-oxo-C$_6$-HSL signal has been shown to freely diffuse across the bacterial cell membrane to allow for an internal-external concentration equilibrium.

The general model for quorum sensing control of specific genes is diagrammed in Fig. 1. The quorum sensing mechanism starts with both the autoinducer and the LuxR-type protein being produced at a basal level while the bacteria are at a low cell density. As the bacterial population grows within a confined area (such as a light organ or a macrocolony), the cell density continues to increase. When this happens, more individual cells are present to produce autoinducer and the concentration of the signal increases. Upon attainment of a threshold level of autoinducer, the autoinducer will bind to and thereby activate a LuxR-type protein. Once activated, the LuxR-type protein will bind to a specific DNA sequence within the promoter region of quorum sensing-controlled genes. These genes will then be transcribed as part of the bacterial population's response to its own increased cell density. It is interesting to note that quorum sensing systems usually regulate themselves. This occurs because one of the genes that is often induced by quorum sensing is the gene encoding the LuxI-type autoinducer synthase. Because of this, the increase in autoinducer concentration causes the production of more autoinducer to generate a positive feedback loop.

Since the identification of 3-oxo-C$_6$-HSL and LuxR, numerous gram-negative bacteria have been found to contain similar quorum sensing systems. At least 22 gram-negative species have been shown to utilize an acyl-homoserine lactone-based quorum sensing system to control various genes (4), and more than 50 different species have been shown to produce an acyl-homoserine lactone type of cell-to-cell signal (5). One of the more well-studied pathogens in this group is *P. aeruginosa*, which contains two separate quorum sensing systems.

P. AERUGINOSA QUORUM SENSING

The *las* Quorum Sensing System

The study of quorum sensing in *P. aeruginosa* began when it was discovered that the production of the virulence factor elastase was controlled by LasR, a homolog

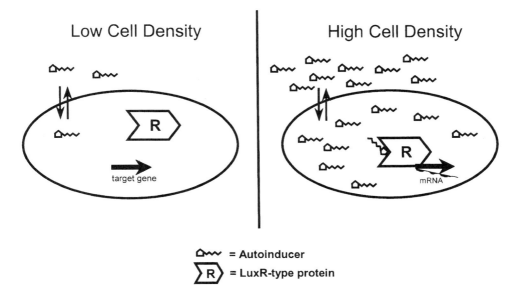

Figure 1 General model for gram-negative quorum sensing. On the left side of the figure, a cell from a dilute bacterial population is shown. Both the autoinducer and the R-protein are produced at a basal level. The autoinducer freely diffuses across the cell envelope to reach a concentration equilibrium between the internal and external environments. If a population of bacteria is growing within a defined space, such as a light organ of a fish or a macro-colony in a host lung, then the autoinducer concentration will increase with cell density. This situation is depicted on the right side of the figure. When the autoinducer reaches a threshold level, it will interact with the R-protein. When this occurs, the R-protein/autoinducer complex will then bind to and thereby activate specific gene promoters.

of LuxR (see references 4, 9, and 10 for complete reviews of *P. aeruginosa* quorum sensing). In that study, an elastase-deficient *P. aeruginosa* strain was analyzed and found to not contain a mutation in or near the *lasB* gene, which encodes LasB elastase. This curious result led to a search for an elastase regulator. A subsequent genetic complementation was successful, and the gene responsible was determined to encode a homolog of the *V. fischeri* LuxR protein. This novel gene was referred to as *lasR*. This gene encodes a 239-amino-acid protein that is well conserved in its autoinducer-binding region (36% identical to the corresponding LuxR region) and its DNA-binding domain (53% identical to the corresponding LuxR region). [Note: To help the reader follow the subsequent discussion, a schematic representing the current model of *P. aeruginosa* quorum sensing is presented in Fig. 2.]

Like most other members of this family, LasR acts as a transcriptional activator only in the presence of its cognate signal. This signal was determined to be *N*-(3-oxododecanoyl)-L-homoserine lactone (3-oxo-C_{12}-HSL) (Fig. 3), which is produced by the enzyme LasI. LasI, which encodes a 201-amino-acid homolog of LuxI (35% identical), is capable of producing 3-oxo-C_{12}-HSL when the enzyme

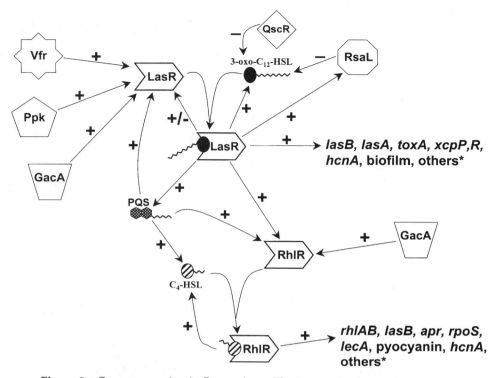

Figure 2 Quorum sensing in *P. aeruginosa*. The *P. aeruginosa* signaling cascade begins with positive control of LasR by Vfr, GacA, and PPK. This presumably occurs at a time when 3-oxo-C_{12}-HSL has accumulated to a level where it will bind LasR. Upon this happening, numerous genes are induced by LasR-3-oxo-C_{12}-HSL. Among these are those that encode for RhlR and an enzyme responsible for PQS production. The production of RhlR and PQS leads to an up-regulation of the *rhl* quorum sensing system, which functions through the signal C_4-HSL. Upon the interaction of RhlR and C_4-HSL, the genes controlled by *rhl* quorum sensing are induced and the cascade is complete. The asterisk in the gene list indicates that an estimated 3 to 4% of the genes in the genome are believed to be controlled by quorum sensing (11).

is expressed in *E. coli*. The *lasI* gene is located just downstream from *lasR* on the same strand of DNA. The two genes do not overlap and are transcribed separately to form the *las* regulon. The signal 3-oxo-C_{12}-HSL has been indirectly shown to bind LasR, and these two components are sufficient to induce genes controlled by the *las* quorum sensing system.

The regulation of the *las* regulon is complex. First, the expression of *lasR* is dependent on multiple factors, including the *vfr* gene. This gene, which encodes a homolog of the *E. coli* cyclic AMP receptor protein (CRP), is absolutely essential for the transcription of *lasR*. There is a CRP-binding consensus sequence in the promoter region of *lasR*, and Vfr has been shown to bind to *lasR* promoter DNA. It is also important to note that the transcription of *lasR* occurs from two transcriptional start sites separated by 30 base pairs. This probably explains some of

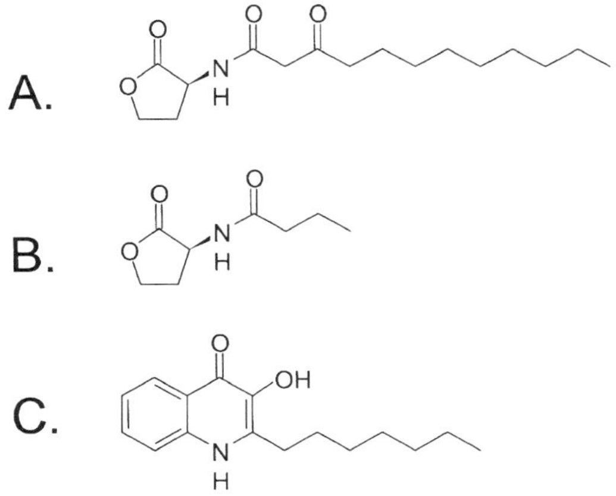

Figure 3 The intercellular signals of *P. aeruginosa*. The chemical structures of the three major signals utilized by *P. aeruginosa* are shown. (A) 3-oxo-C_{12}-HSL; (B) C_4-HSL; (C) PQS.

the confusion surrounding *lasR* autoregulation. Conflicting reports have shown both positive and negative autoregulation for *lasR*, as well as a constitutive type of expression. It is likely that *lasR* is transcribed from different promoters under different growth conditions and this would allow *P. aeruginosa* to finely regulate *lasR* transcription in response to environmental changes. In addition to Vfr regulation, the global regulators GacA and PPK have been shown to effect *lasR*, indicating again that *lasR* is regulated by multiple environmental signals.

Adding to the complexities surrounding *lasR*, the *lasI* gene is also regulated from multiple fronts. The *lasI* gene contains a *lux* box consensus sequence in its promoter region that allows this gene to be up-regulated by LasR and 3-oxo-C_{12}-HSL. There is also a very interesting negative regulator involved in *lasI* transcription. The *rsaL* gene is located between *lasR* and *lasI* and is transcribed in the opposite direction of these genes. This gene was found to be positively regulated by LasR and 3-oxo-C_{12}-HSL, but its product, RsaL, was determined to be a strong repressor of *lasI*. This curious result caused the authors to speculate that *lasI* is tightly controlled to prevent early activation of the quorum sensing cascade. This negative *lasI* regulation is also bolstered by a second protein referred to as QscR. QscR is a LuxR homolog for which the cognate signal has not been identified. The *qscR* gene does not have an autoinducer synthase gene located near it, and aside from *rhlI* (see below) there are no other LuxI homologs encoded by *P. aeruginosa*. Despite this, QscR has been shown to act as a *lasI* repressor, indicating that the regulation of the *las* regulon is multifactorial and complex.

The *rhl* Quorum Sensing System

Shortly after the *las* quorum sensing system was discovered, a second complete quorum sensing system was found in *P. aeruginosa*. The *rhl* quorum sensing system (named for its ability to control rhamnolipid) was found to encode LuxR and LuxI homologs, referred to as RhlR and RhlI, respectively. RhlR is 31% identical to LasR, with highest identity in the autoinducer and DNA-binding regions. RhlI is an autoinducer synthase that produces the signal *N*-butyryl-L-homoserine lactone (C_4-HSL) (Fig. 3). The *rhl* quorum sensing system functions similarly to the *las* quorum sensing system, with C_4-HSL binding to and thereby activating RhlR to allow for the induction of RhlR-controlled genes.

The genetic organization of the *rhl* quorum sensing system is also similar to the *las* quorum sensing system. The *rhlR* and *rhlI* genes form a regulon of two independently transcribed genes on the same DNA strand, with *rhlI* just downstream from *rhlR*. One difference from the *las* regulon is that a primary target of the *rhl* quorum sensing system, *rhlAB*, is located directly upstream from *rhlR*. The *rhlAB* operon encodes a rhamnosyltransferase required for rhamnolipid production and is activated by RhlR and C_4-HSL.

The *P. aeruginosa* Quorum Sensing Hierarchy

The relationship between the *las* and *rhl* quorum sensing systems is interesting. The components of the two systems are not interchangeable, and there was relatively high specificity with regard to the genes that each system could activate. However, it was shown that the systems were linked to each other. LasR and 3-oxo-C_{12}-HSL activated the transcription of *rhlR*, indicating that the *las* quorum sensing system was dominant over the *rhl* quorum sensing system in a quorum sensing hierarchy. It was also shown that 3-oxo-C_{12}-HSL could block C_4-HSL from binding to cells expressing RhlR and that 3-oxo-C_{12}-HSL significantly inhibited the ability of RhlR and C_4-HSL to activate *rhlA*. These data suggested that RhlR was also controlled at the posttranslational level by 3-oxo-C_{12}-HSL, indicating that the *las* quorum sensing system exerts multiple levels of control to finely tune the regulation of the *rhl* quorum sensing system. The significance of this is not known, but it could be suggested that only the *las* quorum sensing system truly senses cell density and triggers the signaling cascade that affects a large portion of the *P. aeruginosa* genome.

P. aeruginosa Quorum Sensing-Controlled Genes

Together, the *las* and *rhl* quorum sensing systems have been shown to control numerous genes. At this point, it is estimated that approximately 4% of the genes of *P. aeruginosa* are regulated by quorum sensing (11). For the sake of brevity, a summary of *P. aeruginosa* quorum sensing-controlled genes is presented in Table 1.

A Third *P. aeruginosa* Intercellular Signal

As the quorum sensing hierarchy of *P. aeruginosa* was being examined, it became apparent that a third cell-to-cell signal was being produced (see reference 2 for a summary of this subject). Culture supernatants from wild-type *P. aeruginosa* contained a signal that had properties different from those of 3-oxo-C_{12}-HSL and C_4-HSL. This novel signal was purified and identified as 2-heptyl-3-hydroxy-4-quinolone (Fig. 3). Because of its quinolone base structure, the signal was named the *Pseudomonas* quinolone signal (PQS). The genetics behind PQS production are not defined, but it has been shown that anthranilate is a precursor of PQS.

PQS is unlike any known cell-to-cell signal, and in fact, the 4-quinolone family of compounds includes antipseudomonal antibiotics such as ciprofloxacin.

Table 1   Genes controlled by quorum sensing in *P. aeruginosa*

Genome No.[a]	Gene and/or protein	QS control[b]
PA0051	Probable glutamine amidotransferase	*las* and *rhl*
PA0109	Hypothetical protein	*las* and *rhl*
PA0852	*cbpD*—chitin-binding protein CbpD	*las* and *rhl*
PA0855	Hypothetical protein	*las*
PA1148	*toxA*—exotoxin A	*las*
PA1129	Probable fosfomycin resistance protein	*las* and *rhl*
PA1249	*aprA*—alkaline metalloproteinase	*las* and *rhl*
PA1430	*lasR*—quorum sensing transcriptional regulator	*las*
PA1431	*rsaL*—repressor protein	*las*
PA1432	*lasI*—3-oxo-C_{12}-HSL synthase	*las*
PA1869	Probable acyl carrier protein	*las* and *rhl*
PA1871	*lasA*—LasA protease	*las*
PA1894	Hypothetical protein	*las*
PA1896	Hypothetical protein	*las*
PA2193	*hcnA*—hydrogen cyanide synthase component	*las* and *rhl*
PA2194	*hcnB*—hydrogen cyanide synthase component	*las* and *rhl*
PA2302	Probable nonribosomal peptide synthetase	*las*
PA2385	Probable acylase	*las*
PA2401	Probable nonribosomal peptide synthetase	*las*
PA2402	Probable nonribosomal peptide synthetase	*las*
PA2424	Probable nonribosomal peptide synthetase	*las*
PA2570	*pa1L (lecA)*—lectin	*rhl*
PA2587	Probable FAD-dependent monooxygenase	*las*
PA2592	Probable periplasmic spermidine-binding protein	*las*
PA2770	Hypothetical protein	*las*
PA3032	*snr1*—cytochrome *c* precursor	*las* and *rhl*
PA3103	*xcpR*—general secretion pathway protein	*las* and *rhl*
PA3104	*xcpP*—general secretion pathway protein	*las* and *rhl*
PA3325	Hypothetical protein	*las* and *rhl*
PA3327	Probable nonribosomal peptide synthetase	*las* and *rhl*
PA3328	Probable FAD-dependent monooxygenase	*las* and *rhl*
PA3329	Hypothetical protein	*las* and *rhl*
PA3330	Probable short-chain dehydrogenase	*las* and *rhl*
PA3331	Probable cytochrome P450	*las* and *rhl*
PA3333	*fabH2*—3-oxoacyl-acyl-carrier-protein synthase III	*las* and *rhl*
PA3336	Probable small molecule transporter (MFS class)	*las* and *rhl*
PA3476	*rhlI*—C_4-HSL synthase	*las* and *rhl*
PA3477	*rhlR*—quorum sensing transcriptional regulator	*las*
PA3478/9	*rhlAB*—rhamnosyltransferase	*las* and *rhl*
PA3724	*lasB*—LasB elastase	*las* and *rhl*
PA3907	Hypothetical protein	*las*
PA4078	Probable nonribosomal peptide synthetase	*las* and *rhl*
PA4084	Probable fimbrial biogenesis usher protein	*las*
PA4207	Probable RND efflux transporter	*las* and *rhl*
PA4212	*phzC*—phenazine biosynthesis protein PhzC	*las* and *rhl*
PA4217	Probable FAD-dependent monooxygenase	*las* and *rhl*
PA4869	Hypothetical protein	*las*
PA5356	*glcC*—probable transcriptional regulator GlcC	*las* and *rhl*

[a] Genome number is from the *Pseudomonas* Genome Project website at Pseudomonas.com.
[b] QS control indicates which quorum sensing system has been shown to control each gene.

The discovery of PQS was allowed because of its ability to induce the *lasB* gene, making this the third intercellular signal that affects the expression of this crucial virulence factor. PQS falls under the regulation of the *las* quorum sensing system as it is only produced in the presence of activated LasR. However, the bioactivity of PQS requires at least RhlR, indicating that this signal is in the middle of the *P. aeruginosa* quorum sensing hierarchy. Whether PQS and RhlR directly interact is currently being studied. In addition to *lasB*, it was also shown that PQS can induce the expression of *rhlI*, which encodes the C_4-HSL synthase. This led to the conclusion that PQS serves as some sort of connector between the *las* and *rhl* quorum sensing systems to provide an additional layer of control over the *rhl* quorum sensing system (see Fig. 2). The reason for this is not obvious, but it was shown that PQS bioactivity is not produced by *P. aeruginosa* until late in the stationary phase of growth. It was also found that PQS and C_4-HSL had a synergistic effect on *lasB* induction. Taken together, these data led to the speculation that PQS may be important for survival at a time when bacterial cells are under stress. Theoretically, PQS could up-regulate *rhlI* to increase the expression of *lasB*. The production of the LasB elastase could then lead to the degradation of host or environmental proteins to provide nutrients for *P. aeruginosa*. While its role is still undetermined, PQS was shown to be required for virulence in a nematode killing assay, indicating that this signal is most likely a *P. aeruginosa* virulence determinant (7).

Quorum Sensing and *P. aeruginosa* Virulence

The exact role of quorum sensing in the infectious process has not been defined, but there is evidence that supports its importance for virulence. A positive correlation was found between *lasR* transcription and the transcription of *lasB*, *lasA*, and *toxA* in infected lungs. More important, a *lasR* mutant was significantly less virulent in a neonatal mouse model of pneumonia. In a follow-up to those studies with the same model, it was also shown that a *lasI* mutant, a *rhlI* mutant, and a *lasI*, *rhlI* double mutant were all significantly less virulent than their appropriate parental strain. Evidence for the importance of quorum sensing in infections was also found by randomly mutagenizing the *P. aeruginosa* wild-type strain PA14 in search of virulence factors. It was found that a *lasR* mutant, along with several other mutated strains, was less virulent in pathogenicity screens in which *Caenorhabditis elegans* and *Arabidopsis thaliana* were used as eukaryotic hosts. This same *lasR* mutant was further tested in a burned mouse model of infection, which showed that it was significantly less virulent than its parent strain. This indicated that quorum sensing was critical for virulence in several different infection models. Finally, an exciting report has shown that in the lungs of infected mice, *P. aeruginosa* produced intercellular signals that were capable of activating an acyl-homoserine lactone bioassay. Taken together, these studies provide strong evidence to support the argument that quorum sensing is an important part of *P. aeruginosa* virulence.

Recent evidence also indicates that the autoinducers themselves may be involved in pathogenesis. The 3-oxo-C_{12}-HSL signal has been shown to induce interleukin-8 production by human airway epithelial cells and human lung cells in a dose-dependent manner. It was also shown that 3-oxo-C_{12}-HSL is capable of modulating several additional immunological activities.

The reason that *P. aeruginosa* regulates numerous factors in response to cell density is not clear. A popular theory is that delaying the production of certain virulence factors may allow *P. aeruginosa* to face a lesser immune response while its population builds. Whatever the case, quorum sensing controls numerous virulence factors, and functional quorum sensing is required in every *P. aeruginosa* model of virulence tested so far.

Quorum Sensing and Biofilms

Biofilm formation is believed to be a critical step in the disease produced when *P. aeruginosa* chronically infects the lungs of patients with cystic fibrosis. The requirement for functional quorum sensing in biofilm formation was shown when it was found that a *lasI* mutant, but not a *rhlI* mutant, was incapable of forming a biofilm. A link to the infectious process was then made with indirect data that suggested that *P. aeruginosa* forms biofilms within the lungs of patients with cystic fibrosis. They showed that the ratio of 3-oxo-C_{12}-HSL to C_4-HSL produced by *P. aeruginosa* growing ex vivo within sputum from patients with cystic fibrosis was indicative of the biofilm growth state. Finally, it has been reported that the expression of *lasI* and *rhlI* within a biofilm is not uniform but occurs within regions of local intensity. This study also showed that these genes were expressed in a dynamic fashion over several days of biofilm growth, suggesting that differential quorum sensing-controlled gene expression occurs during biofilm growth.

A NEW PLAYER IN THE CELL-TO-CELL SIGNALING GAME

Recently, a new class of cell-to-cell signal has been discovered. The story of this signal began in the marine bacterium *Vibrio harveyi*, in which a case of convergent evolution led to the development of a unique quorum sensing system (see references 1 and 8 for a review of *V. harveyi* quorum sensing). This system functions through an acyl-homoserine lactone signal (AI-1) and a second signal (AI-2) that is different from other known signals. The identification of AI-2 was difficult, but it finally was reported to be a furanosyl borate diester (3). This signal exerts its activity by forming a complex with a periplasmic protein, LuxP, and this complex is sensed by LuxQ. LuxQ is the membrane-spanning protein of a two-component regulatory system that also includes LuxU. When the signal from LuxQ is transduced to LuxU, this phosphotransferase passes the signal on to LuxO. LuxO is a repressor that controls the *luxCDABEGH* operon that is responsible for light production. The discovery of a furanosyl borate diester controlling light production in *V. harveyi* was somewhat interesting, but the story took on a broader scope when it was shown that an AI-2 activity was also produced by *E. coli* and *Salmonella enterica* serovar Typhimurium. After this, a gene responsible for AI-2 was found in *V. harveyi*, *E. coli*, and *S. enterica* serovar Typhimurium. This gene encoded a 172-amino-acid protein with no homology to any protein with a known function. A search of protein databases also produced a very exciting result. The *luxS* gene was highly conserved in over 30 species of both gram-positive and gram-negative bacteria. So far, AI-2 has been shown to control virulence in *E. coli* and *Streptococcus pyogenes*. At this time, the studies of AI-2 are still relatively young and we expect that the list of bacteria and genes that depend on AI-2 will continue to grow as we learn more about this new signal that is used by so many evolutionarily distant bacteria.

CONCLUSION

Bacteria have developed multiple ways to communicate with each other. As we learn more about these talkative creatures, it becomes apparent that their lifestyle is not as simple as it once appeared. The discovery of each new intercellular communication method and the elucidation of the involved molecular biology lead us closer to understanding the complexities of bacterial life. With the understanding of quorum sensing, we will begin to know how bacteria coordinate specific responses. It is our hope that this understanding will provide the background for the development of new and effective antimicrobial therapies that will provide much needed options for the treatment of bacterial infections.

Acknowledgments
E. C. Pesci was supported by research grants from the Cystic Fibrosis Foundation (grant PESCI99I0) and the National Institutes of Health (grant R01-AI46682). B. H. Iglewski was supported by NIH grant R01-AI33713 and a grant from the Cystic Fibrosis Foundation (IGLEW S00G0).

REFERENCES
1. **Bassler, B. L.** 1999. A multichannel two-component signaling relay controls quorum sensing in *Vibrio harveyi*, p. 259–273. *In* G. M. Dunny and S. C. Winans (ed.), *Cell-Cell Signaling in Bacteria*. ASM Press, Washington, D.C.
2. **Calfee, M. W., J. P. Coleman, and E. C. Pesci.** 2001. Interference with *Pseudomonas* quinolone signal synthesis inhibits virulence factor expression by *Pseudomonas aeruginosa*. *Proc. Natl. Acad. Sci. USA* **98:**11633–11637.
3. **Chen, X., S. Schauder, N. Potier, A. Van Dorsselaer, I. Pelczer, B. L. Bassler, and F. M. Hughson.** 2002. Structural identification of a bacterial quorum-sensing signal containing boron. *Nature* **415:**545–549.
4. **de Kievit, T. R., and B. H. Iglewski.** 2000. Bacterial quorum sensing in pathogenic relationships. *Infect. Immun.* **68:**4839–4849.
5. **Fuqua, C., M. R. Parsek, and E. P. Greenberg.** 2001. Regulation of gene expression by cell to cell communication: acyl-homoserine lactone quorum sensing. *Annu. Rev. Genet.* **35:**439–468.
6. **Fuqua, W. C., S. C. Winans, and E. P. Greenberg.** 1996. Census and consensus in bacterial ecosystems: the LuxR-LuxI family of quorum-sensing transcriptional regulators. *Annu. Rev. Microbiol.* **50:**727–751.
7. **Gallagher, L. A., and C. Manoil.** 2001. *Pseudomonas aeruginosa* PAO1 kills *Caenorhabditis elegans* by cyanide poisoning. *J. Bacteriol.* **183:**6207–6214.
8. **Miller, M. B., and B. L. Bassler.** 2001. Quorum sensing in bacteria. *Annu. Rev. Microbiol.* **55:**165–199.
9. **Pesci, E. C., and B. H. Iglewski.** 1999. Quorum sensing in *Pseudomonas aeruginosa*, p. 147–155. *In* G. M. Dunny and S. C. Winans (ed.), *Cell-Cell Signaling in Bacteria*. ASM Press, Washington, D.C.

10. **Van Delden, C., and B. H. Iglewski.** 1998. Cell-to-cell signaling and *Pseudomonas aeruginosa* infections. *Emerg. Infect. Dis.* **4:**551–560.
11. **Whiteley, M., K. M. Lee, and E. P. Greenberg.** 1999. Identification of genes controlled by quorum sensing in *Pseudomonas aeruginosa. Proc. Natl. Acad. Sci. USA* **96:**13904–13909.

SECTION II
TOXIN BIOGENESIS: CROSSING BACTERIAL MEMBRANE BARRIERS

Between the time that a bacterial protein toxin is synthesized and the time that the toxin gains access to the eukaryotic cell, a complex series of events must take place. The protein toxin must fold to attain its correct tertiary structure and, if the toxin is oligomeric, the protein subunits must associate. The toxin must also cross the bacterial membrane(s). For toxins produced by gram-negative bacteria, exiting the bacterial cell is particularly difficult since synthesis of the toxin protein occurs in the cytoplasm of the cell, and both inner and outer membrane barriers must be crossed.

Toxin biogenesis and secretion occur in different ways for different toxins. For example, certain toxins are transported across bacterial membranes in an unfolded or partially folded state and attain their final structure only after the transport process is complete. Other toxins fold and assemble within the bacterial cell and only then are transported across the bacterial membrane barrier. Whichever pathway is utilized, the process is complex and energy-consuming. Most often, toxins cannot cross bacterial membranes by themselves, but rather, they utilize sets of accessory proteins to help them out of the cell.

The complexity of the export process is reflected in the fact that only a few types of secretion systems have evolved and these systems are used to transport diverse types of toxins. Secretion systems are classified according to their protein sequence and their general structural architecture. Five general families of transport systems, known as types I to V, exist (Table 1). Types I to IV are discussed in detail in the chapters of this section of the book. Type V transporters are described in Box 1.

The different families of transport systems show significant variation in their structures and the mechanisms by which they export toxins. Some secretion systems, such as type I, type II, and certain type IV secretion systems, export the protein toxins directly to the extracellular milieu. In this case, the toxin then finds its own way to the eukaryotic cell, and the ability to cross the eukaryotic cell membrane is a property that is inherent to the toxin molecule. Toxins secreted in this manner are thought of as "classic" bacterial toxins because the toxin molecule itself contains all of the information necessary to seek out the eukaryotic

Table 1 Toxin secretion systems

Export system	Organism	Example of toxin or effector substrate	Toxin activity
Type I	*Escherichia coli*	Hemolysin	Pore formation
	Bordetella pertussis	Adenylate cyclase toxin	Increases intracellular cyclic AMP levels
Type II	*Vibrio cholerae*	Cholera toxin	ADP-ribosylates G_s
	Aeromonas hydrophila	Aerolysin	Pore formation
Type III	*Yersinia* spp.	YopH	Phosphatase
		YopE	Modulates G proteins
	Pseudomonas aeruginosa	ExoS	ADP-ribosylates Ras and modulates G proteins
	Salmonella spp.	SptP	Phosphatase activity; modulates G proteins
Type IV	*Bordetella pertussis*	Pertussis toxin	ADP-ribosylates G_i
	Helicobacter pylori	CagA	Stimulates phosphatase activity
Type V	*Helicobacter pylori*	VacA	Induces structural and functional alterations in cells

---

BOX 1

Type V export systems: proteins that export themselves

Although many toxins require elaborate and complicated transport systems for their export across the outer membrane of gram-negative bacteria, some proteins have the ability to export themselves. These proteins belong to the type V family of exporters. Members of this family of exporters share several common characteristics. As shown in Fig. 1, autotransporters are synthesized as preproteins in the cytoplasm of the bacterium. Each is synthesized as a precursor form in which the mature protein is preceded by a signal sequence and followed at its C-terminal end by an extension that has the capability of forming an amphipathic β-barrel structure in membranes. The signal sequence directs export of the protein across the inner membrane,

most likely by the general Sec system. Once the protein crosses the inner membrane, the signal sequence of the protein is cleaved and the C-terminal end inserts into the outer membrane, forming a porelike structure in which hydrophilic amino acid side chains face inward into the lumen of the pore and hydrophobic amino acid side chains face outward into the membrane. The N-terminal portion of the protein or "passenger domain" is then threaded through the pore and emerges at the surface of the cell. Finally, the protein is cleaved, releasing the mature form of the protein into the extracellular space. Within a subset of the type V family of transporters, the proteins are actually synthesized as two distinct proteins, a passenger or functional protein and a protein that is capable of forming a β-barrel structure in the outer membrane. After the passenger protein is secreted across the inner cell, cross the membrane barriers of that cell, and damage the cell by manifestation of a toxic activity. Other transport systems require contact of the bacterium with the eukaryotic target cell before they allow export of the protein toxins. For example, type III transport systems and certain type IV systems inject proteins directly from the bacterium into the eukaryotic cell. The proteins delivered by these secretion systems differ from classic bacterial toxins in that they are not able to cross the eukaryotic cell membranes by themselves; rather, they rely on the transporter to get them across the eukaryotic cell membranes. Once inside the eukaryotic cell, these proteins manifest their toxic properties by interrupting critical biochemical pathways of the host cell. Protein toxins delivered directly into the eukaryotic cell by bacterial transporters are known as "effector proteins."

Interestingly, several secretion systems used for transport of toxins appear to have evolved from transport systems used for export of other macromolecules. For example, type II systems are related to transport systems that are used for the export of pili involved in bacterial adherence. Type III secretion systems are related to the flagella export system, a system that transports the flagellar proteins that are essential for bacterial motility. Type IV systems are related to conjugation machinery responsible for transfer of plasmids from one bacterium to another and thus the transfer of genetic information.

BOX 1 (*Continued*)

membrane, it inserts into the pore formed by its partner protein and passes through this pore to the extracellular milieu.

A number of virulence factors have been shown to belong to the type V family of transporters, including many adhesins and several toxins. Toxins that belong to this family of transporters include the VacA protein of *Helicobacter pylori* and the hemolysins of *Proteus mirabilis* and *Serratia marcescens*.

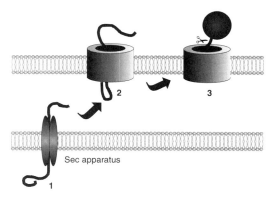

Figure 1 Type V transport. The autotransporter protein is synthesized in the cytoplasm of the bacterial cell and is then exported across the inner membrane by the general secretion pathway (1). A portion of the autotransporter forms a β-barrel structure in the membrane through which the remainder of the protein is transported (2). The passenger domain is then cleaved, releasing the mature form of the protein into the extracellular space (3).

Most export systems consist of proteins distinct from the toxins that are transported, as is the case for type I, II, III, and IV export systems, each of which comprises a set of accessory proteins that aid in the export process. However, in the case of type V secretion, the toxin exports itself across the outer membrane of the bacterium!

The complexities of these systems and their varying mechanisms are fascinating and provide scientists with the difficult challenge of deciphering how these systems work. While a tremendous amount of information on these transport systems has been elucidated in recent years, we clearly have a lot more to learn.

Bacterial Protein Toxins
D. Burns et al., Editors
©2003 ASM Press, Washington, DC 20036

Chapter 5

The Type I Export Mechanism

Vassilis Koronakis, Jeyanthy Eswaran, and Colin Hughes

In gram-negative bacteria such as *Escherichia coli,* proteins destined for the cell surface or surrounding medium must cross the cytoplasmic (inner) and outer membranes and the periplasm between them. Several mechanisms employ periplasmic intermediates, e.g., for assembly of adhesion pili, or utilize a large number of proteins to span the envelope, e.g., in the assembly of flagella. The type I export mechanism contrasts with these as it does not generate periplasmic intermediates and employs a dedicated secretory apparatus of just three proteins. These are a traffic ATPase and an adaptor protein, which together form an energized substrate-specific complex in the inner membrane, and an outer membrane protein of the TolC family.

DIVERSE SUBSTRATES FOR TYPE I EXPORT

Gram-negative pathogens of humans, animals, and plants export a range of virulence proteins by the type I process (Table 1). These export substrates range in size from 11 kDa (colicin V) to 110 to 178 kDa (cytolytic toxins). Type I export is best characterized in the biogenesis of the 110-kDa hemolysin secreted by uropathogenic, enterohemorrhagic, and other disease-causing *E. coli*. It is a member of a membrane-targeted toxin family that includes the hemolysins and leukotoxins of *Actinobacillus*, *Pasteurella*, *Proteus*, *Morganella*, and *Moraxella* species and the bifunctional adenylate cyclase-hemolysin of *Bordetella pertussis*. These Ca^{2+}-requiring toxins are closely related and have a comparable structure (Fig. 1), including a C-terminal domain of acidic glycine-rich repeats (11 to 17 in HlyA) that binds Ca^{2+} to form a β-superhelix (see chapter 14 for more information about these toxins). All are made as inactive protoxins that must undergo a fatty acylation maturation step directed by the coexpressed HlyC, a novel and specific homodimeric lysyl-acyltransferase that uses acyl-acyl carrier protein as fatty acid donor. Although acylation is essential for toxin interaction with mammalian cell target membranes, it is not required for export across the prokaryotic envelope.

Vassilis Koronakis, Jeyanthy Eswaran, and Colin Hughes • Department of Pathology, Cambridge University, Tennis Court Road, Cambridge CB2 1QP, United Kingdom.

Table 1   The substrates of type I export systems

Bacterium	Substrate
Escherichia coli	HlyA hemolysin
	ColV (colicin V)
Pasteurella haemolytica	LktA leukotoxin
Actinobacillus actinomycetemcomitas	AaltA leukotoxin
Actinobacillus pleuropneumoniae	ApxI/II/III hemolysin/leukotoxin
Bordetella pertussis	CyaA adenylate cyclase/hemolysin
Erwinia chrysanthemi	PrtC protease
Serratia marcescens	HasA hemoprotein
	Metalloprotease, lipase, SlaA (S-layer protein)
Pseudomonas aeruginosa	Apr alkaline protease
Pseudomonas fluorescens	TliA thermostable lipase
Campylobacter fetus	SapA (S-layer protein)
Caulobacter crescentus	RasA (S-layer protein)
Anabaena sp.	DevA (heterocyst development protein)
Rhizobium meliloti	ExsH (exopolysaccharide glycanase)
Rhizobium leguminosarum	PlyA, PlyB (exopolysaccharide glycanase)
	NodO (plant nodulation protein)

Other extracellular proteins translocated by the type I mechanism include hydrolytic enzymes (metalloproteases and lipases) and heme-binding proteins, while substrates destined for the bacterial cell surface include glycanases and the S-layer protein that makes up the paracrystalline surface layer of many bacteria (e.g., *Campylobacter fetus* SapA). Type I systems are also involved in cell development and motility in diverse bacteria, e.g., the DevBCA exporter of *Anabaena* species, which is necessary for the assembly of the cell surface laminated layer, and oscillin, which has a role in locomotion of the cyanobacterium *Phomidium* species.

Genes encoding type I export substrates and the three corresponding export machinery components are typically found in the same operon, with the gene for the substrate followed immediately by those encoding the traffic ATPase, the

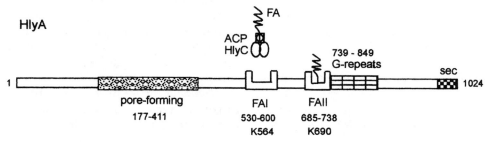

Figure 1   Representation of the 1,024-residue hemolysin HlyA. HlyC fatty acylation recognition domains FAI and FAII span the target lysines K564 and K690, upstream of the glycine-rich Ca^{2+}-binding repeats (G-repeats) and the C-terminal secretion signal (sec).

adaptor protein, and the outer membrane TolC homolog. Nevertheless, the *E. coli* system is atypical in that the *hlyCABD* operon is not closely linked to the *tolC* gene. The export proteins HlyB and HlyD are expressed at a lower level than the HlyC and proHlyA proteins, which determine synthesis and maturation of the toxin. This is due to transcription polarity within the *hlyCABD* operon, which can be suppressed by the elongation protein RfaH, a positive regulator of other virulence and fertility operons, including those involved in the assembly of lipopolysaccharide. The *hlyCABD* operon is generally linked closely to other virulence genes on a pathogenicity island and its 39% G+C content suggests that it been acquired only recently by *E. coli*.

A C-TERMINAL SIGNAL TARGETS SUBSTRATES TO THE EXPORT APPARATUS

Type I export substrates do not have a conventional N-terminal secretion signal and are therefore not subject to N-terminal processing. The secretion signal of these substrates is uncleaved and located at the extreme C terminus, although exceptions are the *Serratia marcescens* hemoprotein HasA, reported to have a cleaved C-terminal signal, and colicin V, observed to possess an uncleaved N-terminal signal. Despite a lack of identity between C-terminal signal primary sequences, interchangeability of type I exporter genes suggests that higher-order structures are shared by the substrates. Three contiguous features are suggested to be conserved in the final 50 residues, namely, a potential amphipathic helix, a cluster of charged residues, and a weakly hydrophobic terminal sequence rich in hydroxylated residues. A C-terminal 53-amino-acid peptide can be exported, albeit weakly, and must therefore contain the features critical to interaction with two inner membrane transporter components.

THE MEMBRANE COMPONENTS DIRECTING TYPE I EXPORT AND ALSO MULTIDRUG EFFLUX

The type I export machinery requires only three export components. These are all integral membrane proteins, a traffic ATPase (in the hemolysin system HlyB), an accessory or "adaptor" protein (HlyD), and the outer membrane protein TolC. HlyB and HlyD are inner membrane proteins, which together form an energized substrate-specific complex, but this complex does not function alone. Without TolC there is no export to the periplasm, i.e., TolC is an integral part of the type I machinery.

A closely related machinery effects the efflux of small noxious compounds from gram-negative bacteria. These include detergents, organic solvents, and antibacterial drugs such as nalidixic acid and antibiotics like tetracycline, erythromycin, and chloramphenicol. These "efflux pumps" are also characterized by an outer membrane TolC homolog cooperating with a substrate-specific inner membrane complex that contains an HlyD-like adaptor protein. However, in efflux pumps the adaptor protein interacts in the inner membrane complex with a protein providing energy from proton antiport, not as in type I protein export ATP hydrolysis.

The HlyB Traffic ATPase: Energy and Substrate Engagement

Cell membrane traffic ATPases are members of the superfamily ubiquitous throughout prokaryotes and eukaryotes, including mammals. They provide energy from ATP hydrolysis for movement of various molecules, large polypeptides to small ions, across membranes. All have a large cytosolic nucleotide-binding domain (the crystal structure of the ATP-binding domain from both the histidine permease HisP and the sugar importer MalK is now available) and a transmembrane domain. These two domains can be encoded separately as distinct proteins or fused together as in the hemolysin export protein HlyB. The fusion may also be tandemly duplicated within a single polypeptide to form a dimer, e.g., in eukaryotic proteins such as the major histocompatibility complex-linked TAP peptide transporter or multidrug resistance proteins.

The hemolysin export ATPase HlyB has 707 residues and is assumed to function as a homodimer. Putatively, six transmembrane helices between amino acids 158 and 432 interact with the bacterial membrane, whereas the C-terminal ca. 200 residues form the ATPase domain located in the cytoplasm (Fig. 2). The HlyB cytosolic domain has an in vitro V_{max} of 1 μmol of ATP per min per mg and a K_m of 0.2 mM of ATP. As in other transporters, the glycine-rich Walker A and Walker B motifs are central to HlyB nucleotide binding and hydrolysis, and mutations in these highly conserved sequences cause a complete loss of hemolysin export and ATPase activity, even though ATP is still bound. HlyB can interact independently with the adaptor protein HlyD and the export substrate HlyA. Mutations in the HlyA C-terminal secretion signal are partially compensated by suppressor mutations in HlyB, compatible with the possibility that HlyB acts as the initial "signal receptor protein" in the inner membrane complex.

The Adaptor Protein: Substrate-Triggered Recruitment of TolC

The inner membrane adaptor protein HlyD (478 residues) has a large periplasmic domain (amino acids 81 to 478) connected by a single transmembrane helix to a small N-terminal cytosolic domain (the residues 1 to 59) (Fig. 3). HlyD family members have similar sizes and hydropathy profiles, and all are embedded in the inner membrane or anchored to it by an N-terminal lipid modification. Cross-linking experiments have shown that HlyD is a trimeric membrane protein and

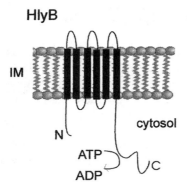

Figure 2 Topology of the *E. coli* traffic ATPase HlyB in the inner membrane.

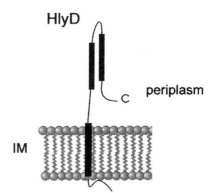

Figure 3 Topology of the *E. coli* adaptor protein HlyD in the inner membrane.

forms a complex with the inner membrane ATPase HlyB, independent of the substrate HlyA and the outer membrane component TolC. The hemolysin HlyA independently interacts with both components of the inner membrane complex. The small N-terminal cytosolic domain is central to substrate engagement as its deletion abolishes hemolysin export. This domain has a conserved potential amphipathic helix (residues 2 to 26) followed by a group of charged residues (amino acids 34 to 38). Specific deletion of the charged cluster disables export and TolC recruitment without affecting substrate engagement, i.e., uncouples the two critical events. As a result, the HlyD cytosolic domain is posulated to mediate transduction of the substrate-binding signal to the HlyD periplasmic domain. This large coiled-coil-containing domain is in turn thought to bind TolC to effect recruitment, and as expected, small deletions or masking of the periplasmic C-terminal domain causes strong attenuation of HlyA export.

The TolC Channel Tunnel: Exit from the Cell

The 471-residue TolC assembles as a trimer in the *E. coli* outer membrane, and two-dimensional crystallization in phospholipid bilayers indicated a trimeric structure of 58 Å outer diameter. Although this was reminiscent of the porins, the TolC images showed a central pool of stain, hinting at a single pore rather than three. Side views revealed a further difference, an additional domain outside the membrane, which was interesting as it seemed such a domain could form part of the periplasmic bypass. Subsequent X-ray crystallography revealed that TolC is indeed fundamentally different from other outer membrane proteins (Fig. 4) (see Color Plates following p. 256). At 2.1 Å resolution, TolC is seen as a homotrimeric tapered cylinder 140 Å in length. This comprises a 40-Å-long outer membrane β-barrel (the channel domain) anchoring a contiguous 100-Å-long α-helical barrel (the tunnel domain) that projects across the periplasmic space. A third domain, a mixed α/β-structure, forms a strap around the equator of the tunnel.

Each of the three TolC monomers contributes four antiparallel β-strands and four antiparallel α-helical strands to form the channel and tunnel domains, respectively. Whereas a β-barrel is a typical feature of outer membrane proteins, the TolC channel domain is different in that the three monomers form a single

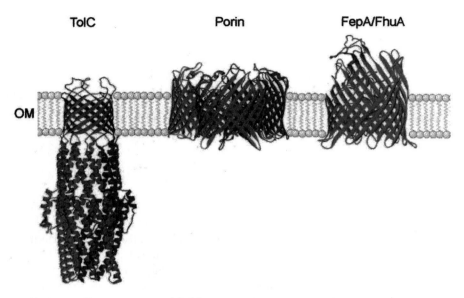

Figure 4 The structures of TolC porins (e.g., OmpF and SecY) and the siderophore transporters FhuA and FepA. (See Color Plates following p. 256.)

β-barrel. All other known outer membrane proteins, including the trimeric porins, form one barrel per protein monomer (Fig. 4), in which each monomer forms a β-barrel of 16 or 18 β-strands. TolC also lacks a common structural element of these channel-forming proteins, an inward folded loop that constricts the internal diameter of the β-barrel, and it does not have the plug domain of the larger β-barrels, e.g., in the iron transporters FhuA and FepA. The unique TolC α-barrel is assembled by each helix, packing laterally with two neighboring helices to generate "knobs-into-holes" interactions at the two interfaces.

The TolC structure provides a large water-filled exit duct with an internal diameter of about 35 Å and cross-sectional area of 960 Å², 15-fold larger than that of the general diffusion pore OmpF. The outer membrane β-barrel channel is constitutively open to the cell exterior, but the tunnel diameter decreases to a virtual close at its periplasmic entrance, which is reflected by the low conductance of TolC reconstituted in artificial membranes. Clearly, during protein export this entrance must be opened to allow substrate to transit out of the cell.

Channel Tunnels: Ubiquitous Exit Ducts for Gram-Negative Bacteria

TolC homologs are seemingly used by all gram-negative bacteria and have an important role in bacterial survival, most notably conferring virulence and multidrug resistance during infection of mammalian hosts. There are currently more than 70 TolC homologs known throughout more than 30 species, and these can be divided into three groups in which sequence relatedness correlates with function in protein export, cation efflux, or multidrug efflux. Remarkably, the sequences of the N- and C-terminal halves of the TolC monomer are similar,

generating an internal structural duplication. Both sequence and structural du-
plications are evident throughout the family, with the strongest intramolecular
identity seen in *Bordetella pertussis* CyaE, which is involved in toxin-adenylate
cyclase export. CyaE is also nearest to the root of the tree, suggesting that it is
closest to the family progenitor. Although TolC homologs vary in length, this
variation is due primarily to variable extensions at the periplasmic N and C
termini, and significant gaps or insertions occur only in the extracellular loops
and the equatorial domain. Sequences determining the α-helices and β-strands
of the channel tunnel do not vary substantially in length, compatible with conservation of the basic fold of the structure throughout the family.

Despite the overall conservation, few amino acids are very well conserved
in all the TolC homologs, but these seem structurally significant. In particular, at
the periplasmic entrance, glycines facilitate a tight turn between the helices forming the periplasmic tunnel entrance, while small residues like alanine and serine
at the interface of tunnel-forming helices allow the very dense packing that de-

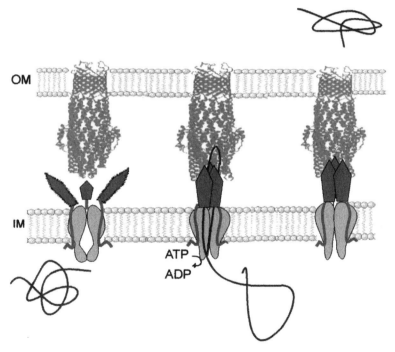

Figure 5 Schematic presentation of the proposed interaction of trimeric TolC
(blue) with substrate-specific inner membrane complexes containing an adaptor protein (red) and a traffic ATPase (green). When the substrate binds to the
inner membrane complex, the trimeric adaptor protein contacts the periplasmic tunnel, possibly via the predicted coiled-coil structures, triggering
the conformational change that opens the entrance and presents the exit duct.
Following export, the components revert to resting state. An animated model
of protein export is available at http://archive.bmn.com/supp/ceb/ain1.
html. (See Color Plates following p. 256.)

TolC closed open

Figure 6 The TolC tunnel entrance closed and modeled open state showing the four monomers colored in green, blue, and red. (See Color Plates following p. 256.)

termines tapering and entrance closure. In addition, a ring of aspartic acid residues maintains an electronegative inner surface at the entrance, which may influence substrate movement and also account for the cation selectivity of TolC in planar lipid bilayers. The aspartate ring acts as a binding site for divalent and trivalent cations, suggesting that the entrance could be a target for blocking by positively charged drugs.

THE MECHANISM OF TYPE I EXPORT

TolC is an integral part of the type I protein export machinery; without it there is no "halfway" export to the periplasm. Cross-linking has defined the sequence of protein-protein interactions underlying export of the hemolysin toxin, in particular showing that a defining event is the recruitment of TolC by inner membrane complex (Fig. 5) (see Color Plates following p. 256). This recruitment is triggered by engagement of the HlyA export substrate with the inner membrane complex, in particular with the N-terminal cytosolic domain of the trimeric adaptor protein HlyD. The active type I export complex contains the substrate and all three export proteins, each of which undergoes conformational change. This complex is transient; once the substrate passes out of the cell, the inner membrane and outer membrane components disengage.

A key requirement is that the periplasmic entrance of TolC must open, i.e., undergo conformational change to allow passage of substrate (Fig. 6) (see Color Plates following p. 256). This is envisaged to occur by an allosteric mechanism in which the inner pair of helices of each TolC monomer realigns with respect to the outer pair to open the entrance, like an iris. In support of this view, elimination of the circular network of intra- and intermonomer salt bridges and hydrogen bonds linking the coiled coils of the α-helical domain causes destabilization of the TolC entrance and a substantial increase in the size of the aperture. It is proposed that this transition to the open state of TolC is triggered and stabilized by interaction with the coiled coils of the large periplasmic domain of the adaptor protein. In this way a contiguous apparatus, which spans the cell envelope, is assembled and allows direct exit of hemolysin and other proteins from the cytosol to the extracellular medium.

Acknowledgments
Work is supported by a Medical Research Council Programme grant.

REFERENCES

1. **Balakrishnan, L., C. Hughes, and V. Koronakis.** 2001. Substrate-triggered recruitment of the TolC channel-tunnel during type I export of hemolysin by *Escherichia coli. J. Mol. Biol.* **313**:501–510.

2. **Johnson, J. M., and G. M. Church.** 1999. Alignment and structure prediction of divergent protein families: periplasmic and outer membrane proteins of bacterial efflux pumps. *J. Mol. Biol.* **287**:695–715.

3. **Koronakis, V., C. Andersen, and C. Hughes.** 2001. Channel-tunnels. *Curr. Opin. Struct. Biol.* **4**:403–407.

4. **Koronakis, V., A. Sharff, E. Koronakis, B. Luisi, and C. Hughes.** 2000. Crystal structure of the bacterial membrane protein TolC central to multidrug efflux and protein export. *Nature* **405**:914–919.

5. **Nikaido, H.** 1998. Multiple antibiotic resistance and efflux. *Curr. Opin. Microbiol.* **1**:516–523.

6. **Stanley, P., V. Koronakis, and C. Hughes.** 1998. Acylation of *Escherichia coli* hemolysin: a unique protein lipidation mechanism underlying toxin function. *Microbiol. Mol. Biol. Rev.* **62**:309–333.

7. **Stanley, P., L. C. Packman, V. Koronakis, and C. Hughes.** 1994. Fatty acylation of two internal lysine residues required for the toxic activity of *Escherichia coli* hemolysin. *Science* **266**:1992–1996.

8. **Thanabalu, T., E. Koronakis, C. Hughes, and V. Koronakis.** 1998. Substrate-induced assembly of a contiguous channel for protein export from *E. coli*: reversible bridging of an inner-membrane translocase to an outer membrane exit pore. *EMBO J.* **17**:6487–6496.

Bacterial Protein Toxins
D. Burns et al., Editors
©2003 ASM Press, Washington, DC 20036

Chapter 6

Toxins and Type II Secretion Systems

Maria E. Scott and Maria Sandkvist

DISTRIBUTION OF THE TYPE II SECRETION PATHWAY

A number of highly specialized pathways have evolved to overcome the outer membrane barrier and to allow for extracellular protein secretion in gram-negative bacteria. One such pathway is the type II secretion pathway or the main terminal branch of the general secretion pathway, which is widely distributed and generally present in species that can be isolated from a diverse range of habitats. The type II secretion pathway has been discovered in species that include human pathogens such as *Vibrio cholerae*, *Pseudomonas aeruginosa*, and *Legionella pneumophila*, fish pathogens such as *Aeromonas hydrophila*, and the plant pathogens *Erwinia chrysanthemi*, *Erwinia carotovora*, and *Xanthomonas campestris*. This pathway is responsible for secretion of toxins and a variety of hydrolytic enzymes that include proteases, lipases, phospholipases, cellulases, and pectinases, which contribute to tissue damage and disease of animals and plants (11). Only a single protein is secreted via the type II pathway in *Klebsiella oxytoca* and *Escherichia coli* K-12. The starch-hydrolyzing enzyme pullulanase is the only known substrate for the Pul secretion pathway in *K. oxytoca*. The expression of the genes that encode pullulanase and the type II secretion components is dependent on MalT and is induced by the addition of maltose to the growth medium. The only known substrate for the *E. coli* type II secretion pathway is chitinase. However, the system has to be genetically modified to promote chitinase secretion. Secretion of chitinase can only be detected when the *E. coli gsp* genes are expressed from a high-copy-number plasmid in an H-NS (histone-like nucleoid structuring protein) mutant.

The tight regulation of genes that encode secreted proteins and the type II secretion genes themselves, in some cases, is a feature common to other species as well. Expression appears to be growth-phase dependent and may be induced by environmental conditions present at the site of colonization. Strict regulation ensures that secretion will occur only when the bacteria have reached their cor-

Maria E. Scott and Maria Sandkvist • Department of Biochemistry, American Red Cross, Jerome H. Holland Laboratory, 15601 Crabbs Branch Way, Rockville, MD 20855.

rect location or target, have multiplied to a critical number, and have received the appropriate environmental signal(s).

In addition to the above-mentioned species, genome sequencing has identified other species with putative type II secretion genes. These species include *E. coli* O157, *Xylella fastidiosa, Caulobacter crescentus, Yersina pestis, Shewanella putrefaciens,* and *Geobacter sulfurreducens* (11). *E. coli* O157 is the only species known to date to harbor a plasmid-encoded type II secretion operon. Furthermore, in contrast to *E. coli* K-12, novel chromosomal DNA in this species has replaced the region of DNA that contains the chitinase and type II secretion genes.

TWO-STEP SECRETION PROCESS

Cholera toxin, exotoxin A, and aerolysin produced by *V. cholerae, P. aeruginosa,* and *A. hydrophila,* respectively, are examples of three toxins that utilize the type II pathway for extracellular secretion. The biogenesis of all three toxins has been analyzed, and their crystal structures have been solved. We will use cholera toxin (CT) as the prototype to describe the type II secretion process. CT is the most intensely studied protein of all factors that are secreted via the type II pathway. The structure of CT has been determined to 2.5 Å, and its folding and assembly have been analyzed in great detail. Its activity is also very well characterized, and the site of receptor binding is known. Furthermore, CT-producing *V. cholerae* remains a significant health threat in several parts of the world, where it causes a severe diarrheal disease. A critical facet of *V. cholerae* pathogenesis is its ability to actively secrete CT to the extracellular environment. The symptoms of cholera are in large part a result of activation of the adenylate cyclase complex and disregulation of ion channels within intestinal epithelial cells by the CT. The multisubunit CT is composed of a doughnut-shaped structure of five identical B subunits of 11.6 kDa each and one A subunit of 28 kDa.

The secretion of CT across the bacterial cell envelope occurs in two distinct steps that have different requirements; transport across the cytoplasmic membrane is followed by outer membrane translocation (3). First, the toxin subunits are produced as precursor proteins with N-terminal signal peptides, which aid in the transport through the cytoplasmic membrane (Fig. 1). Following cytoplasmic membrane transport, the signal peptides are removed and the subunits undergo folding and are released into the periplasmic compartment. Here, the subunits assemble noncovalently into an 85-kDa AB_5 holotoxin complex (Fig. 1). The folding and assembly of the subunits are assisted by the disulfide isomerase DsbA (TcpG). When the *dsbA* gene is inactivated, the subunits do not assemble and are rapidly degraded. Similar effects are observed when the reducing agent dithiothreitol is included in the growth medium, thus indicating that disulfide bond formation is required for toxin biogenesis.

When assembled, the AB_5 complex is secreted, in a second step, across the outer membrane into the extracellular milieu via the type II secretion pathway encoded by the extracellular protein secretion (*eps*) genes and *vcpD* (Fig. 1). The *vcpD* and *eps* genes are specifically required for the outer membrane translocation step, since inactivation of these genes results in accumulation of the assembled toxin in the periplasmic compartment. All type II secretion mutants of *V. cholerae* and other species characterized to date exhibit the same phenotype; secreted

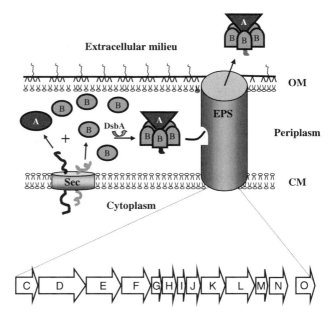

Figure 1 Assembly and secretion of CT. The individual A and B subunits of CT are transported as precursor proteins across the cytoplasmic membrane via a Sec-like mechanism. The signal peptides are removed, and the subunits fold and assemble with the assistance of disulfide isomerase DsbA into the AB_5 complex in the periplasmic compartment. The assembled toxin is then translocated across the outer membrane via the type II secretion apparatus that is encoded by the *eps* genes (C to N) and *vcpD* (O).

proteins accumulate in the periplasm. The suggestion is that unless all components are produced, a functional complex cannot be formed, and as a consequence, no transport across the outer membrane can occur. Also, the implication is that no intermediates exist in the outer membrane translocation process. However, there may be intermediates we cannot detect with the limited cell fractionation techniques available.

SECRETION SIGNAL

The type II secretion gene products are believed to form an envelope-spanning multiprotein secretion complex (secreton) (2, 10) (Fig. 1). However, the proteins to be secreted are thought to engage this complex first when they have reached the periplasmic compartment. The current hypothesis postulates that molecules secreted by the type II system may encode information critical to their secretion within their tertiary and quarternary structures (3). It is believed that recognition and outer membrane translocation of the secreted proteins occur once they have folded into a secretion-competent conformation (2, 10). This may expose a specific recognition sequence or a structural motif that has yet to be identified. The B subunit pentamer, in the case of CT, has been suggested to carry the information necessary for outer membrane translocation. When B subunits are produced in the absence of A subunit expression, they assemble into a pentamer and translocate across the outer membrane. The A subunit, on the other hand, remains in the periplasm and is degraded when B subunits are not produced. Thus, the A subunit is secreted by virtue of its association with the B subunit pentamer. The periplasmic and the secreted forms of CT are very similar; therefore, it is theorized that the substrates for the type II secretion pathway are fully or nearly fully folded proteins. Moreover, several other proteins secreted via the type II pathway have also been shown to undergo folding and disulfide bond formation prior to outer membrane translocation. Proaerolysin, for instance, assembles into a dimer in the periplasmic compartment prior to secretion. Some proteins undergo further modification either during or immediately after outer membrane translocation. These proteins include, among others, the elastase of *P. aeruginosa* and HA/protease of *V. cholerae*, which are both secreted as proforms. In the case of elastase, it has been shown that the propeptide is cleaved in the periplasmic compartment, but the propeptide remains associated with the rest of the molecule until it reaches the extracellular milieu, where it is removed.

It comes as no great surprise that the *E. coli* heat-labile enterotoxin B subunit (LT-IB) is secreted via the Eps pathway when expressed in *V. cholerae*. After all, the amino acid sequences of the B subunits of CT and LT-I are 83% identical. However, what may be surprising is that the B subunit of *E. coli* LT-II shares only 11% identity with CT-B but is nevertheless secreted when produced in *V. cholerae*. An explanation for this may be revealed in the structures of these proteins, which are very similar and superimposable. Hence, a structural motif, instead of specific amino acid content, may constitute the secretion signal.

The putative secretion signal may be continuous but is only recognized by the secretion apparatus when it is correctly presented on the surface of the secretion-competent protein. Alternatively, it may be discontinuous and composed of residues from different positions in the linear polypeptide. Folding of the protein will bring these residues together to form the secretion signal. A consensus secretion signal or structural motif required for secretion has not been identified by sequence or structure comparisons, nor has such a signal been identified by mutagenesis. However, protein fusion technology has been utilized to identify larger domains required for secretion. For example, different portions of exotoxin A were fused to the normally periplasmic enzyme β-lactamase and analyzed for extracellular secretion. A minimal region spanning amino acids 60 to 120 was sufficient to target β-lactamase, which is not a secreted protein, to the extracellular environment. This suggested that the secretion signal is present within the N-terminal region containing residues 60 to 120 of exotoxin A. However, this conclusion was not completely compatible with the results obtained in another study, where a deletion of residues 31 to 243 from exotoxin A had no adverse effects on secretion. Instead, it was suggested that the C-terminal half carries the information required for secretion. One possible interpretation is that both regions may be important for secretion of the intact exotoxin A molecule and may act sequentially. A third study showed that correct positioning of these two domains is important for secretion, because deletions of regions between these domains prevented secretion of exotoxin A.

TYPE II SECRETION APPARATUS

Studies are under way to dissect the relevant interactions between individual components of the type II secretion apparatus in an effort to determine the organization of this multiprotein complex and to understand the mechanism of secretion via the type II pathway. Mutations have been introduced into several genes to determine the contribution of individual components to the activity, stability, and subcellular location of other proteins of the secretion apparatus (Table 1) (2, 10). These analyses have allowed the identification of a number of interacting partners within the apparatus (Table 2). They have also demonstrated interactions between some components known to be present in separate subcellular compartments, which has led to the conclusion that the type II apparatus very likely spans the periplasmic space and both the inner and outer membranes (Figs. 1 and 2). However, it is not known whether this is a permanent or transient envelope-spanning structure. The function of the individual type II components is, with few exceptions, not known, but several of them are assumed to play structural roles.

Sequence analysis of type II secretion components has revealed homology with several components involved in type IV pilus biogenesis, suggesting that the type II apparatus may assemble or function in a manner similar to the pilus biogenesis machinery. Type IV pili are surface appendages, which are several microns long and are composed of polymerized type IV pilin subunits. Generally, they are arranged in a polar fashion and are associated with virulence. Type IV pili support adherence, bacterium-bacterium interactions, and biofilm formation.

Table 1 Subcellular location and putative function of the components of the type II secretion apparatus

Type II secretion component[a]	Size (kDa)[b]	Subcellular location[c]	Putative function[d]
A	60	im	ATPase/kinase?
B	22	im	?
S	14	om	Stabilizes and promotes outer membrane insertion of D
C$_{(P)}$	30	im/om	
D$_{(Q)}$	73	om	Channel
E$_{(R)}$	55	c	ATPase, kinase?
F$_{(S)}$	44	im	?
G$_{(T)}$	16	im/om	Forms a pilus-like structure
H$_{(U)}$	20	im/om	?
I$_{(V)}$	14	im/om	?
J$_{(W)}$	24	im/om	?
K$_{(X)}$	37	im/om	?
L$_{(Y)}$	44	im	Stabilizes and promotes membrane-association of E
M$_{(Z)}$	18	im	Polar targeting
N	27	im	?
O$_{(A)}$	30	im	Peptidase/methyltransferase Substrates: G, H, I, J, and K

[a] The *P. aeruginosa* components are listed in parentheses.
[b] Average size of secretion components in kDa.
[c] Determined by subcellular fractionation and sodium dodecyl sulfate-polyacrylamide gel electrophoresis and immunoblot analysis or by enzymatic activity determination of components fused to either alkaline phosphatase or β-lactamase. im, inner membrane; om, outer membrane; c, cytoplasm.
[d] Based on experimental data or sequence homology.

Table 2 Interacting partners within secretion apparatus

Type II component	Direct or indirect interaction with:[a]
A	B
B	A, D
S	D
C	D, L, M
D	S, C
E	F, G,[b] L, M
F	E
G	E,[b] H, I, J
H	G
I	G
J	G
K	
L	E, M
M	C, E, L
N	
O	G[c], H[c], I[c], J[c], K[c]

[a] Based on coimmunoprecipitation or stabilization toward proteolysis.
[b] Suppressor mutation in *xcpT* (G gene) complements secretion defect in an *xcpR* (E gene) mutant of *P. aeruginosa*.
[c] The G, H, I, J, and K components are cleaved by protein O.

They are also involved in twitching motility of *P. aeruginosa* and *Neisseria gonorrhoeae* and social gliding of *Myxococcus xanthus*, two forms of surface movements that are results of pilus extension and retraction (8). Proteins G, H, I, J, and K of the type II system display some similarity to type IV pilin subunits. These proteins are approximately of the same size and share homologous N-terminal domains that contain the leader peptide required for cytoplasmic membrane translocation and pilin subunit assembly. Furthermore, protein O, which is the most well-characterized protein of all type II secretion components, shares a high degree of identity with prepilin peptidase, a bifunctional enzyme. Prepilin peptidase, often called PilD, is a novel aspartic acid protease that cleaves the leader peptides of type IV pilin subunits and *N*-methylates the newly generated N termini (4). Protein O performs the same task on the prepilin-like components G, H, I, J, and K of the type II secretion apparatus. Prepilin peptidase, in some species, acts on both the precursor forms of type IV pilin subunits and the prepilin-like proteins of the type II system, and in these cases, there is no gene O present in the type II secretion operon.

Components D, E, and F exhibit homology with proteins that are required for the assembly and export of the type IV pilus. Protein D is a member of the secretin family. This family is composed of proteins required not only for type II secretion and type IV pilus biogenesis but also for type III secretion and filamentous phage extrusion. The secretins are located in the outer membrane, where they form very stable oligomeric structures composed of 12 to 14 individual subunits. Protein S, a small lipoprotein of 14 kDa, promotes outer membrane insertion and oligomerization of protein D in a subset of species. Species that

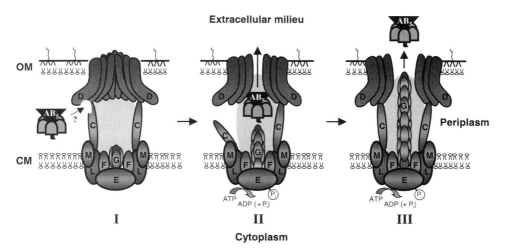

Figure 2 Model of pilus-mediated secretion by the type II apparatus. Assembled CT (AB_5) is targeted to the secretion apparatus via specific recognition of B_5 in a process that may involve components C and/or D (I). Upon binding, conformational changes in the type II apparatus may result in accommodation of the AB_5 complex and lead to polymerization of component G and the other pilin-like proteins into a piston-like structure that extends from the E-F-L-M platform in the cytoplasmic membrane (II). Through the active process of pilin subunit polymerization, AB_5 is then pushed through the channel to the extracellular environment in a process that is thought to be regulated and/or energized by the E component (III). Proteins A, B, N, and S are not shown in this model since they are not present in all organisms that encode and assemble a type II pathway and these proteins may not be required for secretion in every case.

lack the protein S homolog may or may not produce another protein that assists in the outer membrane localization of protein D. Two D components, PulD and $XcpQ_D$ from *K. oxytoca* and *P. aeruginosa*, respectively, and pIV, a protein D homolog required for f1 phage secretion in *E. coli*, have been purified and analyzed in detail. When inserted into planar lipid bilayers, they conduct ions. Moreover, electron microscopic analyses show that these proteins are present in ring-like structures that contain 75- to 95-Å-wide openings in the center that could accommodate folded proteins during their transport across the outer membrane. Taken together, these studies show that protein D, with most certainty, forms an outer membrane pore or channel. A closer analysis of the D structures revealed an electron-dense mass in the center of the cavities, suggesting that these pores may be gated. A gating mechanism must exist; to have such a large open channel would make the cells vulnerable to the outside environment and allow the periplasmic contents to leak out. How the pore is gated and what components are involved in the opening of the closed channel are not known. It is postulated that the protein D oligomer serves as the export channel for secreted proteins, but this has not been visualized yet. However, with the protein D homolog pIV, phage particles have been shown to extrude through the pore during transport from the *E. coli* cell (6). It is believed that in the case of the type IV pilus system, pilin subunits that polymerize into the approximately 60-Å-wide pilus structure are also extruded through the outer membrane D oligomer. The situation is less clear for the type II system since it is not known whether the pilus-like structure and the secreted proteins extrude the pore separately or simultaneously. Size restriction of the pore dictates that both structures cannot occupy the channel simultaneously.

Protein E and the homologous components of the type IV pilus assembly apparatus constitute a family of putative ATPases that are peripherally associated with the cytoplasmic side of the inner membrane. They have typical ATP-binding sites (Walker box motifs) commonly present in ATPases, ATP synthases, and kinases. No ATPase activity has so far been detected for this class of proteins. However, the less related plasmid R64-encoded PilQ, a protein required for formation of thin flexible pilus, which is essential for transfer of the R64 plasmid from one bacterium to another via conjugation, demonstrates ATP hydrolysis (9). PilQ is 30% identical to the E component of the type II secretion system. The homology spans approximately 400 out of 500 residues and covers the region containing the Walker A and B boxes, an aspartic acid-rich region, and two CysXXCys motifs. Hence, it is likely that the E component is also an ATPase that provides energy to either the assembly of the secretion apparatus or for the actual secretion process. The only protein E with detectable enzymatic activity is EpsE from *V. cholerae*, which has been found to exhibit weak autophosphorylation activity. This may suggest that, in addition to being an ATPase, the E component may also function as a kinase that regulates the secretion process.

The last of the type II secretion components to show homology with a type IV pilus assembly protein is protein F. No activity has been demonstrated for this component. However, protein F is an inner membrane protein with three predicted membrane-spanning regions and has been coimmunoprecipitated with component E.

Besides the shared homology with type IV pilus components, the type II secretion proteins, when overexpressed, can assemble protein G into pilus-like structures on the cell surface (12). The type II apparatus can also assemble bona fide pilin subunits into bundled pili. Normal-level expression of these components from their chromosomally located genes does not give rise to detectable appendages on the cell surfaces. However, analogous to the type IV pilus system, there is a possibility that a pilus-like structure that extends from the cytoplasmic membrane and spans the periplasmic compartment is formed (Fig. 2). It has been hypothesized that with overproduction of the type II secretion components, there may be a shift in the equilibrium toward the extended surface-exposed form, while under normal conditions the pilus-like structure may not extend much past the outer membrane and cell surface (Fig. 2). Type IV pilus causes twitching motility in *N. gonorrhoeae* and *P. aeruginosa* as a result of pilin subunit extension and retraction (8). A similar mechanism might support type II-dependent secretion (2). Pilus extension from the cytoplasmic membrane may push the secreted proteins through the pore or may actually push open the pore (Fig. 2).

Components C, L, M, and N appear to be unique to the type II secretion system. Proteins homologous to these components are not present in the type IV pilus biogenesis system. However, this does not rule out the presence of components with similar functions. There is, in fact, precedence for the existence of proteins with similar functions. Protein E, for instance, is dependent on the cytoplasmic membrane protein L for membrane association. The E protein is completely soluble and remains in the cytoplasm in mutants lacking the L protein. Yet, the protein E (such as *P. aeruginosa* PilB) of the type IV pilus system is associated with the cytoplasmic membrane even though no obvious L homolog exists. Protein L, in turn, interacts with the cytoplasmic membrane protein M of the type II system. These two components stabilize each other from proteolytic degradation. A trimolecular complex formed by the three components E, L, and M can form in the absence of other proteins of the secretion machinery when they are expressed in *E. coli*. These three components stabilize each other and can be coimmunoprecipitated. A fourth component of the cytoplasmic membrane complex is protein F. The E, F, L, and M proteins may form a platform in the cytoplasmic membrane upon which the rest of the machinery is assembled (Fig. 2). Genetic analysis has shown that E also interacts with G, one of the pilin-like components. This interaction may be transient because mutual stabilization or coimmunoprecipitation of these proteins has not been observed. Protein C may be one of the components that connects the inner membrane platform composed of E, F, L, and M with the gated pore. Protein C appears to interact with L and/or M present in the inner membrane and protein D in the outer membrane (Fig. 2).

Finally, proteins A and B are present in only a handful of species and have only been shown to be required for extracellular secretion in *A. hydrophila* and *E. coli*. While there is no protein A in *Erwinia*, protein B is required for secretion in this species. Protein A may be an ATPase or kinase. Mutations in the Walker A motif of this protein result in inhibition of secretion in *A. hydrophila*. Proteins A and B stabilize each other and can be coimmunoprecipitated. It is possible that protein B also interacts with protein D, since inactivation of the B gene in *Erwinia* species results in proteolytic degradation of the D protein.

On the basis of the homology with type IV pilus components and available information on the various interactions between proteins of the type II secretion apparatus obtained through experimental analysis, a model for pilus-mediated secretion is presented in Fig. 2. Several of the interactions are hypothetical and some are likely to be transient. The number of individual components within the apparatus is not known, with the possible exception of protein D.

THE Eps COMPLEX IS RESTRICTED TO THE POLE OF *V. CHOLERAE*

Studies of *E. coli, Bacillus subtilis, C. crescentus,* and *P. aeruginosa* have shown that various proteins important in chemotaxis, cell division, development, motility, and adhesion are localized to distinct sites within the bacterial cell. More important, however, it is evident that the placement of proteins at the correct location within the cell appears to be critical for the proper function of many of these vital cellular components. For example, studies of the *E. coli* MinC, MinD, and MinE proteins indicate that cooperative interactions and proper distribution of these proteins during cell division are critical for formation of the septum at the midpoint of the cell. Coordinated and rapid movement of these proteins back and forth between the cell poles ensures that the septum is assembled only at the midpoint of the cell to yield daughter cells of equal size (5). Another example is the polar pilus produced by *P. aeruginosa*, which is important for adherence to tracheal epithelium. Attachment to the epithelium occurs via the polar pilus, followed by unidirectional penetration of the submucosal epithelial layer. The latter example also implies that distinct compartmentalization of proteins within the cell envelope may enhance virulence.

Two recent studies demonstrate that yet another multiprotein complex, the type II secretion apparatus, is localized to discrete sites within the cell. First, proaerolysin was overexpressed from a plasmid that had been genetically modified by the removal of a regulatory secondary structure that controls the rate of production of proaerolysin in *Aeromonas salmonicida* (1). Electron microscopy revealed that overexpression resulted in accumulation of proaerolysin within the periplasm, which consequently caused an increase in cellular osmotic pressure and a corresponding increase in the size of the periplasmic compartment. Enlargement of the periplasmic compartment, which was believed to be due to proaerolysin overproduction, was restricted to the cell poles. Furthermore, immunogold electron microscopy demonstrated the proaerolysin was predominantly limited to the pole, exclusive of the lateral portions of the cell (1). The authors suggested that restriction of proaerolysin to the poles could be evidence of periplasmic compartmentalization with concomitant secretion at the poles. Second, the green fluorescent protein (GFP) was fused with Eps proteins to determine the cellular location, in living cells, of the type II secretion apparatus in *V. cholerae* (13). A biologically active GFP-EpsL fusion that complemented the secretion defect in a *V. cholerae epsL* mutant was detected at the pole of the *V. cholerae* cell (Fig. 3B). Likewise, fluorescent microscopy revealed that functional GFP-EpsM was also restricted to the pole (Fig. 3C). Immunofluorescence with

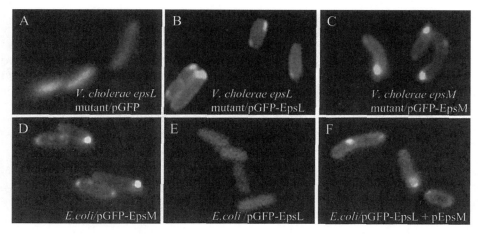

Figure 3   Polar localization of GFP-EpsL and GFP-EpsM protein fusions. Fluorescent microscopy, with the aid of an FITC filter, was used to determine the location within the cell of different GFP constructs. (A) GFP in *V. cholerae epsL* mutant. (B) GFP-EpsL in *V. cholerae epsL* mutant. (C) GFP-EpsM in *V. cholerae epsM* mutant. (D) GFP-EpsM in *E. coli* MC1061. (E) GFP-EpsL in *E. coli* MC1061. (F) GFP-EpsL and EpsM in *E. coli* MC1061.

anti-EpsL or anti-EpsG antibodies supported the GFP protein fusion data and confirmed that native Eps proteins are confined to the cell poles (13). Furthermore, the site of Eps-dependent protease secretion in single *V. cholerae* cells, like the Eps apparatus itself, was confined to the pole of the cell (Fig. 4). This study also indicated that the Eps apparatus and active HA/protease secretion were colocalized to the same pole (13).

Restriction of the type II apparatus to the pole could serve to cluster the relatively low number of complexes, estimated to be 50 to 100 per cell in *P. aeruginosa*. Grouping of the complexes may provide a means to deliver a concentrated quantity of secreted material directly to one site. Therefore, polar restriction of the Eps complex coupled with directed secretion could provide *V. cholerae* with a significant biological advantage. Furthermore, conservation of secreted material and cellular energy may be accomplished by directed secretion of chitinase during nutrient acquisition when *V. cholerae* is associated with chitinous particles in the aquatic environment. Likewise, if the Eps machinery is located at the same pole used by *V. cholerae* to attach to the epithelial cell surface during colonization of the small intestine, this would provide for targeted delivery of virulence factors such as CT. Moreover, HA/protease, which is also secreted by the Eps apparatus, may be responsible for detachment of *V. cholerae*

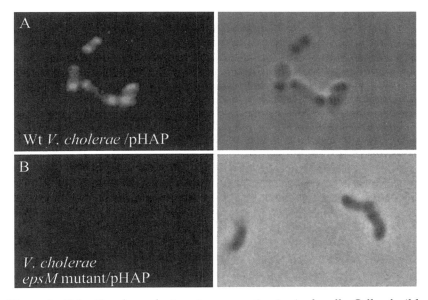

Figure 4 Polar Eps-dependent protease secretion in single cells. Cells of wild-type *V. cholerae* TRH7000 (A) and *epsM* mutant PU3 (B) that expressed IPTG-inducible HA/protease from plasmid (pHAP) were embedded in agarose supplemented with M9 salts, amino acids, IPTG, and intramolecularly quenched casein. Casein became highly fluorescent when cleaved by secreted HA/protease. The location of HA/protease secreted from single cells via the Eps apparatus was determined by fluorescence microscopy with the use of a TRITC filter after overnight incubation at 37°C. Next to panels (A) and (B) are the corresponding phase-contrast images.

from the epithelial cell surface following digestion of putative cell surface receptors. The directed secretion of HA/protease at the site of bacterial adherence, if the Eps apparatus is located at the site of attachment, could result in rapid release and augment dissemination of the bacteria. Thus, the inference is that polar localization of the secretion apparatus may play a key role in the pathogenesis of *V. cholerae*.

TARGETING OF THE Eps APPARATUS TO THE POLE

Preliminary data demonstrated that EpsM is targeted to the pole in the absence of other Eps components when expressed in *E. coli* (Fig. 3D) and that detection of EpsL at the pole required EpsM expression (Fig. 3F). Hence, not all Eps proteins possess the necessary information for polar localization. Therefore, EpsM alone or in combination with other Eps proteins may promote polar clustering of other Eps components.

Understanding the spatial distribution of specific proteins and the exact mechanisms used to direct these proteins to distinct regions of the cell envelope continues to elude investigators. Several hypotheses have been put forth to explain how proteins are targeted to specific destinations within the bacterial cell. One theory suggests that, in comparison to the lateral wall, the presence or absence of specific components, such as proteins and lipids, at the poles may designate these sites for delivery of polar complexes (5). Other studies indicate that new glycan strands are incorporated into the elongating sacculus along the lateral membrane and the septum while the "old" poles remain static (5). Consequently, the stable environment that exists at the old poles (as opposed to the newly generated poles formed at the septum after cell division) may provide a mechanism to anchor complexes at this location. Results of time-lapse fluorescent microscopy of *V. cholerae* that expresses GFP-EpsM indicates that the Eps apparatus is assembled at the old pole (13). However, during the experiment, as the cells stopped dividing and "aged," a shift from unipolar to bipolar distribution of the Eps complex occurred. This may reflect the maturation of the new poles and development of properties typical of the old pole.

Recent intriguing data based on experiments accomplished with an active blue fluorescent protein (BFP) fused with ribosomal protein L1 in *B. subtilis* showed that ribosomes are located around the periphery of nucleoids, predominantly at the cell poles (7). Location of ribosomes to the poles appears to be a dynamic process, dependent on transcription, and not simply the result of nucleoid exclusion. Additionally, cold shock proteins (CSPs) ensure proper translation initiation of nascent mRNA as they are expelled from the border of the nucleoid following transcription. Extension of these experiments illustrated that GFP-CSP and L1-BFP colocalized to the cell pole (7). Taken together, the evidence suggests that CSPs serve to couple transcription and translation in the bacterial cell. The implication is that proteins may simply localize to the pole by default through the expression and translation of their genes at this site. Thus, a retention signal, instead of a targeting signal, may be present on these proteins to prevent their diffusion and maintain them at the pole. Future in situ experiments in which mRNAs, specific for polar proteins, are efficiently labeled and detected by high-resolution microscopy may help resolve the dilemma of how some polar proteins arrive at their destinations.

FUTURE DIRECTIONS

Future investigations seek to comprehend how the individual Eps components are targeted and retained at the pole and how they assemble to form the highly organized multiprotein secretion complex. Experiments will also address the mechanism of secretion used by proteins to gain access to the extracellular environment. Type II-dependent secretion is complex because the proteins that are secreted are in the native conformation or nearly folded state and display no obvious sequence homology with each other. What structural commonality or specific secretion signal designates these proteins for type II-dependent secretion? These diverse proteins are first transported by way of the Sec pathway to the periplasmic compartment and must be directed to the pole before they are translocated across the outer membrane. Does this step require a Sec machinery located at the pole to preferentially transport these proteins? Conversely, are these proteins randomly transported across the cytoplasmic membrane, uniformly distributed within the periplasm, and then directed to the polar Eps apparatus by diffusion or via a special Eps targeting protein? Does polar retention of the Eps apparatus augment pathogenesis? Is there an unequivocal link between the location of the Eps apparatus, directed secretion of putative virulence factors, and the pathophysiology of *V. cholerae* in animal models or the human host?

Finally, the 21st century presents new therapeutic challenges to be met by the medical community given that many bacteria, both opportunistic and frank pathogens, have become antibiotic resistant. Yet, many important pathogens that represent a significant health risk, such as *V. cholerae*, *P. aeruginosa*, and *E. coli* O157, encode the type II secretion pathway, which could serve as an Achilles' heel against which treatment can be targeted. Therefore, elucidation of the type II system may be pivotal in the identification of novel therapeutic agents that can disrupt the function of the secretion apparatus and thereby block secretion of important colonization or potent virulence factors.

REFERENCES

1. **Burr, S. E., D. B. Diep, and J. T. Buckley.** 2001. Type II secretion by *Aeromonas salmonicida*: evidence for two periplasmic pools of proaerolysin. *J. Bacteriol.* **183:**5956–5963.
2. **Filloux, A., G. Michel, and M. Bally.** 1998. GSP-dependent protein secretion in gram-negative bacteria: the Xcp system of *Pseudomonas aeruginosa*. *FEMS Microbiol. Rev.* **22:** 177–198.
3. **Hirst, T. R., and J. Holmgren.** 1987. Conformation of protein secreted across bacterial outer membranes: a study of enterotoxin translocation from *Vibrio cholerae*. *Proc. Natl. Acad. Sci. USA* **84:**7418–7422.
4. **LaPointe, C. F., and R. K. Taylor.** 2000. The type 4 prepilin peptidases comprise a novel family of aspartic acid proteases. *J. Biol. Chem.* **275:**1502–1510.
5. **Lybarger, S. R., and J. R. Maddock.** 2001. Polarity in action: asymmetric protein localization in bacteria. *J. Bacteriol.* **183:**3261–3267.
6. **Marciano, D. K., M. Russel, and S. M. Simon.** 2001. Assembling filamentous phage occlude pIV channels. *Proc. Natl. Acad. Sci. USA* **98:**9359–9364.

7. **Mascarenhas, J., M. H. Weber, and P. L. Graumann.** 2001. Specific polar localization of ribosomes in *Bacillus subtilis* depends on active transcription. *EMBO Rep.* **2:**685–689.

8. **Merz, A. J., M. So, and M. P. Sheetz.** 2000. Pilus retraction powers bacterial twitching motility. *Nature* **407:**98–102.

9. **Sakai, D., T. Horiuchi, and T. Komano.** 2001. ATPase activity and multimer formation of Pilq protein are required for thin pilus biogenesis in plasmid R64. *J. Biol. Chem.* **276:**17968–17975.

10. **Sandkvist, M.** 2001. Biology of type II secretion. *Mol. Microbiol.* **40:**271–283.

11. **Sandkvist, M.** 2001. Type II secretion and pathogenesis. *Infect. Immun.* **69:**3523–3535.

12. **Sauvonnet, N., G. Vignon, A. P. Pugsley, and P. Gounon.** 2000. Pilus formation and protein secretion by the same machinery in *Escherichia coli. EMBO J.* **19:**2221–2228.

13. **Scott, M. E., Z. Y. Dossani, and M. Sandkvist.** 2001. Directed polar secretion of protease from single cells of *Vibrio cholerae* via the type II secretion pathway. *Proc. Natl. Acad. Sci. USA* **98:**13978–13983.

Bacterial Protein Toxins
D. Burns et al., Editors
©2003 ASM Press, Washington, DC 20036

Chapter 7

Type III Secretion Systems

Gregory V. Plano, Kurt Schesser, and Matthew L. Nilles

Various types of bacteria have evolved the capacity to survive and proliferate while closely associated with eukaryotic cells. In some cases this ability is dependent on the bacterium injecting proteins directly into the eukaryotic cell cytoplasm. One such injection system, designated as the type III secretion system or TTSS, is believed to have originally evolved from the flagellar export system and is now dispersed among a number of both animal- and plant-interacting gram-negative bacteria. Although our ignorance is vast, we are slowly gaining ground in understanding both the mechanism of TTSS-mediated protein transfer and what the transferred proteins, which are called effectors, are doing once they are inside the eukaryotic cell. What is emerging from a variety of experimental work is that the TTSS and its effectors play an important role in shaping the outcome of the microbe-host interaction.

The TTSS was initially described and characterized in *Yersinia* species. In the 1980s it was found that a 70-kb "virulence plasmid" (so called since it was required to cause disease in mice) directed the massive secretion of approximately 10 proteins into the culture supernatant when a *Yersinia* cell culture was shifted from 26° to 37°C. At first, the function of these proteins was unknown, and many believed at the time that their observed "secretion" was simply due to the proteins being nonspecifically released from the bacterial membrane under laboratory culture conditions. With the development of molecular genetic techniques that made it possible to correlate genes with functions, three different types of genes on the virulence plasmid were identified: (i) those encoding the structural components of the secretion/injection apparatus, (ii) those encoding the effector proteins, and (iii) those encoding regulatory factors controlling the temporal expression of the structural and effector proteins (Fig. 1).

The secretion/injection apparatus of TTSSs of other gram-negative bacteria turned out to be fairly similar to that of *Yersinia* species. This was remarkable

Gregory V. Plano and Kurt Schesser • Department of Microbiology and Immunology, University of Miami, School of Medicine, 1600 NW 10th Ave., Miami, FL 33136. *Matthew L. Nilles* • Department of Microbiology and Immunology, University of North Dakota, School of Medicine and Health Sciences, Grand Forks, ND 58202.

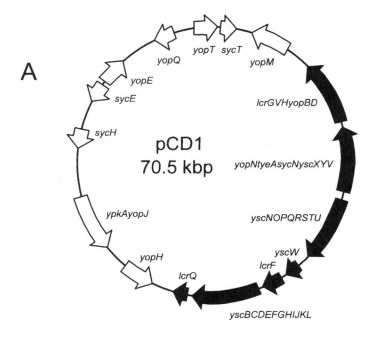

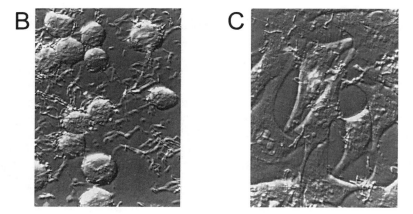

Figure 1 Genetic organization of the *Yersinia pestis* virulence plasmid pCD1.
(A) Genes encoding the components of the *Yersinia* TTSS are depicted as arrows showing the direction of transcription. Genes encoding proteins involved in the assembly, regulation, and function of the type III secretion apparatus are arranged in polycistronic operons within an approximately 25-kb region of the plasmid (filled-in arrows), suggesting that they were inherited collectively. In contrast, genes encoding the secreted effector proteins and their chaperones (open arrows) are found in mono- or bicistronic operons scattered around the remainder of the plasmid, suggesting that they were inherited independently. (B) HeLa cells infected with *Y. pestis*-carrying plasmid pCD1 showing YopE-dependent cytotoxicity (from reference 6 with permission). (C) HeLa cells infected with a *Y. pestis* strain with a defective TTSS (from reference 6 with permission).

given that these various species occupy disparate ecological niches (Table 1). On the other hand, the TTSS effectors are for the most part unique to each species, likely reflecting the various strategies and challenges these species face during their encounters with eukaryotic organisms. The "conserved delivery systems/ unique effectors" paradigm is reflected in their genetic organization. The 20 to 30 genes (depending on the species) encoding the secretion/injection apparatus are arranged in closely linked polycistronic operons within pathogenicity islands (see Box 2). In contrast, the majority of the TTSS effectors are encoded by un-linked genes dispersed throughout the genome. Since there can be variability in the composition of TTSS effector-encoding genes even within a given species, it is believed that these genes enter and exit the genome independently of the secretion/injection TTSS components. Although it may be easier to understand

Table 1 Bacterial virulence-associated type III secretion systems

Organism	Location[a]	Effector proteins	Function in virulence
Yersinia spp.	70-kb plasmid	YopE, YopH, YopJ, YopM, YopT, LcrV, YpkA	Inhibition of phagocytosis; induction of apoptosis; suppression of the in-flammatory response
Pseudomonas aeruginosa	Chromosome	ExoS, ExoT, ExoU, PcrV	Inhibition of phagocytosis; induction of apoptosis
Salmonella enterica; SPI-1	Chromosome, centisome 63	AvrA, SipA, SlrP, SopA, SopB, SopD, SopE, SopE2, SptP	Invasion of epithelial cells; induction of apoptosis; cytoskeletal rearrange-ments; induction of the inflammatory response
Salmonella enterica; SPI-2	Chromosome, centisome 31	SpiC, SifA	Required for systemic in-fection, intracellular growth in macrophages; avoidance of NADPH oxidase-dependent kill-ing
Shigella flexneri	220-kb plasmid	IpaA, IpaB, IpaC, IpgD	Invasion of epithelial cells; induction of apoptosis
Enteropathogenic *Escherichia coli*	Chromosome, centisome 82	EspF, Tir (EspE)	Adherence to intestinal mucosa; effacement of intestinal microvilli; ped-estal formation
Chlamydia spp.	Chromosomal	Inc proteins[b]	Unknown
Plant pathogens	Chromosomal	Harpin, Avr proteins[b]	Induction of necrotic lesions in susceptible plants; induction of the hypersensitive response in resistant plants

[a] Location of TTSS gene clusters or pathogenicity islands; genes encoding individual effector proteins may be located outside these regions; specific strains may carry different combinations of effector protein genes.

[b] Inc proteins, inclusion membrane proteins; Avr proteins, avirulence proteins.

BOX 2

Pathogenicity islands

General structure of pathogenicity islands (PAIs)

Many years ago the question arose why bacteria belonging to the same or related species may significantly differ from each other in their pathogenic potency. Through molecular techniques it quickly became clear that many pathogens carry "additional" pieces of DNA, which may be part of plasmids, phages, or the chromosome. These DNA regions have been termed PAIs. Since the first description of PAIs as particular genomic regions of different *Escherichia coli* pathotypes, these genetic elements have been described in more than 20 bacterial species. As indicated in Fig. 1, PAIs may cover the following characteristics: They are present in the genome of pathogens but absent from the genome of the majority of non-pathogens and they encode often

more than one virulence factor. PAIs may comprise large genomic regions, most likely on the chromosomes, and are often flanked by directly repeated DNA segments, which may form the basis for their instability. They are often located next to tRNA genes, carry additional (often cryptic) "mobility genes," and often exhibit an altered G+C content and differences in the codon usage, compared to the "core" genome. Many PAIs may have evolved in a multistep process, a phenomenon that is reflected by their mosaic-like structure.

Virulence factors encoded by PAIs

As PAIs may be part of the genome of many gram-negative and gram-positive pathogens, different types of virulence factors may be encoded by these elements. Thus, adherence factors such as P fimbriae of uropathogenic *E. coli* or iron-uptake systems such as yersiniabactin or aerobactin, produced by many enterobacteria, are

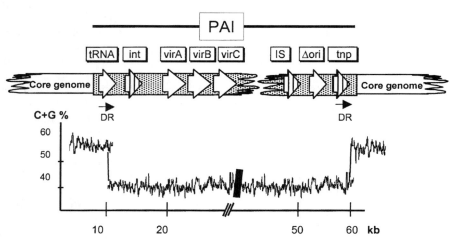

Figure 1 Model of a bacterial PAI. The boxes represent genes, the arrows indicate the presence of direct repeats at the ends of the PAI. DR, direct repeats; *int*, integrase gene; *vir*, virulence-associated gene; IS, insertion sequence; Δori, pseudo-origin of replication; *tnp*, transposase gene. At the lower part of the figure, the estimated G+C content is indicated.

BOX 2 (Continued)

PAI encoded. Furthermore, the genes responsible for capsule antigens and toxins such as α-hemolysin of *E. coli* or staphylococcal superantigens may be parts of PAIs. In addition, genes coding for secretion systems and/or their effector molecules may be located on PAIs. This is true for genes of the type I secretion systems (which, in the case of α-hemolysin, are part of the hemolysin operon) and also of the type III and type IV systems. The type III and IV secretion systems are of particular interest as these systems are often encoded by pathogens, where they deliver effector molecules, which interact with host cell structures. Genes encoding secretion systems as well as other virulence determinants may be located on plasmids and on chromosomes as well, suggesting a horizontal transfer of the respective gene clusters. Type III systems are expressed not only by pathogens of humans and animals but also by plant pathogens.

Other PAI-encoded properties

In addition to structural genes encoding virulence factors, genes necessary for other functions may be part of PAIs. Thus, regulatory elements may be important for the regulation of PAI-encoded genes and also for genes of the core genome. The presence of different mobility genes encoding integrases or transposases, origins of replication or insertion sequence elements, reflects that PAIs represent formally transferred DNA regions and that many of these elements are instable or even mobilizable by helper phages and plasmids. tRNA genes often act as targets for the integration of foreign DNA, especially following phage transfer. That tRNA genes often flank PAIs is a further indication that they were horizontally transferred in the past. Many of the mobil-

ity genes, however, are destroyed by point mutations or small deletions, which may indicate that successful genomic variants have been fixed in the genomes of the respective carrier organisms.

PAIs and their impact on microbial evolution

Many PAIs have been selected not only on the basis of their pathogenic potency but on their contribution to adaptation of microbes to certain ecological niches. Therefore, it is not surprising that genomic regions that are considered as PAIs in pathogenic organisms were also detected in the genomes of nonpathogenic organisms where they may play a role as "fitness" or "ecological islands." In general, DNA regions with PAI-like structures are located in the genomes of many organisms; thus, they have been termed "genomic islands" (GIs). PAIs form a particular subgroup of GIs. GIs in general and PAIs in particular represent DNA fragments that are part of the so-called "flexible gene pool," which has been horizontally transferred from one strain or species to another via conjugation or transduction and whose products may significantly contribute to the adaptation of microbes. As such, events of gene transfer may alter the properties of microbes dramatically. Accordingly, the occurrence of such islands reflects microbial evolution "in quantum leaps." Such processes seem to be still ongoing, and it is tempting to speculate that new variants of pathogens will arise in the future as a result of such evolutionary processes. Therefore, new PAIs will certainly be discovered in the next decades.

1. **Hacker, J., and J. B. Kaper.** 2000. Pathogenicity islands and the evolution of microbes. *Annu. Rev. Microbiol.* **54:** 641–679.

Jörg Hacker

the advantages of acquiring TTSS effectors through horizontal genetic transfer, it may be equally important to "lose" the ability to secrete and inject a particular TTSS effector. This is clearly the case in some plant-microbe relationships in which a plant defense response (analogous to an animal inflammatory response) can be activated by a bacterial TTSS effector. If a given plant acquires such an ability to recognize a bacterial "invader," then it clearly would be to the bacterium's advantage to cease expression of that particular TTSS effector.

Adding even another level of complexity to the roles TTSSs play in influencing the host-microbe interaction, several species of bacteria have been shown to possess two separate TTSSs. For example, enteropathogenic *Salmonella* spp. possess two genetically unlinked TTSSs that appear to operate independently of one another during different stages of interaction of *Salmonella* cells with host cells. The first discovered TTSS of *Salmonella* spp., that encoded within *Salmonella* pathogenicity island 1 (SPI-1), is required for *Salmonella* spp. to invade host cells. Following invasion of the host cell, the second TTSS of *Salmonella* spp., SPI-2, is expressed and required for intracellular proliferation. Apparently SPI-1 and SPI-2 TTSSs secrete and inject different effectors, but how the effectors are distinguished by the two different systems is unknown.

The aim of much current research is to understand the mechanisms involved in effector secretion/injection and what the effectors are doing inside the host cell. In this chapter the secretion/injection issue will be addressed.

RECOGNITION OF EXPORTED PROTEINS

Bacterial pathogens use several different protein secretion pathways to export virulence proteins from the bacterial cytoplasm to their site of action (see section overview). Proteins destined for secretion by each of these pathways must possess unique features or distinct secretion signals that target them to the correct membrane secretion complex. System-specific accessory proteins or chaperones often assist in the sorting and export of secretion substrates through specific recognition of substrates and/or by maintaining substrates in an unfolded and secretion-competent state prior to their export.

Proteins exported by TTSSs are not cleaved and have no classical *sec*-dependent signal sequence; however, the information required to direct proteins into the type III pathway is contained within the N-terminal coding region of each respective gene. For example, gene fusion experiments carried out with YopE, YopH, and YopQ of *Yersinia enterocolitica* and *Yersinia pseudotuberculosis* determined that the N-terminal 15 to 17 residues of these proteins were sufficient to direct the secretion of heterologous reporter proteins. Interestingly, no primary sequence homology or common secondary structure could be detected between the N-terminal regions of the various secreted Yops. Furthermore, frameshift mutations that completely altered the amino acid sequences within the identified regions failed to prevent secretion. These observations indicate that proteins secreted via the type III pathway are recognized in a manner that is fundamentally different from that described for the classical signal-peptide-dependent secretion systems of bacteria and eukaryotes.

mRNA or Amino Acid Secretion Signals

Two hypotheses have been proposed to explain how N-terminal secretion signals are recognized by the type III secretion machinery (Fig. 2). One theory proposes that the 5′-coding regions of the *yop* mRNAs serve as secretion signals, suggesting that Yops can be secreted in a cotranslational manner (1). Indeed, the 5′ ends of several *yop* mRNAs are predicted to form stem-loop structures that could conceal translational start codons and thus prevent Yop translation until the transcripts and associated ribosomes can interact with the secretion machinery. Analysis of YopQ expression and secretion confirmed that *yopQ* mRNA translation and secretion are coupled; however, experiments with YopE failed to demonstrate a similar relationship. Several proteins from the plant pathogen *Pseudomonas syringae* also appear to utilize mRNA secretion signals, indicating that such signals may be a general feature of TTSSs.

Conversely, experiments with YopE of *Y. pseudotuberculosis* suggest that N-terminal amino acid residues, and not mRNA signals, are involved in the recognition and export of YopE (3). A comparison of the N-terminal amino acid

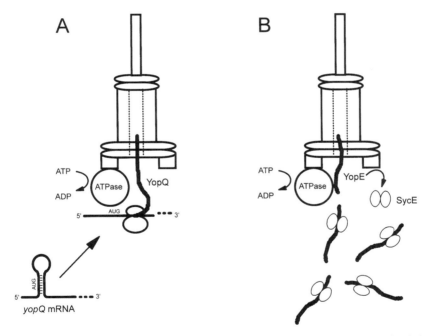

Figure 2 Models for recognition of *Yersinia* type III secretion signals. (A) mRNA secretion signal hypothesis. Translation of *yopQ* mRNA is proposed to be inhibited by its own RNA structure or by a component of the type III secretion machinery. The inhibition of translation is relieved upon interaction of the mRNA with the type III secretion apparatus, triggering cotranslational secretion of YopQ. (B) Amino acid secretion signal hypothesis. Translated YopE is recognized by the SycE chaperone, forming a stable cytosolic YopE/SycE complex. Recognition of an N-terminal amino acid sequence by a component of the secretion apparatus triggers YopE secretion and release of SycE.

sequences of several Yops led investigators to suggest that an N-terminal amphipathic sequence may be involved in Yop export. In support of this hypothesis,
a sequence consisting of alternating polar (serine) and hydrophobic (isoleucine)
residues was shown to function as an N-terminal secretion signal in *Y. pseudotuberculosis*. N-terminal amino acid secretion signals have also been shown to be
present in several exported flagellar proteins, including the anti-sigma factor
FlgM. Structural analyses of several flagellar substrates indicate that the N termini of these proteins are unstructured in solution, suggesting that the type III
secretion machinery is capable of recognizing a variety of unstructured and/or
amphipathic amino acid sequences.

Chaperone-Dependent Secretion Signals

A number of type III secretion substrates have a second amino acid targeting
domain that is dependent on the binding of a specific type III chaperone (see
below). For example, YopE is recognized by a type III chaperone termed SycE.
SycE is a small homodimeric cytoplasmic protein that specifically binds within
amino acid residues 15 to 50 of YopE and functions to solubilize and stabilize
YopE in the bacterial cytosol. *Yersinia* YopE mutants with defective N-terminal
secretion signals are absolutely dependent on SycE binding for type III export.
Furthermore, strains lacking SycE secrete and translocate reduced levels of YopE
or YopE-hybrid proteins that carry the chaperone-binding domain. SycE binding
appears to facilitate YopE export in part by countering the effect of a YopE domain that is inhibitory to export in the absence of the chaperone and in part by
preventing aggregation and degradation of YopE. SycE expression also enables
Yersinia spp. to accumulate or store YopE within the bacterial cell. The cytosolic
YopE/SycE complexes provide the bacterium a means to maintain a pool of YopE
that can be mobilized upon contact with a eukaryotic cell. Indeed, a SycE-
dependent burst of YopE secretion and translocation sufficient to cause cytotoxic
effects occurs immediately following cell contact.

The different targeting pathways and chaperones discussed above provide
bacteria a means to organize their assault on eukaryotic cells. In *Yersinia* spp.,
for example, Yops involved in antiphagocytosis (YopE, YopH, and YopT) utilize
type III chaperones, whereas the other effector Yops (YopJ, YopM, and YpkA) are
targeted exclusively via N-terminal or mRNA secretion signals. The chaperone-
dependent targeting of the antiphagocytic Yops enables these bacteria to prevent
engulfment by delivering a burst of the appropriate effectors immediately following contact with a phagocytic cell. Safe from engulfment, bacteria then deliver
other effector Yops that modulate the host's response to the infection. In the end,
the ability to order or time the delivery of effector proteins enables pathogens to
fine-tune the delivery of effector proteins according to their role in the infectious
process.

The high level of similarity shared between many of the components from
the various TTSSs suggests that these systems recognize similar secretion signals.
In fact, it has been shown that secreted proteins from one TTSS can often be
secreted from a heterologous TTSS. For example, IpaB, a *Shigella* sp. type III
secretion substrate, is secreted via the *Yersinia* spp. TTSS in the presence, but not
in the absence, of its chaperone, IpgC. Likewise, YopE is secreted via the *Sal-*

monella spp. SPI-1 TTSS in a SycE-dependent manner. Similarly, *P. syringae* export substrates AvrB and AvrPto were secreted by the *Yersinia* spp. TTSS and via the *Erwinia chrysanthemi* TTSS cloned in *Escherichia coli*. These findings confirm that the various TTSSs use a relatively well-conserved delivery apparatus to dispense a wide variety of effector proteins by recognizing common secretion signals (mRNA, amino acid, and type III chaperone-dependent).

TYPE III SECRETION CHAPERONES

Type III secretion chaperones are found in almost all bacteria that use TTSSs; however, not all type III secretion substrates utilize chaperones. Chaperones bind to effector proteins in the bacterial cytosol and remain cytosolic following export of their cognate substrate. Each chaperone recognizes and binds to one or two effector proteins that are generally encoded just upstream or downstream of that chaperone's structural gene. In virulence-associated systems, type III chaperones function as homodimers that bind to an N-terminal region of their substrate, whereas in the flagellar system the chaperone-binding site is localized at the C terminus.

Although there is only limited amino acid sequence homology among these chaperones, they share common features, including small size (12 to 18 kDa), an acidic pI, and an overall α-helical character. Recent structural studies of several chaperones, including SigE and SicP of *Salmonella enterica* and CesT of *E. coli*, demonstrate that these proteins function as homodimers that share a common overall fold (4, 7). In addition to a shared structure, each chaperone homodimer exhibits an extensive amount of surface-localized negative charges that are interspersed with hydrophobic patches. Structural analysis of SicP complexed with the chaperone-binding domain of SptP indicates that the substrate-interacting surfaces of these chaperones are primarily hydrophobic in nature. Interestingly, the chaperone-binding domain of SptP bound by SicP is maintained in an extended, unfolded conformation, suggesting that type III chaperones maintain their substrates in an unfolded, secretion-competent conformation. However, studies involving the *E. coli* Tir protein and its cognate chaperone CesT have shown that CesT binding has no effect on the ability of Tir to function as a receptor for intimin, suggesting that the unfolding associated with the binding of type III chaperones may be localized to the chaperone-binding region of the protein.

In the absence of a chaperone, secretion and translocation of the cognate protein are prevented or dramatically reduced (see "Recognition of Exported Proteins"); however, the exact function of these chaperones in the secretion process remains elusive. Type III chaperones may directly participate in the delivery of the protein to the type III secretion complex. In the flagellar secretion system, the ATPase FliI has been shown to interact directly with the flagellar chaperone FliJ and several secreted proteins. On the other hand, type III chaperones may play an indirect role in the secretion process, functioning primarily to maintain secretion substrates in a secretion-competent state by preventing their folding, aggregation, and degradation prior to their export. In addition to their role in secretion, several type III chaperones appear to perform functions that are not directly related to the type III export process. For example, SicA, a type III chaperone of *Salmonella enterica*, directly interacts with the AraC-like transcriptional regulator InvF and functions with InvF to regulate the expression of effector proteins and chaperones. Thus, type III chaperones have apparently evolved to perform a variety of different functions, including functions that are not directly related to the secretion process.

TRANSLOCATION OF PROTEINS ACROSS THE EUKARYOTIC MEMBRANE

Virulence-associated TTSSs export proteins from the bacterial cytosol directly into the cytosol of a eukaryotic cell. In general, the process of moving a protein across the bacterial membranes is referred to as secretion or export, whereas the process of moving a protein across the eukaryotic cell membrane is termed translocation or targeting. Although these two processes are closely coupled, the secretion process can be examined independently under conditions that induce secretion in the absence of eukaryotic cells. On the other hand, translocation of substrates across the eukaryotic cell membrane can only be measured in the presence of eukaryotic cells. Translocation of proteins across eukaryotic cell membranes requires specific secreted accessory proteins that insert into the eukaryotic membrane and assemble into pore-forming multiprotein complexes termed bacterial translocons. Translocons are thought to facilitate the entry of secreted effector proteins into the eukaryotic cytosol (Fig. 3).

Proteins secreted and translocated via the type III secretion pathway carry multiple signals that direct the proteins to the type III membrane secretion complex and into the eukaryotic cell via the assembled type III translocon. This process is best understood in the pathogenic yersiniae. These organisms are capable of transporting several toxins into the cytosol of eukaryotic cells. The immediate effect of this targeted secretion of toxins is a blockage of phagocytosis that allows invading *Yersinia* cells to remain in an extracellular environment. The directional secretion of toxins into eukaryotic cells is accomplished using a type III secretion apparatus and several secreted accessory proteins that insert into the eukaryotic membrane and facilitate intoxication of the cell. One of the hallmarks of type III intoxication is the directional nature of the process. The translocation process results in intracellular delivery of most of the Yops without significant release into the culture medium, resulting in the characterization of the process as being directional or vectorial.

The Bacterial Translocon

Three proteins (YopB, YopD, and LcrV) secreted by the *Yersinia* spp. type III secretion apparatus play clear roles in the translocation of the effector Yops into eukaryotic cells. The majority of current data support a role for all three proteins in the insertion of a 16- to 23-Å pore in the cytoplasmic membrane of eukaryotic cells. YopB, which contains two central hydrophobic domains, has a demonstrated ability to disrupt membranes and is required to lyse sheep red blood cells. YopD has a central hydrophobic domain and is also required for lysis of sheep red blood cells. Both YopB and YopD can be recovered from membranes of *Y. enterocolitica*-infected cells, suggesting that these proteins are directly in-

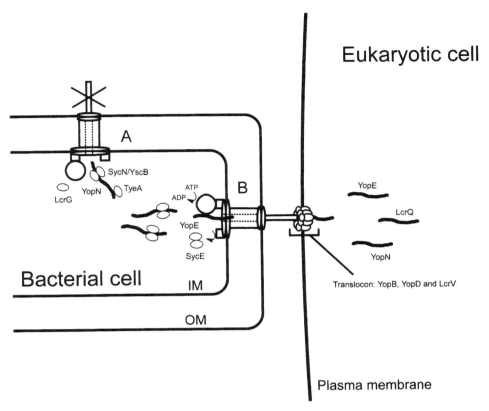

Figure 3 Cell contact-dependent secretion and translocation of Yops. Before contact with a eukaryotic cell, YopN, TyeA, SycN, YscB, and LcrG prevent Yop secretion, possibly by physically blocking the secretion channel. Contact with a eukaryotic cell relieves the block in secretion at the point of cell-to-cell contact. Secreted YopB, YopD, and LcrV assemble into a pore-forming translocon within the eukaryotic membrane, facilitating the translocation of effector proteins into the eukaryotic cell. Secretion of LcrQ, a negative regulatory protein, triggers increased transcription of genes encoding the Yop effector proteins, ensuring the availability of effector proteins for delivery. Once inside the eukaryotic cell, the Yop effector proteins block phagocytosis and subvert the normal signaling pathways of the cell.

serted into eukaryotic membranes. In addition to YopB and YopD, LcrV (the classic V antigen) is also required for secretion and targeting of effector Yops. LcrV has a demonstrated role in the lysis of sheep red blood cells and in the formation of pores in eukaryotic membranes. Moreover, LcrV is exposed on the surface of the bacterium in discrete foci prior to cell contact, and some antibodies directed against LcrV can block translocation of Yops into eukaryotic cells. Taken together, all of the evidence strongly suggests roles for YopB, YopD, and LcrV in the translocation of Yops into eukaryotic cells.

Secreted proteins exhibiting significant structural similarities to YopB and YopD are present in all TTSSs of animal pathogens but not plant pathogens.

Although the amino acid sequences of these proteins are not highly conserved, their role in translocation has been confirmed in several other pathogens, including *Salmonella enterica, Shigella flexneri, Pseudomonas aeruginosa,* and enteropathogenic *E. coli* (Table 2). The only known type III homolog of LcrV is the *P. aeruginosa* protein PcrV. Expression of PcrV in a *Y. pseudotuberculosis lcrV* mutant restored Yop translocation, confirming a role for both proteins in the translocation process. Accumulating evidence suggests that YopB, YopD, LcrV, and related proteins in other bacterial pathogens represent the individual components of channel-forming bacterial translocons, which facilitate the entry of toxins into eukaryotic cells.

Role of the Needle and Associated Structures in Translocation

In enteropathogenic *E. coli,* type III-mediated translocation occurs through a pilus-like structure composed of the protein EspA. The EspA pilus is formed on the tip of the putative type III needle structure that is composed of polymerized EscF. EscF is required for effector protein secretion, EspA filament assembly, and translocation of Esp proteins, whereas EspA is only required for the translocation process. Furthermore, EspA has been shown to interact directly with both EscF and EspB. Thus, the EspA pilus-like structure likely represents the physical bridge between the type III secretion apparatus and the eukaryotic membrane-embedded EspB- and EspD-containing translocon.

Table 2 Secreted proteins required for translocation of effector proteins into eukaryotic cells[a]

Salmonella (SPI-1)[b]	Shigella[b]	Enteropathogenic E. coli[b]	Yersinia[b]	Pseudomonas[b]	Function
SipB[c] (2)	IpaB (2)	EspB (1); EspD (2)	YopB (2)	PopB (2)	Translocation, pore formation
SipC[c] (1)	IpaC (2)	–		–	Translocation, pore formation
–	–	–	YopD[c] (1)	PopD (1)	Translocation, pore formation
SipD[c] (0)	IpaD (0)	–	–	–	Translocation, regulation of secretion
–	–	–	LcrV[c] (0)	PcrV (0)	Translocation, pore formation
–	–	EspA[c] (0)	–	–	Translocation, pilus subunit

[a] Secreted proteins specifically required to translocate effector proteins across the eukaryotic cell membrane.
[b] Numbers in parentheses indicate the number of predicted transmembrane domains in each protein.
[c] Based on amino acid identity and similarity, proteins were grouped into SipB/YopB, SipC/IpaC, YopD/PopD, SipD/IpaD, LcrV/PcrV, and EspA families.

REGULATION OF TTSSs

Expression and secretion of type III secretion components and substrates in the various bacterial pathogens that employ TTSSs are regulated by a multiplicity of signals and signal-response systems. In general, TTSS components are expressed under conditions similar to those the pathogen encounters during the infectious process. For example, expression of type III secretion components and substrates in *Yersinia* spp. is controlled by several known environmental signals, including temperature, contact with eukaryotic cells, and the removal of Ca^{2+} from the growth environment. Although the first two signals are most likely to be relevant during an infection, study of the latter signal has been extremely useful as an investigative tool. Control of Yop expression and secretion in the yersiniae is complex and occurs at multiple levels. The best-characterized control mechanisms exist at the levels of transcription and secretion. These control circuits are interconnected, but the mechanisms connecting them remain elusive, and progress is being made at unraveling the details.

Regulation at the Level of Secretion

Secretion of proteins by the type III pathway is not a constitutive process; instead, export is triggered by exogenous signals. For example, initiation of Yop secretion in *Yersinia* spp. is triggered by contact between the bacterium and a eukaryotic cell. Secreted effector Yops are targeted directly into the eukaryotic cell and are not found in the extracellular milieu. *Salmonella* and *Shigella* species also initiate effector protein secretion after contact with a eukaryotic cell; however, in these cases, effector proteins are found both in the extracellular medium and within the targeted cell, suggesting that secretion and translocation are not tightly coupled.

Secretion of Yops by *Yersinia* spp. is blocked in the presence of Ca^{2+} prior to contact with the surface of a eukaryotic cell. The block in Yop secretion is dependent on the secreted YopN proteins, SycN, YscB, TyeA, and LcrG. Mutational inactivation of *yopN*, *sycN*, *yscB*, *tyeA*, or *lcrG* results in uncontrolled secretion prior to host cell contact and in a loss of polarized translocation after host cell contact. SycN and YscB form a SycN-YscB heterodimeric chaperone that specifically binds to YopN and facilitates entry of YopN into the type III secretory pathway. In contrast, TyeA binds to a C-terminal domain of YopN and prevents secretion of YopN from the bacterial cell. Contact with a eukaryotic cell is hypothesized to relieve the TyeA-mediated arrest in YopN secretion, triggering YopN export and activating type III secretion (Fig. 3).

LcrG is a small, apparently cytoplasmic protein that directly binds to a C-terminal domain of intracellular LcrV, an essential secreted component of the *Yersinia* translocon. LcrV and LcrG appear to work together to control the activity of the type III secretion apparatus. Evidence for regulatory roles of LcrG and LcrV comes primarily from studies on mutant strains of *Yersinia* lacking LcrG, LcrV, or both proteins. *lcrG* strains secrete Yops constitutively, while *lcrV* strains fail to secrete Yops. Strains lacking both *lcrG* and *lcrV* behave like *lcrG* strains, suggesting that the *lcrV* strains are not defective in the type III secretion process, but in control of the type III secretion process. A current model suggests that the activity of the type III secretion apparatus is controlled by varying the relative levels of LcrG and LcrV. The "LcrG-titration" model hypothesizes that activation of type III secretion involves inactivation of LcrG's secretion-blocking function by its interaction with LcrV. This model suggests that as the type III pathway is activated, perhaps via export of YopN, levels of LcrV increase relative to LcrG levels and that the excess LcrV titrates LcrG's secretion-blocking activity. The removal of LcrG from its secretion-blocking function triggers full activation of type III secretion. The "LcrG-titration" model is supported by the fact that mutant forms of LcrG that are incapable of interaction with LcrV abolish the secretion of Yops in vitro and block the translocation of Yops into cultured eukaryotic cells regardless of the level of LcrV present.

Regulation at the Level of Transcription

Genes that encode type III secretion components and substrates are regulated by system-specific transcriptional regulators and by components of global regulatory networks. For example, *Salmonella* invasion genes encoded within the SPI-1 pathogenicity island are regulated by a complex network of regulators that responds to a variety of environmental conditions. Regulators of SPI-1 genes are encoded both within SPI-1 (HilA, HilC, HilD, and InvF) and outside SPI-1 (PhoPQ, BarA/SirA, flagellar regulatory genes, and nucleoid-binding proteins). The complex regulatory network and wide variety of environmental signals involved in the transcriptional regulation of SPI-1 genes presumably reflect the importance of expressing these genes at the appropriate time during an infection.

   Yersinia type III genes are regulated in response to environmental signals (temperature and extracellular Ca^{2+}) and according to the status of the type III secretion apparatus. Transcriptional control is primarily exerted by two proteins, LcrF and LcrQ. LcrF, a member of the AraC family of transcriptional regulators (see chapter 8), activates transcription of type III secretion genes at temperatures approximating mammalian body temperature ($\geq 34°C$). An additional level of LcrF-dependent transcriptional activation is directly coupled to the activation state of the Yop export apparatus through LcrQ, a secreted negative regulatory protein. Secretion of LcrQ upon cell contact depletes LcrQ from the cytoplasmic compartment and triggers increased transcription of type III genes. The mechanism by which cytoplasmic LcrQ represses type III gene transcription is unknown; however, LcrQ-dependent repression also requires functional YopD and the SycD chaperone. The coupling of transcriptional regulation to the status of the export apparatus allows *Yersinia* spp. to coordinate effector protein expression and secretion. Interestingly, LcrQ shares sequence homology with the chaperone-binding domain of YopH. Indeed, the SycH chaperone is required for LcrQ secretion and translocation, thereby linking effector Yop secretion to negative regulation. The involvement of SycH and other secretion chaperones in the regulation of TTSSs may be a generalized phenomenon. For example, the *Salmonella enterica* SicA chaperone directly interacts with the InvF transcriptional regulator and functions with InvF to regulate the expression of type III chaperones and their substrates.

THE TYPE III SECRETION MACHINERY

Proteins that assemble to form the secretion apparatus are among the most highly conserved components of TTSSs. All virulence-associated TTSSs share at least 11 essential core structural components, 10 of which also share significant amino acid similarity with components of the flagellar export apparatus (Table 3). These components can be divided into three groups based on specific features and predicted subcellular location. One group consists of three predicted cytosolic or peripheral membrane proteins, including a highly conserved ATPase. The second group includes six proteins that have been demonstrated, or are predicted, to be integral inner membrane proteins. The last group consists of one outer membrane protein with sequence similarity to the secretin family of channel-forming proteins and one lipoprotein that may link the inner and outer membranes. In addition, there appear to be several less-conserved components that assist in the assembly of the secretion apparatus and/or are secreted by the TTSS.

Table 3   Broadly conserved and system-specific type III secretion components[a]

Salmonella (SPI-1)[b]	Yersinia[b]	Flagellar biosynthesis[b]	Cellular location[c]	Function
InvC (432)	YscN (439)	FliI (456)	C/P	ATPase
OrgB (227)	YscL (221)	FliH (235)	C/P	Regulates ATPase
SpaO (303)	YscQ (307)	FliN (137)	C/P	Export apparatus
SpaP (224)	YscR (217)	FliP (245)	IM	Export apparatus
SpaQ (86)	YscS (88)	FliQ (89)	IM	Export apparatus
SpaR (263)	YscT (261)	FliR (264)	IM	Export apparatus
SpaS (356)	YscU (354)	FlhB (383)	IM	Export apparatus
InvA (685)	YscV (704)	FlhA (692)	IM	Export apparatus
PrgH[d] (392)	YscD (419)	FliG (331)	IM	Needle complex
PrgK (252)	YscJ (224)	FliF (560)	IM/OM	Lipoprotein, needle complex
InvG (563)	YscC (607)	–	OM	Secretin, needle complex
InvH (147)	YscW (131)	–	OM	Lipoprotein, secretin pilot
PrgI (80)	YscF (87)	–	Secreted	Needle subunit
PrgJ (101)	YscI (115)	–	Secreted	Unknown, needle complex
OrgA (200)	YscK (209)	–	C/P	Unknown
InvI (147)	YscO (154)	–	Secreted	Unknown
InvJ (336)	YscP (455)	FliK (405)	Secreted	Regulates needle or hook length
–	YscE (66)	–	C/P	Unknown
–	YscG (115)	–	C/P	Chaperone for YscG
–	YscX (122)	–	Secreted	Unknown
–	YscY (114)	–	C/P	Chaperone for YscX

[a] Proteins required to form a functional TTS apparatus in *Yersinia*, *Salmonella* (SPI-1), and flagellar TTSSs.

[b] Numbers in parentheses indicate the length of the protein in amino acids.

[c] Locations are indicated by IM for inner membrane, OM for outer membrane, and C/P for cytoplasmic or peripheral membrane protein.

[d] PrgH shows structural similarities but no significant amino acid similarities to YscD and FliG.

Cytoplasmic ATPase Complex

An essential component of all TTSSs is a cytosolic ATPase related to the catalytic subunits of bacterial F_0F_1 ATPases. The role of these proteins in the export process is unknown; however, it has been suggested that they may play a part in targeting proteins to the membrane secretion complex and/or energizing the export process, functions similar to those performed by SecA in the general secretory pathway (5). The flagellar export system ATPase FliI has been shown to form a soluble complex with a homodimer of FliH. The interaction of FliH with FliI inhibited FliI ATPase activity, suggesting that FliH family members may regulate this activity. FliI has also been shown to interact with the flagellar secretion chaperone FliJ and with a number of secreted proteins, including flagellin and hook protein. The FliI-FliH complex may directly interact with secretion substrates and facilitate delivery to the secretion machinery. Interaction with the secretion machinery may remove FliH, thus stimulating ATPase activity and protein export.

Inner Membrane Secretion Components

Several conserved inner membrane proteins that are predicted to possess multiple transmembrane domains are found in all TTSSs, including the flagellar export system. In *Yersinia* spp. these include YscD, YscR, YscS, YscT, YscU, and YscV. Four of these proteins (YscD, YscR, YscU, and YscV) have been shown to span the inner membrane, whereas YscS and YscT are predicted to be integral membrane proteins. Homologs of these proteins in the flagellar system are thought to be located within the central pore of the flagellar MS-ring complex. Intergenic suppression analyses suggest that the flagellar YscV homolog FlhA directly interacts with FliF, the MS-ring component. Membrane topology models of YscR, YscU, and YscV predict these proteins to possess large hydrophilic cytoplasmic domains that could interact with secretion substrates and with more peripherally associated secretion components, such as the ATPase complex.

Outer Membrane Secretion Components

Secretin superfamily members are outer membrane proteins that multimerize to form channels for transport of molecules across the bacterial outer membrane. The *Yersinia* spp. secretin superfamily member YscC forms ring-shaped structures with an external diameter of about 200 Å and apparent central pores of about 50 Å. These structures, which require an outer membrane lipoprotein (YscW in *Yersinia* spp.) for proper insertion, are hypothesized to form the channel for secretion across the outer membrane. The flagellar export apparatus lacks this ring-like structure and homologs of the proteins required for its formation; instead, the flagellar rod and hook structures are thought to form the conduit for export across the outer membrane. Penetration of the outer membrane by the flagellar rod is facilitated by the L-ring lipoprotein FlgH, which some consider the flagellar equivalent of the virulence-associated secretin family.

Exported Secretion Components

Proteins secreted by the type III secretion pathway appear to be exported via flagellar hook-filament, pilin-like, or needle-like structures (see "Molecular Architecture of the Assembled Type III Secretion Machinery"). Structural components of the bacterial flagellum are exported through the flagellar rod-hook-filament structure to assemble at its distal end. Similarly, several structural components of the virulence-associated TTSSs are also export substrates. For example, the *Yersinia* YscO and YscP proteins are members of two poorly conserved families of secretion components that are exported to an extracellular location. Another secretion component, termed YscX, directly interacts with a cytosolic chaperone, YscY, prior to its export via the *Yersinia* TTSS. Likewise, the structural proteins that polymerize to form the type III needle structures in *Salmonella*, *Shigella*, and *Yersinia* species, termed PrgI, MxiH, and YscF, respectively, are exported via the type III pathway. The exported secretion components are thought to be required to complete the assembly of the external structures of the secretion complex. Proper assembly of external structures, including the needle, must be completed before the secretion apparatus can become competent for effector protein secretion, confirming that a hierarchy exists for type III substrates.

Molecular Architecture of the Assembled Type III Secretion Machinery

The bacterial flagellum consists of three structural components: a basal body, a hook, and a thin helical filament (Fig. 4A). The basal body is embedded in the bacterial envelope and serves to anchor the flagellar filament to the cell via the hook structure. The basal body consists of a hollow rod-like structure surrounded by several ring structures. The L-ring (L for lipopolysaccharide) resides in the outer membrane, whereas the P-ring (P for peptidoglycan) is embedded in the peptidoglycan layer. The MS-ring (MS for membrane/supramembrane) is located in the inner membrane, and the C-ring (C for cytoplasmic) extends into the cytoplasmic compartment. Flagellar assembly is thought to initiate with MS-ring assembly (FliF), followed closely by incorporation of proteins that form the C-ring (FliG, FliM, and FliN). Integral membrane proteins directly involved in flagellar export (FlhA, FlhB, FliO, FliP, FliQ, and FliR) are thought to reside in a specialized patch of membrane within the MS-ring structure (5). Completion of a functional flagellar export system allows export of flagellar rod-type substrates and initiates assembly of the rod, hook, and filament in that order. Exported rod-, hook-, and filament-type substrates diffuse down the hollow rod, hook, and filament and polymerize at the distal end of the growing structure. Thus, flagellar assembly begins with the basal body, proceeds with the hook, and finishes with the filament.

The recently visualized *Salmonella* SPI-1 virulence-associated type III needle complex (Fig. 4B to D) bears a striking resemblance to the flagellar hook-basal body structure discussed above, except that the extracellular hook-filament structure is replaced by a hollow needle-like structure (2). The assembled *Salmonella* needle complex consists of two pairs of rings connected by a central rod. The lower pair of rings, which are embedded in the inner membrane, are approximately 30 to 40 nm in diameter and 10 to 20 nm in height. The upper rings, which appear to interact with the outer membrane and the peptidoglycan layer, are 15 to 20 nm in diameter and 10 to 18 nm in height. A hollow straight-needle structure 50 to 80 nm in length and 8 to 13 nm wide extends outward from the

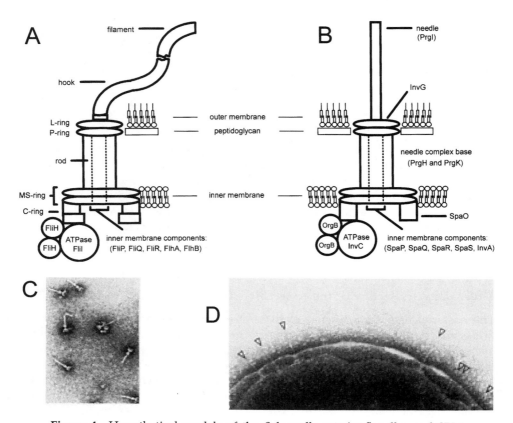

Figure 4 Hypothetical models of the *Salmonella enterica* flagellar and SPI-1 type III secretion machines. The proposed locations of type III secretion components within the flagellar basal body-hook-filament structure (A) and the needle complex structures (B) are shown. The flagellar FliI-FliH ATPase complex may deliver substrates and energize export via an inner membrane secretory complex contained within a patch of membrane within the MS-ring structure. Secreted substrates are delivered into a central channel within the flagellar basal body-hook-filament structure, where they eventually assemble at the distal end of this structure. A similar targeting pathway may function to deliver proteins to the SPI-1 type III secretion apparatus. Effector proteins exported across the inner membrane may enter a channel within the needle complex base and eventually be targeted to the extracellular environment via the needle structure or into a eukaryotic cell with the help of the SipBCD translocon. Electron micrographs of purified *Salmonella* SPI-1 needle complexes (C) and surface-localized needle complexes (D) visualized on osmotically shocked bacteria (from reference 2 with permission).

base structure. A supramolecular structure similar in size and appearance to the *Salmonella* needle complex has also been isolated from *S. flexneri* and enteropathogenic *E. coli*.

The remarkable structural similarities between the flagellar basal body-hook complex and the needle complex are not reflected at the amino acid sequence level. For example, two of the three predominant constituents of the *Salmonella* needle complex, PrgH and InvG, have no counterparts in the flagellar system, and the third major constituent, PrgK, shows only limited amino acid similarity to the flagellar FliF protein. Interestingly, the majority of the structural components that are conserved between the flagellar and virulence-associated type III export systems are localized to the innermost part of the flagellar basal body, the MS- and C-ring structure. The assembled needle complex base is likely the functional equivalent of the flagellar MS- and C-ring structure and presumably functions to house the essential integral membrane components of the type III export system (SpaP, SpaQ, SpaR, SpaS, and InvA).

Assembly of the virulence-associated TTSSs is thought to proceed in a manner that parallels flagellar basal body-hook structure assembly. The major constituents of the *Salmonella* needle complex base (InvG, PrgH, and PrgK) carry *sec*-dependent secretion signals, and assembly of this substructure, therefore, does not require a functional TTS apparatus. On the other hand, assembly of the external needle requires type III export, and the major constituent of this structure (PrgI) is a secreted protein. PrgH and PrgK, in the absence of all other secretion components, assemble into ring-shaped structures that closely resemble the base of the isolated needle complex. Assembly of the integral inner membrane secretion components (SpaP, SpaQ, SpaR, SpaS, and InvA) must also be completed prior to PrgI export and needle assembly. Secretion and polymerization of PrgI are required to complete the assembly of the needle complex and type III export apparatus. The role of the other exported secretion components is unknown; however, they likely function in the assembly of external structures, including needle assembly.

Substrate Specificity Switching in TTSSs

The question of how the secretion apparatus selects the appropriate secretion substrate at the appropriate time has recently been addressed in the flagellar export system (5). The flagellar export apparatus switches substrate specificity from rod- and hook-type substrates during hook assembly to filament-type substrates upon completion of the hook structure. Two flagellar components, FlhB, an inner membrane protein with a substantial C-terminal cytoplasmic domain, and FliK, the secreted hook-length control protein, are proposed to be involved in this process. An absence of FliK results in abnormally long hooks, presumably due to a failure to switch from rod- and hook-type substrates to filament-type substrates. The cytoplasmic C-terminal domain of FlhB is proposed to exist in two conformational states that correspond to the two substrate specificity states. A conformational change in this domain is thought to be responsible for the switch in substrate specificity. The mechanism by which FliK senses hook length and communicates with FlhB is unknown. Similar mechanisms of substrate selection also appear to exist in the virulence-associated TTSSs. For example, the length of the external needle in *Salmonella* spp. is negatively regulated by the secreted InvJ protein, such that an *invJ* null mutant expresses abnormally long needle segments. Thus, InvJ may regulate needle length in a manner similar to that of FliK in the flagellar system.

CONCLUSION

The assembly and function of the type III secretion apparatus appear to be complex, with many questions remaining to be answered. Fortunately, a number of them, especially in regard to structure and assembly, have at least begun to be addressed in the flagellar system. An important question is how far knowledge of flagellar biogenesis can be extrapolated to understand the structure and function of the type III export apparatus. It is obvious that there will be features unique to each system; however, the basic process of transporting substrates across the bacterial membranes appears to be relatively well conserved. Still, we should not lose sight of the fundamental differences between these two systems or among the different virulence-associated TTSSs. Clearly, there are a number of gene products that are present in some TTSSs but not in others. These may represent minor differences in the secretion machinery or adaptations to unique obstacles. For example, bacterial plant pathogens produce a long pilus-like structure instead of a short needle, which presumably facilitates effector protein delivery through the thick plant cell wall. An appreciation of the similarities and differences among the different type III export systems will be invaluable for their complete understanding.

REFERENCES
1. **Anderson, D. M., and O. Schneewind.** 1997. A mRNA signal for the type III secretion of Yop proteins by *Yersinia enterocolitica. Science* **278:**1140–1143.
2. **Kubori, T., Y. Matsushima, D. Nakamura, J. Uralil, M. Lara-Tejero, A. Sukhan, J. E. Galán, and S. Aizawa.** 1998. Supramolecular structure of the *Salmonella typhimurium* type III protein secretion system. *Science* **280:**602–605.
3. **Lloyd, S. A., M. Norman, R. Rosqvist, and H. Wolf-Watz.** 2001. *Yersinia* YopE is targeted for type III secretion by N-terminal, not mRNA, signals. *Mol. Microbiol.* **39:** 520–531.
4. **Luo, Y., M. G. Bertero, E. A. Frey, R. A. Pfuetzner, M. R. Wenk, L. Creagh, S. L. Marcus, D. Lim, F. Sicheri, C. Kay, C. Haynes, B. B. Finlay, and N. C. J. Strynadka.** 2001. Structural and biochemical characterization of the type III secretion chaperones CesT and SigE. *Nat. Struct. Biol.* **8:**1031–1036.
5. **Macnab, R. M.** 1999. The bacterial flagellum: reversible rotary propellor and type III export apparatus. *J. Bacteriol.* **181:**7149–7153.
6. **Matson, J. S., and M. L. Nilles.** 2001. LcrG-LcrV interaction is required for control of Yop secretion in *Yersinia pestis. J. Bacteriol.* **183:**5082–5091.
7. **Stebbins, C. E., and J. E. Galán.** 2001. Maintenance of an unfolded polypeptide by a cognate chaperone in bacterial type III secretion. *Nature* **414:**77–81.

SUGGESTED READING
1. **Cornelis, G. R., and F. Van Gijsegem.** 2000. Assembly and function of type III secretory systems. *Annu. Rev. Microbiol.* **54:**735–774.
2. **Galán, J. E.** 2001. *Salmonella* interactions with host cells: type III secretion at work. *Annu. Rev. Cell Dev. Biol.* **17:**53–86.
3. **Kimbrough, T. G., and S. I. Miller.** 2001. Assembly of the type III secretion needle complex of *Salmonella typhimurium. Microb. Infect.* **4:**75–82.
4. **Lloyd, S. A., Å. Forsberg, H. Wolf-Watz, and M. S. Francis.** 2001. Targeting exported substrates to the *Yersinia* TTSS: different functions for different signals. *Trends Microbiol.* **9:**367–371.

Bacterial Protein Toxins
D. Burns et al., Editors
©2003 ASM Press, Washington, DC 20036

Chapter 8

Type IV Secretion Systems

Drusilla L. Burns

Transport of macromolecules across bacterial membranes is a difficult, energy-consuming process. Probably for this reason, only a handful of distinct types of bacterial transporters have evolved that are capable of exporting large proteins across the inner and outer membranes of gram-negative bacteria. One family of transporters, termed type IV transporters, has members that are utilized by a variety of gram-negative bacteria to transport large macromolecules, including protein toxins, across bacterial membranes. Type IV transporters differ from other transporter systems in that type IV systems are used not only to transport proteins but also to mobilize DNA. In fact, it is likely that type IV transporters first evolved as conjugation systems that functioned to transfer genetic information between bacteria and only later were these systems modified by pathogenic bacteria to transfer critical virulence factors across the bacterial membranes and into the host eukaryotic cell. Any transport system that is related to conjugation systems is considered a member of the family of type IV transporters. This transporter family is ubiquitous in gram-negative bacteria; members are found in nonpathogenic bacteria as well as in pathogens that infect both animals and plants.

Type IV transporters can be divided into three categories: (i) those that function to transport DNA from one bacterium to another; (ii) those that function to transport DNA from a pathogen into a target host cell; and (iii) those that transport protein virulence factors across the membranes of bacterial pathogens. Although these functions appear, on the surface, to be disparate, in reality, they have striking similarities because DNA transferred by type IV systems is covalently attached to a protein, which is the entity that is actively transported. Thus, all type IV transporters are truly protein transporters.

THE ANCESTRAL TYPE IV TRANSPORTER: BACTERIAL CONJUGATION MACHINERY

Because transfer of genetic information from one bacterium to another via bacterial conjugation is an ancient system that likely predates the evolution of path-

---

Drusilla L. Burns • Food and Drug Administration, 8800 Rockville Pike, Bethesda, MD 20892.

ogens and therefore predates the necessity to transport virulence factors across bacterial membranes, it seems likely that the first type IV transporter was a conjugation system that transferred DNA from one bacterium to another. The other type IV systems then evolved from conjugation machinery and represent modifications of this basic machinery. Thus, to best understand type IV transporters and how they function, it is important to be familiar with the conjugation process.

During conjugation, DNA is transferred from a donor bacterium to a recipient bacterium. The genes encoding the transport apparatus or conjugation machinery are most often located on an extrachromosomal element that is self-replicating and self-transmissible. This element is termed a conjugative plasmid. One of the best-studied conjugative plasmids is the F plasmid of *Escherichia coli* K-12. The series of events that occur during transfer of the F plasmid from one cell to another have been elucidated and are likely prototypic of conjugation events in general (Fig. 1). The conjugation machinery encoded by the F plasmid in the donor cell first forms a pilus, i.e., a long, thin macromolecular structure that can extend 1 to 2 μm from the surface of the donor cell. The pilus is composed of many individual protein subunits; the major subunit type is known as pilin. The tip of the F pilus interacts with the recipient cell, grabbing onto it tightly. The pilus then retracts by depolymerizing, bringing the surfaces of the donor and recipient cells together. A conjugation junction is formed between the two cells, through which the DNA will pass. An enzyme then nicks the circular, double-stranded F plasmid and becomes covalently attached to the 5' end of the DNA. A single strand of the DNA is unwound from the remainder of the plasmid in the 5' → 3' direction and is translocated into the recipient cell through the conjugation junction. Replacement strand synthesis in both the donor and recipient cell occurs to regenerate copies of the original plasmid.

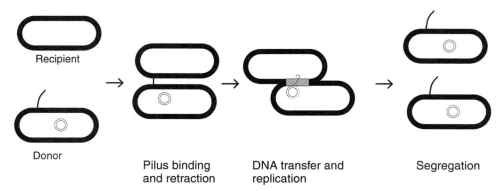

Figure 1 Schematic diagram of the events that occur during bacterial conjugation. The donor bacterium attaches to the recipient cell by its pilus. The pilus retracts and a mating channel is formed through which the DNA passes. After replication of the DNA is complete, the two bacteria separate, each now carrying its own copy of the plasmid. Adapted from reference 2 with permission.

Some basic questions remain unanswered concerning the conjugation process. First and foremost is the question of how the DNA is transferred. The possibility exists that the pilus acts as a channel through which the DNA is transferred to the recipient cell. Measurements of the lumen of the pilus indicate that it is large enough to allow passage of single-stranded DNA covalently attached to its pilot protein. Alternatively, the pilus might simply act to grab onto the recipient cell and bring it close enough to the donor cell such that the DNA can be transported from one cell to the next via a mating channel that has yet to be defined. Extended conjugation junctions between the outer membranes of donor cells carrying the conjugative plasmid RP4 and recipient cells have been observed that have a mean length of 200 nm. However, clear structures that might correspond to mating channels have not been detected.

MEMBERS OF THE TYPE IV TRANSPORTER FAMILY

By definition, all members of the type IV transporter family are related to DNA conjugation systems. DNA conjugation systems that belong to the type IV family include transfer machineries of the plasmids F, pKM101, R388, and RP4, which belong to the incompatibility groups IncF, IncN, IncW, and IncP, respectively. While these transporters are used to transfer distinct plasmids, they share considerable homology (Fig. 2).

The plant pathogen *Agrobacterium tumefaciens* appears to have usurped the type IV conjugation machinery and tailored it for a slightly different purpose. This pathogen utilizes a well-studied type IV system, known as the VirB system, to transport a piece of bacterial DNA into a plant host cell, where it incorporates into the plant genome. This DNA encodes proteins that induce proliferation of plant cells and tumor formation, processes that are key to the pathogenesis of *A. tumefaciens*. Thus, this pathogen transports an important virulence factor, in this case a DNA molecule, via its type IV system. This system is closely related to the conjugal DNA transport systems discussed above in that DNA is transported from one cell to another; however, instead of transfer between two bacteria, interkingdom transfer occurs between a bacterium and a plant cell. The mechanism by which the DNA is transferred is strikingly similar to that observed during conjugation. After *A. tumefaciens* comes in close contact with the plant cell, the first step in the generation of the piece of single-stranded DNA, known as T-DNA, that will ultimately be transferred is the nicking of the Ti (tumor-inducing) plasmid that is carried by virulent forms of this bacterium. The nicking enzyme becomes covalently attached to the DNA. The DNA-protein complex is then transported into the plant cell via the VirB system.

A slight variation of type IV DNA transport systems appears to have occurred when type IV systems evolved that solely transport proteins without any DNA attached. Such a modification is not a large evolutionary leap, since the DNA is not believed to be actively transported by conjugation systems. Rather, a protein covalently attached to the DNA is actively transported, with the DNA simply coming along for the ride. We now know that a number of pathogenic bacteria have adapted type IV secretion systems to export critical protein virulence factors, including toxins and effector proteins, from the bacterium either into the extracellular milieu or directly into the host cell. Although the substrate,

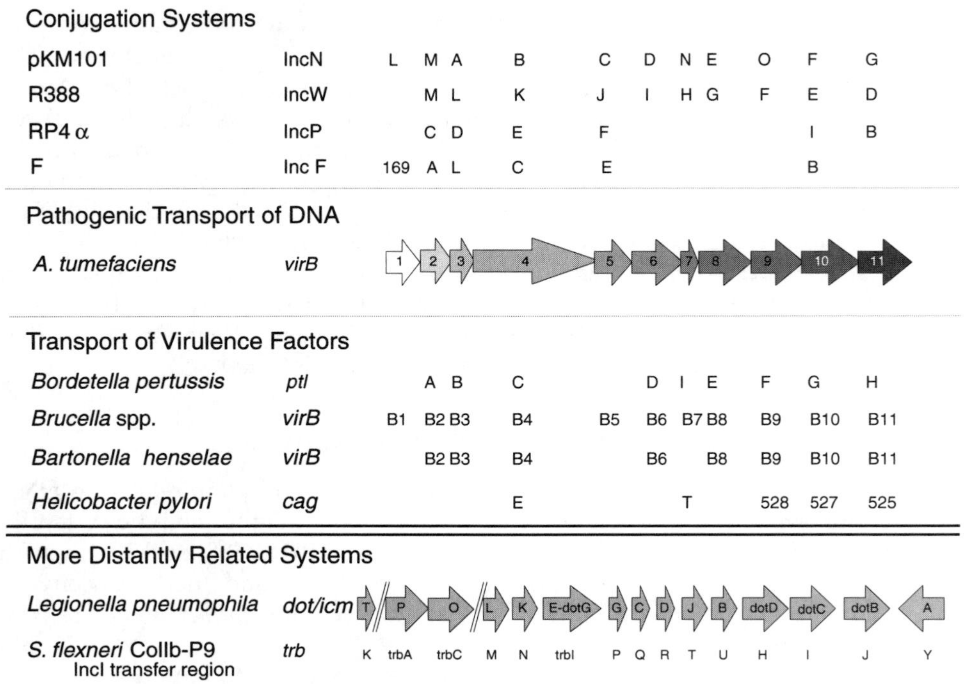

Conjugation Systems

pKM101	IncN		L	M	A		B		C	D	N	E		O	F	G
R388	IncW			M	L		K		J	I	H	G		F	E	D
RP4 α	IncP			C	D		E		F					I		B
F	Inc F	169	A	L			C		E					B		

Pathogenic Transport of DNA

A. tumefaciens   *virB*   [gene arrows: 1 2 3 4 5 6 7 8 9 10 11]

Transport of Virulence Factors

Bordetella pertussis	*ptl*		A	B		C			D	I	E		F	G	H
Brucella spp.	*virB*	B1	B2	B3		B4		B5	B6	B7	B8		B9	B10	B11
Bartonella henselae	*virB*		B2	B3		B4			B6		B8		B9	B10	B11
Helicobacter pylori	*cag*					E					T		528	527	525

More Distantly Related Systems

Legionella pneumophila   *dot/icm*   [gene arrows: T P O L K E-dotG G C D J B dotD dotC dotB A]

S. flexneri ColIb-P9   *trb*   K trbA trbC M N trbI P Q R T U H I J Y
Incl transfer region

Figure 2 The family of type IV transporters. Shown are homologies between the genes encoding the VirB transporter of *A. tumefaciens*, closely related genes of transfer regions of conjugative plasmids, and the genes of transport systems of pathogenic bacteria. Also shown is the relationship between the genes of the transfer regions of the ColIb-P9 plasmid and the genes of the *Legionella dot/icm* system.

in these instances, is exclusively protein, the structure of the transporter is likely similar to those of type IV DNA transporters because of the striking sequence homologies between the various systems (Fig. 2). Pathogens that produce type IV transporters include *Bordetella pertussis*, *Brucella* spp., *Bartonella henselae*, *Helicobacter pylori*, *Rickettsia prowazekii*, and *Legionella pneumophila*.

Type IV DNA transporters share homologous proteins (Fig. 2). The core set of type IV proteins is perhaps best represented by the VirB system of *A. tumefaciens*. This type IV system is encoded by 11 genes, *virB1* to *virB11*. Although some type IV DNA transporters may be missing certain of these genes and/or have additional genes, the core set of these genes is present in a number of members of this family and all members have at least a subset of these genes. For example, the conjugation system of pKM101 is encoded by homologs to each of the *virB* genes and two additional genes that have no homolog in the *virB* system. The F plasmid system has homologs to only six of the *virB* genes and has a number of additional genes that encode proteins important to the functioning of this system. *Bordetella pertussis*, *Brucella* spp., and *Bartonella henselae* have homologs to most or all of the *virB* genes whereas *H. pylori* and *R. prowazekii* contain only a subset of genes homologous to the *virB* genes. Whether the transport systems of these latter two pathogens contain additional proteins encoded by as yet to be discovered genes or whether the transport systems of these pathogens resemble just a part of the VirB apparatus remains to be determined.

The *dot/icm* system of *L. pneumophila* consists of a number of genes, only two of which display weak homology to the *virB* genes (*virB10* and *virB11*); however, at least 15 genes of this system are homologous to genes encoding a conjugation system of the IncI CollIb-P9 plasmid of *Shigella flexneri*. Thus, this protein transport system, while not closely related to many of the type IV transporters, is closely related to a DNA conjugation system and therefore has been included in the type IV family.

STRUCTURE OF TYPE IV TRANSPORTERS

At the present time, the best-studied type IV transporter is the VirB system of *A. tumefaciens*. Because of the striking homologies between the genes encoding this system and those of the other type IV systems, it seems reasonable to assume that many of the notable architectural features of the VirB system will be present in other type IV systems. The VirB transporter is composed of 10 proteins, VirB2 to 11, each of which is essential for DNA transfer since deletion of any of the genes encoding these proteins completely abolishes virulence (1). An 11th protein, VirB1, plays a role in transporter biogenesis; however, it does not appear to be part of the transport apparatus itself. Deletion of *virB1* results in a 100- to 1,000-fold attenuation of virulence of the organism.

Through the intense efforts of a number of research groups, the structure of the VirB transporter as well as the role of the individual VirB proteins in the transport process are becoming better understood. A schematic diagram of the VirB transporter is shown in Fig. 3.

Perhaps the most notable morphological feature of the VirB transport system is its pilus structure (3, 5). Like the type IV conjugation systems, the VirB transporter consists, in part, of a long, thin pilus, known as the T-pilus, which is approximately 10 nm in diameter. The major subunit comprising the pilus is VirB2. The VirB2 subunit undergoes an interesting cyclization whereby the C-terminal end becomes attached to the N-terminal end via an intramolecular covalent head-to-tail peptide bond. This cyclization may contribute to the overall stability of the pilus structure. VirB5 has been identified as a minor pilin subunit. It is not known whether copies of VirB5 are dispersed throughout the pilus or whether they localize to one portion of the pilus such as the tip or the base.

At present, the role of the T-pilus is unknown; however, it may play a role in bringing the bacterial cell in contact with the plant cell, just as in conjugation in which the mating pilus is thought to be important for contact of the donor cell with the receptor cell. Also, it is unknown whether the DNA traverses through a channel in the middle of the pilus or whether the DNA transfer channel is distinct from the pilus structure.

A number of the VirB proteins are believed to be important structural components of the pilus/transfer system. Current evidence suggests that many of the VirB proteins may assemble to form a single structure. VirB2 and VirB5 are the major and minor subunits, respectively, that form the pilus that extends outward from the cell. The pilus appears be anchored to the cell by other VirB

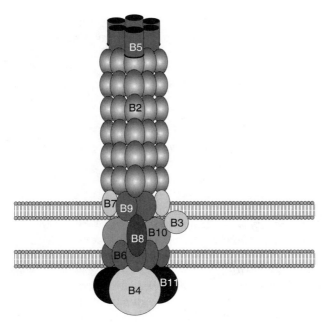

Figure 3 Schematic model of the structure of the VirB transporter. Proteins encoded by the *virB* genes of *A. tumefaciens* form a pilus structure. Most of the VirB proteins are believed to associate to form this structure, which begins in the inner membrane, spans the periplasmic space and the outer membrane, and then extends into the extracellular space.

proteins that assemble and form a bridge spanning from the inner membrane through the outer membrane. VirB7 is a lipoprotein attached to the outer membrane. Recently, evidence has surfaced suggesting that VirB7 may interact with subunits of the T-pilus. VirB7 is linked to VirB9, another outer membrane protein, which in turn associates with VirB10 and VirB8, an inner membrane protein. Other VirB proteins include VirB6, which appears to be an integral membrane protein that may play a role in pore formation, and VirB3, which is thought to be an outer membrane protein. Two VirB proteins, VirB4 and VirB11, have demonstrated ATPase activity. Because ATP is found in the cytoplasm and not the periplasm, these proteins are believed to be inner membrane proteins with their active sites exposed to the cytoplasm of the cell. These critical proteins may provide the energy for biogenesis of the pilus/transfer apparatus or for the transport process itself.

VirB1 has homology with transglycosylases, and it has been postulated that this protein may possess an activity that allows the protein to bore a channel through the peptidoglycan layer found in the periplasm of the cell, permitting assembly of a transporter within this channel. The transporter would then extend from the inner membrane to the outer membrane as a single continuous structure. Such a proposed structure is particularly appealing if the DNA-protein substrate traverses a channel formed by the apparatus, since the DNA substrate of type IV DNA transporters originates in the cytoplasm of the cell. VirB1 is ultimately proteolyzed and the C-terminal 73-amino-acid polypeptide is secreted into the extracellular space. Whether this proteolyzed fragment plays an additional role in the transport process remains to be determined.

TYPE IV SYSTEMS OF PATHOGENIC BACTERIA

Many of the type IV systems that have been discovered recently are now known to transport protein virulence factors across the membranes of medically important pathogens. Some of these proteins are well-known toxins, such as pertussis toxin, and other substrates have not yet been identified. Pathogens that have type IV systems are diverse in nature. Some are respiratory pathogens and certain others are enteric pathogens. Some infect humans and others primarily infect animals. Certain of these pathogens are extracellular whereas others lead an intracellular existence. Because of the diverse nature of the pathogens, each has its own special needs. The type IV system of each of the pathogens appears to be tailored to the needs of that pathogen. For example, *Bordetella pertussis*, a pathogen that infects the human respiratory tract and that remains extracellular, has evolved a type IV system that is critical for the transport of one of its important toxins across the outer membrane and into the extracellular milieu, where the toxin can attack neighboring cells. *H. pylori* is found in the acid environment of the stomach. This pathogen has found a need to deliver important virulence factors from the bacterium directly into the host cell and, therefore, its type IV transporter has evolved to do this. Other pathogens that produce type IV transporters, such as *Brucella, Bartonella,* and *Legionella* species, live in an intracellular environment. Their need is to transport effector molecules into the eukaryotic cell that will make the eukaryotic cell a more hospitable environment for the pathogen.

A Type IV System that Transports a Protein Toxin Substrate to the Extracellular Milieu

Bordetella pertussis is the causative agent of the disease pertussis, or whooping cough. This pathogen secretes one of its important virulence factors, pertussis toxin (PT), using a type IV transporter. The exact role that PT plays in the disease process has not been fully elucidated; however, it is clear that PT is a critical virulence factor, since mutants lacking PT are much less virulent than the wild-type strain in animal models of the disease. Moreover, vaccines composed solely of inactivated PT are capable of providing considerable protection against the disease in humans. Once the toxin is secreted from the bacterium, it travels to its target host cell, binds to that cell, and is internalized into the cell. When inside the cell, it wreaks havoc by ADP-ribosylating a family of GTP-binding regulatory proteins within the cell that are important for signal transduction. PT is capable of ADP-ribosylating both G_i and G_o, which couple hormone receptors to eukaryotic effector proteins within the cell. When these GTP-binding regulatory proteins are ADP-ribosylated, they are no longer capable of transducing signals, resulting in a loss of response to a variety of hormones and cell regulators. PT interacts with cells of many types; however, it is generally believed that the toxin does most of its damage during the disease process by affecting cells of the host immune system. By altering signal transduction, the toxin effectively attenuates the immune system of the host. Although this effect is devastating for the host, who becomes more susceptible to dangerous and life-threatening secondary infections, the effect on the bacterium is quite the opposite. Attenuation of the host immune system gives the bacterium a better chance to thrive and multiply. Thus, the toxin alters the balance in the war between the host and the pathogen in favor of the pathogen.

PT is typical of bacterial toxins in that it has an AB_5 structure (Fig. 4). It is composed of an enzymatically active S1 subunit and a binding component, known as the B oligomer, that is composed of five subunits, one copy each of S2, S3, and S5 and two copies of S4. The crystal structure of the toxin has been solved, and the dimensions of the toxin are known. PT measures approximately 6 nm $\times$ 6 nm $\times$ 10 nm.

Bordetella pertussis secretes PT using a type IV secretion system known as the Ptl system. This was the first type IV secretion system to be discovered that transports only protein and has no role in exporting DNA. The Ptl system appears, on the surface, to be similar to the VirB system. It is encoded by nine genes homologous to *virB* genes (Fig. 2). Only two homologs of *virB* genes are not found among the *ptl* genes; these are homologs to *virB1* and *virB5*. The Ptl system contains a protein, PtlA, that is homologous to the pilin protein of the VirB system. Although no pili have yet been observed that are associated with the Ptl system, the homologies between the two systems suggest that either a yet to be discovered pilus exists or, alternatively, the Ptl system is associated with a very short, stubby pilus-like structure. Many of the other components of the Ptl system are known to have features similar to those of the VirB proteins. For example, the two nucleotide-binding proteins of the Ptl system (PtlC and PtlH,

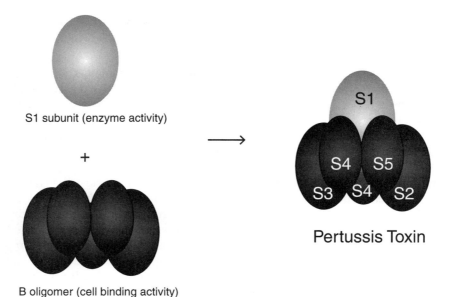

Figure 4   Subunit structure of PT. PT is composed of an enzymatically active S1 subunit and a component, known as the B oligomer, which binds to receptors on eukaryotic cells. The structure of PT is typical of a family of bacterial toxins that have an AB_5 structure, in that it is composed of an enzymatically active subunit and five subunits that form the ring of the binding component.

homologous to VirB4 and Vir11, respectively) are essential for secretion. More-over, the nucleotide-binding region of the proteins was shown to be critical for activity, suggesting that these two proteins are indeed ATPases that are critical for Ptl transporter biogenesis or for the transport process itself.

Although the Ptl and VirB systems appear to be homologous, significant differences exist between the substrates that are transported by these two systems. One striking difference between the Ptl system and the VirB system is that the substrates for the Ptl system (the PT subunits) are each synthesized with their own signal sequence, suggesting that the toxin is exported through the inner membrane of the bacterium via a generic transporter, rather than the Ptl system. The Ptl system is used to transport the toxin across the outer membrane of the bacterium. The protein substrates of the VirB system are not synthesized with their own signal sequences, and it is generally believed that these substrates are transported across both the inner and outer membranes by the VirB transport apparatus. This difference between the two systems has yet to be fully reconciled given the striking similarities of many aspects of the two transport systems.

A considerable body of knowledge about PT is available, including its crystal structure, which allows us to use this system to finely dissect how this substrate is transported and to give us clues as to how other type IV transporters might function. A major question is whether the toxin is secreted piece by piece, i.e., subunit by subunit, or whether the toxin assembles within the bacterial cell and only then is transported as a single entity across the outer membrane via the Ptl system. Substantial evidence now exists to support the idea that PT is assembled into the holotoxin form before it is released from the bacterial cell. This evidence came from experiments in which single toxin components, either the S1 subunit or the B oligomer, were individually expressed in a strain of *Bordetella pertussis* that had a functional Ptl system. Neither the S1 subunit nor the B oligomer was efficiently secreted by the Ptl system when expressed individually. Thus, it appears that the toxin first assembles and then is released into the extracellular milieu by a Ptl-mediated mechanism. Since we know the dimensions of the holo-toxin molecule, any channel through which the toxin might travel would have to be large, having dimensions greater than the 6 nm $\times$ 6 nm $\times$ 10 nm dimensions of the toxin.

If the toxin is exported to the periplasmic space, or possibly even to the outer membrane, before it interacts with the Ptl transport apparatus, the question arises as to how PT enters the transport apparatus. The Ptl system is strikingly homologous to the VirB system, which spans the inner and outer membranes. We have no reason to believe that the homologous Ptl proteins would not localize in a similar manner. In fact, studies that have been conducted indicate that homologous proteins do indeed localize in similar, if not identical, manners. A transport apparatus that spans the inner and outer membranes and a periplasmic or outer membrane location for its substrate present a paradox for researchers in this area. Perhaps the apparatus has an opening to the periplasmic space that would be large enough to accommodate the PT molecule, or the transport apparatus might assemble around the PT molecule. Alternatively, PT may be pushed through the outer membrane of the bacterium by a piston-like action of a growing pilus. Once the toxin traverses the outer membrane of the bacterium, it could dissociate from the pilus and then diffuse through the medium to its target cell.

A Type IV System that Transports an Effector Protein Directly into a Eukaryotic Cell from the Bacterium

The type IV system of *H. pylori* transports the virulence factor known as CagA directly into the eukaryotic cell. Thus, in this case, a type IV secretion system resembles type III systems in that a protein virulence factor is injected directly into the eukaryotic cell from the bacterial cell.

   H. pylori infects about 50% of the world's population and therefore is considered to be a successful pathogen. This pathogen is known to cause chronic gastritis and peptic ulcer disease. Moreover, the pathogen is associated with gastric adenocarcinoma and mucosa-associated lymphoid tissue lymphoma and is therefore considered to be a carcinogen. The mechanism by which *H. pylori* causes gastric disease is not known; however, recent results have implicated the protein CagA in the pathogenesis of this organism.

   The type IV transporter genes involved in CagA transport are located on a pathogenicity island within the *H. pylori* chromosome. This pathogenicity island has been shown to encode virulence factors that are unique to *H. pylori* strains with enhanced virulence. Therefore, it has been suggested that acquisition of this pathogenicity island played a key role in the evolution of *H. pylori* as a pathogen.

   Transport of CagA from the bacterium into the mammalian cell by a type IV secretion system is critical for manifestation of a number of changes in host cell physiology. As indicated earlier, the *H. pylori* type IV transporter is composed of only a subset of the proteins that are found with many of the type IV transporter systems, and the details of the mechanism by which CagA is transported across the bacterial and mammalian cell membranes remain elusive. Once inside the mammalian cell, CagA undergoes phosphorylation on tyrosine residues, which appears to initiate cytoskeletal rearrangements within the cell. A recent study shed light on the molecular events that lead to alterations in cell morphology induced by phosphorylated CagA (4). This study found that phosphorylated CagA is capable of forming a physical complex with the SRC homology 2 domain (SH2)-containing tyrosine phosphatase SHP-2. Formation of this physical complex results in potent stimulation of SHP-2 tyrosine phosphatase activity. Since SHP-2 is involved in the regulation of spreading, migration, and adhesion of cells, it has been postulated (4) that deregulation of SHP-2 activity by CagA may induce abnormal proliferation and movement of gastric epithelial cells, promoting acquisition of a transformed phenotype. CagA thus appears to be an effector protein that is injected directly into the eukaryotic cell by a type IV transporter, just as effector proteins are delivered by type III transporters (see chapter 7). Like other effector proteins produced by bacterial pathogens, CagA may act as a modulating toxin in that it can modulate the activity of critical cellular components that are involved in signal transduction events within the mammalian cell.

Type IV Systems of Intracellular Pathogens

Several intracellular pathogens utilize type IV transporters to export important virulence factors from the bacterium to the cellular milieu of the host cell. These pathogens include *L. pneumophila*, *Brucella* spp., *Bartonella henselae,* and *Rickettsia* spp. Although many of the substrates of these systems have yet to be identified, mutational analysis indicates that these systems each contribute to intracellular survival of the organism.

The type IV system of *L. pneumophila* has been shown to play a key role in ensuring the survival of the bacteria within host cells. *L. pneumophila* is responsible for the bacterial pneumonia known as Legionnaires' disease. This pathogen parasitizes macrophages and monocytes and, to survive, must alter the cellular processes of the eukaryotic cell such that the pathogen is not destroyed in intracellular departments known as phagolysosomes as it enters the cell. Normally, nonpathogens are taken up by macrophages by a process known as phagocytosis in which the macrophage engulfs the bacterium in a membrane-bound vacuole known as a phagosome. The phagosomes fuse with lysosomes, exposing the bacteria to a harsh environment containing a number of degradative enzymes and toxic peptides such that the bacteria are rapidly killed. *Legionella* cells survive by orchestrating their own uptake and altering trafficking patterns within the macrophage to create a vacuole that supports replication. Clues as to how the pathogen accomplishes this came from experiments in which it was shown that high multiplicity infection of cultured macrophages by *L. pneumophila* is cytotoxic because of the formation of pores in the macrophage cell membrane. Mutations in the *dot/icm* genes that comprise a type IV secretion system abolish this cytotoxicity. Thus, it has been hypothesized that certain *dot/icm* genes encode proteins that form this pore structure. This structure is believed to act as a channel for the transport from the bacterium into the macrophage of effector proteins that alter intracellular trafficking. Other *dot/icm* genes may encode effector proteins. The effector(s) appears to be protein in nature rather than nucleic acid because phagosome-lysosome fusion is prevented after less than 30 min, not allowing enough time for DNA to be transcribed and expressed. Recently, one substrate of the *Legionella* type IV transporter was identified as RalF (6), which functions as a guanine nucleotide exchange factor for the ADP-ribosylation factor (ARF) family of small-molecular-weight GTP-binding proteins. RalF promotes the exchange of GDP for GTP in the active site of ARF, resulting in activation of this protein. ARF proteins are key regulators of vesicle traffic, and RalF is required for localization of ARF on phagosomes containing *L. pneumophila*.

Two other types of intracellular pathogens are known to produce typical type IV transporter systems. These pathogens are *Brucella* spp. and *Bartonella henselae*. *Brucella spp.* are facultative intracellular pathogens that infect many mammalian species including humans and produce chronic infections. *Brucella* spp. invade macrophages and have the ability to survive and multiply within these cells. *Bartonella henselae* is the major etiologic agent of cat scratch disease. Both pathogens contain typical type IV secretion systems that share considerable homology with the prototypic type IV system of *A. tumefaciens*. In neither case is the substrate for transfer known; however, these systems are thought to be important for transport of virulence factors that are critical for intracellular survival of these pathogens.

CONCLUSION

Bacteria have modified conjugation systems to perform a variety of transport functions, including interkingdom mobilization of DNA and the transport of pro-

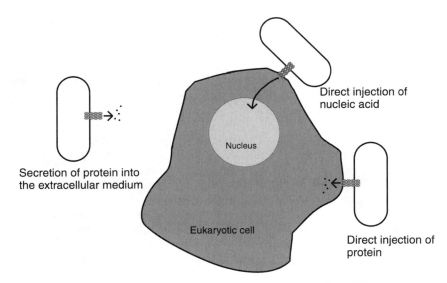

Figure 5 Mechanisms of transport by type IV systems. Type IV transporters exhibit several mechanisms by which they introduce virulence factors into eukaryotic cells. In the case of *Bordetella pertussis*, PT is secreted directly into the extracellular mileu. The toxin then binds to the eukaryotic cell and enters the cell via an endocytic pathway. In contrast, the type IV transporters of *A. tumefaciens* and *H. pylori* are capable of injecting virulence factors directly into the eukaryotic cell. The *A. tumefaciens* type IV transporter injects a nucleic acid substrate whereas the *H. pylori* type IV transporter injects a protein substrate.

tein virulence factors from the bacterium into eukaryotic host cells. Type IV systems can resemble type I and type II transport systems in that they can secrete bacterial virulence factors directly into the extracellular milieu, or they can resemble type III transport systems in that they can inject the virulence factor directly into the eukaryotic cell (Fig. 5). Thus, they exhibit a versatility that allows them to be tailored to the needs of a particular pathogen. While a considerable amount of information concerning these systems has been elucidated, a number of critical issues remain unanswered, foremost of which are the exact architecture of the transport apparatus, how the substrates interact with the transporters, and the mechanism by which the substrates are transported. Understanding these aspects of the transport process will help to elucidate basic mechanisms that are important for transfer of genetic information as well as critical pathogenic mechanisms.

REFERENCES

1. **Berger, B. R., and P. J. Christie.** 1994. Genetic complementation analysis of the *Agrobacterium tumefaciens* operon: *virB2* through *virB11* are essential virulence genes. *J. Bacteriol.* **176:**3646–3660.
2. **Firth, N., K. Ippen-Ihler, and R. A. Skurray.** 1996. Structure and function of the F factor and mechanism of conjugation. *In* F. C. Neidhardt, R. Curtiss III, J. L. Ingraham, E. C. C. Lin, R. B. Low, B. Magasanik, W. S. Reznikoff, M. Riley, M. Schaechter, and

H. E. Umbarger (ed.), Escherichia coli *and* Salmonella, *Cellular and Molecular Biology,* 2nd ed. American Society for Microbiology, Washington, D.C.

3. **Fullner, K. J., J. C. Lara, and E. W. Nester.** 1996. Pilus assembly by *Agrobacterium* T-DNA transfer genes. *Science* **273:**1107–1109.

4. **Higashi, H., R. Tsutsumi, S. Muto, T. Sugiyama, T. Azuma, M. Asaka, and M. Hatakeyama.** 2002. SHP-2 tyrosine phosphatase as an intracelllular target of *Helicobacter pylori* CagA protein. *Science* **295:**683–686.

5. **Lai, E.-M., and C. I. Kado.** 1998. Processed VirB2 is the major subunit of the promiscuous pilus of *Agrobacterium tumefaciens. J. Bacteriol.* **180:**2711–2727.

6. **Nagai, H., J. C. Kagan, X. Zhu, R. A. Kahn, and C. R. Roy.** 2002. A bacterial guanine nucleotide exchange factor activates ARF on *Legionella* phagosomes. *Science* **295:**679–682.

SUGGESTED READING

1. **Lessel, M., D. Balzer, W. Pansegrau, and E. Lanka.** 1992. Sequence similarities between the RP4 Tra2 and the Ti VirB region strongly support the conjugation model for T-DNA transfer. *J. Biol. Chem.* **267:**20471–20480.

2. **Vogel, J. P., and R. R. Isberg.** 1999. Cell biology of *Legionella pneumophila. Curr. Opin. Microbiol.* **2:**30–34.

SECTION III
TOXIN DELIVERY INTO
EUKARYOTIC CELLS

The journey of a toxin from the bacterium to its site of action within the eukaryotic cell involves sequential binding and entry of the toxin into the host cell. Progress has been made in resolving how soluble, hydrophilic protein toxins are transformed into structures that trigger the translocation of their catalytic domain across a cell membrane. Bacterial toxins accomplish this thermodynamically unfavorable translocation process by several mechanisms.

After coming in contact with the eukaryotic cell, certain toxins bind with high affinity to specific protein receptors on the surface of sensitive cells whereas other toxins may interact with glycolipids. Recent identification of cell surface receptors for several bacterial toxins has provided insight into the early stages of the intoxication process. Most protein toxins enter the eukaryotic cell by endocytic processes. One exception to this generalization is *Bordetella* adenylate cyclase toxin, which translocates its catalytic domain directly across the plasma membrane.

A number of protein toxins are transported into the host cell via endocytic processes and become "translocation competent" upon acidification of the endosome. Protein toxins that translocate their catalytic domains across the membrane of an acidified endosome include diphtheria toxin and the protective antigen of anthrax toxin. The crystal structures of these toxins facilitate the modeling of the translocation process. These models involve the partial unfolding of the catalytic domain, allowing its translocation across the membrane through a pore formed by a portion of the toxin.

Other protein toxins are taken up by endocytic processes but continue their intracellular journey into the depths of the host cell, utilizing the host cell's retrograde transport system to migrate through the Golgi apparatus into the endoplasmic reticulum. Within the endoplasmic reticulum these toxins appear to utilize the host cell's sorting machinery to translocate a partially unfolded catalytic domain across the endoplasmic reticulum membrane. In addition to intracellular delivery of the catalytic domain across a cell membrane, some bacterial toxins utilize components of the host cell's retrograde transport system to migrate between the basolateral and apical sides of a polarized cell.

Although the intracellular translocation of the catalytic domain of bacterial toxins is one of the most challenging properties of toxins to study, modern technology and novel approaches to analyze this stage of intoxication have provided insight into the molecular steps that result in successful delivery of the catalytic domain into the host cell cytosol.

Bacterial Protein Toxins
D. Burns et al., Editors
©2003 ASM Press, Washington, DC 20036

Chapter 9

Receptors for Bacterial Toxins

Catharine B. Saelinger

The first step essential to the cellular action of many toxins is binding to structures (receptors) on the target cell surface. The receptors to be considered here are ones involved in interactions with toxins classified as AB_n toxins, i.e., toxins composed of two (or more) domains, where the A domain is the effector moiety and the B domain binds to cell surface receptors (Box 3). To date, the process of identifying receptors for toxins has been laborious, and only a relatively few toxin receptors have been definitely identified. The receptors identified to date belong to diverse groups of cell surface structures.

First, some of the "duties" of a typical receptor will be considered, followed by how one might approach identifying a receptor, general characteristics of toxin receptors, and finally, a review of three representative receptors, namely, the receptor for Shiga toxins, diphtheria toxin, and *Pseudomonas* exotoxin A.

Why would a mammalian cell have a receptor to facilitate the entry of a molecule that would eventually kill, damage, or disrupt it? There is no known answer for this, but it is assumed that the "toxin receptor" or toxin-binding moiety is "tricked" into thinking that it is recognizing a physiologically relevant protein. Thus, the toxin hijacks the receptor molecule and uses the receptor to enter its mammalian target cell.

What evidence is there that mammalian cells have receptors for bacterial toxins? One piece of circumstantial evidence that a receptor is involved in the intoxication process is that some cells are exquisitely sensitive to a toxin, whereas others are exceedingly resistant to the same toxic molecule. Differences in cellular expression levels of receptors could be at least partially responsible for substantial differences in sensitivity of mammalian cells and tissues to different bacterial toxins. Another bit of circumstantial evidence is that treating toxin-sensitive cells with enzymes that remove surface components (e.g., proteases, lipases) often renders the cells transiently resistant to a toxin. Third, incubating medium containing specific toxins with sensitive cells has been shown to reduce or deplete the amount of toxic material in the medium.

Catharine B. Saelinger • Department of Molecular Genetics, Biochemistry and Microbiology, University of Cincinnati College of Medicine, 231 Albert Sabin Way, Cincinnati, OH 45267-0524.

BOX 3

Domain structures of bacterial toxins

Many of the most potent protein toxins are enzymes that catalytically damage essential cytosolic targets. These toxins must accomplish three tasks to be effective. They must bind to receptors on the animal cell surface (the subject of this chapter), they must cross a lipid bilayer membrane to reach the cytosol, and they must perform their destructive catalytic function. It is remarkable that single protein molecules exist that have the ability to perform three completely different, complex functions. However, it becomes easier to understand why such trifunctional proteins have evolved when one recognizes that these potent toxins confer strong selective advantages to bacterial pathogens. Thus, there is strong evolutionary pressure toward solving the problem of how to combine these three functions into a single protein. Nature (i.e., Darwinian selection) has solved this problem in at least three different ways, as illustrated by the three types of toxins discussed below.

AB toxins

In diphtheria toxin, the three functions are performed by separate domains contained within a single polypeptide. Although diphtheria toxin was originally viewed as having an N-terminal A (active) domain joined to the C-terminal B (binding) domain, the crystal structure revealed that the middle portion of the polypeptide constitutes a separate translocation (T) domain. The T domain creates a membrane channel through which the A chain reaches the cytosol. The peptide loop joining the A to the T+B domains is cleaved by a pathogen or host protease, but the A and T+B domains remain associated through a disulfide bond. Only at a late stage in the internalization process is the disulfide bond broken, releasing the A fragment and thereby enhancing its catalytic activity. Diphtheria toxin and other members of this group such as *Pseudomonas* exotoxin A represent a highly evolved toxin design, because the three functional domains have been successfully crowded into a single polypeptide while maintaining high efficiency in each.

AB_n toxins

In this group of toxins, typified by cholera toxin, the B function is performed by a separate polypeptide gene product present in multiple copies. The cholera toxin B subunit pen-

What are the possible contributions of receptor to the intoxication process? First, it is assumed that most AB_n toxins bind to specific (if not yet identified) receptor molecules on the surface of a target cell. This initial binding brings the toxin into close contact with a target cell and also helps determine if the cell is susceptible to the toxin or not. Second, the receptor may moderate toxin entry. That is, if the putative receptor normally is taken into the cell via a clathrin-coated pathway, then the receptor-toxin complex also will enter via this pathway. Third, the receptor may dictate the pathway taken by the toxin after internalization. For example, in the case of *Pseudomonas* exotoxin A, the receptor may facilitate the movement of the toxin to the Golgi apparatus from where it moves more deeply along the retrograde trafficking pathway. Fourth, in some cases (e.g., diphtheria toxin) the receptor may have a direct role in the penetration of toxin

BOX 3 (Continued)

tamer assembles with a single catalytic A chain. (The T function of these toxins is not well understood and cannot be clearly assigned to a particular polypeptide region.) The B subunits bind to sugar residues on target cells with a relatively low intrinsic affinity but achieve high binding due to their multivalency. In these AB_n type toxins, the A chain typically has an internal site that is cleaved by a protease to release the catalytic portion of the A chain. In creating toxins of this type, evolution solved the problem of incorporating three functions into the toxin by having the separate polypeptides bind very tightly to each other while still within the pathogenic bacterium that produces the toxin.

Bipartite toxins

The third group of these toxins, of which anthrax toxin is the most studied, has the B, T, and A functions performed by two entirely distinct, soluble proteins that only associate on the surface of their target cells. The B and T functions of anthrax toxin are performed by the protein designated protective antigen (PA). PA forms a receptor-bound, heptameric ring that serves as a channel for translocation of the other two proteins, an adenylate cyclase designated edema factor (EF) and a protease designated lethal factor (LF). Anthrax toxin is actually two toxins because two different enzymes are delivered into cells. Proteolytic activation plays an essential role here too, but its function is to expose a high-affinity binding site on the PA heptamer, to which the LF and EF then bind. Here the solution to combining the three functions is solved in part by using a cryptic, high-affinity binding site to achieve tight association of the three functional domains, but this occurs only on the target cell surface. This solution to the problem of delivering toxic enzymes into cells is clearly a successful one, because evolution has used this strategy repeatedly. Thus, several pathogenic clostridial species produce toxins having the same domain structure as anthrax toxin but differing in that the catalytic domains of these toxins ADP-ribosylate cytoplasmic actin. The fact that this molecular design has been used by several different pathogens shows that this is a very flexible and effective design for a toxin.

Stephen H. Leppla

through the host cell membrane to access the target of the enzyme-active domain in the cytoplasm.

Toxin receptors come in a variety of flavors. Both cholera toxin and Shiga toxin bind to a membrane glycolipid. Cholera toxin recognizes the ganglioside GM1, whereas Shiga toxin binds to a neutral glycolipid, the globoside globotriaosylceramide (Gb3), not to a ganglioside. Other receptors, such as the receptor for *Pseudomonas* exotoxin A, are glycoproteins, whereas the receptor for diphtheria toxin is a protein (see below). The receptor for the protective antigen component of anthrax toxin is a ubiquitous molecule that is expressed at relatively high levels on cell surfaces. The receptor has recently been shown to be a type I membrane protein having an extracellular von Willebrand factor A domain to which protective antigen (the binding domain of the toxin) binds directly.

Efficient interaction of clostridial neurotoxins with their target cells is complex and may involve two different receptors. These neurotoxins access the nervous system at autonomic nerve endings or at the presynaptic region of the neuromuscular junction. Neuronal tissue is enriched in gangliosides that have more than one neuraminic acid residue, and tetanus and some serotypes of botulinum toxin bind preferentially to this group of gangliosides. Tetanus and botulinum neurotoxins bind to different types of gangliosides and thus do not compete with each other for the same binding sites. Binding efficiencies in artificial systems decorated with gangliosides are low compared to binding of toxin to nerve cells. Therefore, a "double receptor" has been proposed. In this model, one part of the receptor is formed by the ganglioside and the other part is composed of a protein expressed selectively in neuronal tissue. For example, botulinum toxin type B can bind to the vesicular transmembrane protein synaptotagmin. It is hypothesized that other clostridial neurotoxins may bind to different proteins and that these proteins may direct the movement of the toxins in target cells.

BOX 4

Recombinant immunotoxins

Bacterial toxins that exhibit potent cytotoxic activity are attractive agents for the development of novel anticancer drugs called immunotoxins. The goal of immunotoxin development is to harness the toxic activity of these toxins and direct it toward cancer cells and away from normal cells. Diphtheria toxin (DT) and *Pseudomonas* exotoxin A (PEA) kill mammalian cells by gaining access to the cell cytosol, ADP-ribosylating elongation facter 2, inhibiting protein synthesis, and causing cell death. By genetic manipulation, the binding domain of either toxin can be deleted and replaced with a gene encoding a cell-binding antibody fragment or a ligand (such as a growth factor or cytokine). These binding sequences are employed to direct the toxin to bind, enter, and kill cancer cells. This concept is as old as Erhlich's magic bullet, but has taken many decades to develop. Key elements of design include the use of recombinant antibody fragments (or ligands) that retain binding activity for tumor antigens. Antibodies or ligands are best inserted at the location normally occupied by the toxin's own binding domain. For PEA, the tumor-binding antibody is placed at the amino terminus and for DT at the carboxyl terminus. In tissue culture, recombinant immunotoxins exhibit potent cytotoxic activity with 50% inhibitory concentration values for killing cancer cells of 10^{-11} M or better. In the past 5 years recombinant immunotoxins have graduated from the laboratory to the clinic. ONTAK, composed of the first 388 amino acids of DT fused to interleukin 2, is now approved for the treatment of cutaneous T-cell lymphoma. BL22, a disulfide-stabilized antibody Fv fused with PE38 (this truncated form of PEA is composed of amino acids 252 to 365 and 380 to 613), binds, enters, and kills cells expressing CD22 on their cell surface (Fig. 1). BL22 has demonstrated a high complete remission rate in patients with hairy cell leukemia.

**David J. FitzGerald,
Robert J. Kreitman, and Ira Pastan**

Some receptors for bacterial toxins have been isolated, purified, and well characterized. A review of several receptors will illustrate the different ways these molecules are studied and that bacterial toxins have harnessed a variety of processes to enter target cells.

PSEUDOMONAS AERUGINOSA **EXOTOXIN A RECEPTOR**

Pseudomonas exotoxin A (PEA) exerts its toxicity by ADP-ribosylating elongation factor 2 and thus irreversibly stopping protein synthesis. The toxin is internalized by receptor-mediated endocytosis, is cleaved by mammalian cell furin (a pro-protein-converting enzyme that resides in the constitutive secretory pathway), and is translocated to the cytosol, where it inactivates elongation factor 2. The toxin consists of a single polypeptide chain with three functional domains. The N-terminal domain mediates binding to a mammalian target cell, the central domain has the translocation activity and is the substrate for proteolytic cleavage, and the C-terminal domain possesses the enzyme activity. The well-defined structural divisions within the domains of PEA make it an excellent candidate for the development of immunotoxins. These chimeras are formed by replacing the

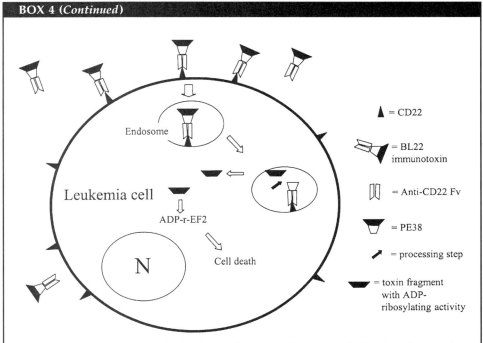

BOX 4 (*Continued*)

▲	= CD22
	= BL22 immunotoxin
	= Anti-CD22 Fv
▽	= PE38
	= processing step
	= toxin fragment with ADP-ribosylating activity

Endosome

Leukemia cell

ADP-r-EF2

N

Cell death

Figure 1 The figure depicts the binding and entry of BL22 into a leukemia cell expressing CD22. Once taken into an endosome, BL22 is processed by the cellular protease, furin. This is followed by the reduction of a key disulfide (Cys-265 to 287), the release of an enzymatically active C-terminal fragment, its translocation to the cytosol, and the cessation of protein synthesis. This causes death of the leukemia cell.

DNA-encoding domain I with DNA that encodes cell-binding domains having different specificities (Box 4).

Early studies provided circumstantial evidence for a PEA-specific receptor on the surface of toxin-sensitive cells such as mouse LM fibroblasts. Mouse LM cells are a cell line of choice in these studies because they are highly sensitive to the biological action of PEA. Initial experiments with electron microscopy have shown that PEA is routed into LM cells via clathrin-coated pits, structures long associated with receptor-mediated endocytosis. Treatment of LM cells with trypsin transiently ablates toxin sensitivity, suggesting that these cells have a proteinaceous PEA-binding moiety on their surface. Sensitivity is recovered rapidly once trypsin is removed. This suggests the presence of an intracellular pool of receptor, capable of recycling to the cell surface. Binding studies demonstrated that toxin binding to LM cells is saturable (K_d = 5.4 nM) and reversible, and indicated that there are approximately 100,000 binding sites per cell. In addition, examination of numerous cell lines showed that there is a several log range of sensitivity to PEA; one explanation is that cells that are resistant to PEA lack functional surface receptors for the toxin whereas sensitive cells would be expected to possess functional receptors.

Initial studies to identify the putative receptor were carried out. PEA-binding glycoprotein was solubilized from mouse LM fibroblasts and purified by single-step affinity chromatography on PEA-Sepharose affinity columns. The purified binding protein migrates on sodium dodecyl sulfate-polyacrylamide gel electrophoresis (SDS-PAGE) as a single high-molecular-weight species. In addition, incubation of PEA with binding protein reduces toxin bioactivity (decreased ability to stop protein synthesis). A similar glycoprotein can be isolated from mouse liver, the primary in vivo target of PEA. This toxin-binding glycoprotein has not been identified in detergent extracts of certain cell lines that exhibit high levels of resistance to the toxin. Further studies identified the receptor for PEA as the low-density lipoprotein receptor-related protein (LRP). Before outlining these studies, it is necessary to characterize the receptor itself.

Characteristics of Low-Density LRP

LRP is a member of the low-density lipoprotein receptor gene family. Other members of this gene family include the low-density lipoprotein receptor, the very-low-density lipoprotein receptor, apolipoprotein E (ApoE) receptor 2, and megalin. All members of this family have five major structural modules: a short cytoplasmic domain, a single transmembrane domain, YWTD domains, epidermal growth factor receptor-like cysteine-rich repeats, and ligand-binding (complement) type cysteine-rich repeats. Receptors in this family recognize extracellular ligands and internalize them, primarily for degradation in lysosomes.

LRP is a heterodimeric protein composed of two noncovalently associated subunits, the α- and β-chains (515 and 85 kDa, respectively) (Fig. 1). LRP is synthesized as a single polypeptide of 4,525 amino acids; it is cleaved by furin in the Golgi compartment to produce the two subunits. The two subunits have independent functions. The α-chain is made up of numerous epidermal growth factor-like domains, complement-like domains, and O-linked glycosylation sites,

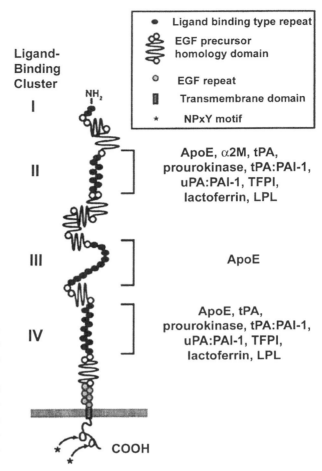

Figure 1 Structure of low-density LRP, the receptor for *Pseudomonas* exotoxin A. LRP is a multifunctional scavenger receptor capable of binding a variety of ligands to different clusters of ligand-binding repeats. tPA, tissue plasminogen activator; LPL, lipoprotein lipase. From reference 2 with permission.

all of which are involved in ligand interactions. These glycosylation sites are rich in serine and threonine residues and are thought to serve to keep the ligand-binding domains away from the cell surface. YWTD repeats separate the growth factor repeats. The β-chain consists of a transmembrane and a cytoplasmic domain that contains two copies of the NPXY sequence involved in targeting receptors to coated pits and their subsequent internalization.

LRP is expressed in a variety of cell types. It is expressed at high levels in liver, where it is involved in removal of circulating ligands such as activated α_2-macroglobulin and lipoproteins; LRP is also relatively abundant in brain and lung. It is found in tissues and cell lines derived from many species, including humans, mice, rats, and chickens. Various signaling molecules including growth factors, hormones, matrix components, and bacterial products such as lipopolysaccharide are known to alter LRP levels in different cell types. This expression is also dependent on oncogenic transformation and cellular differentiation.

LRP is essential for development. Mice deficient in LRP are not viable and die early during embryonic development. The reasons for embryonic lethality have not been determined, but it suggests that LRP is important for blastocyst implantation. Animals that are heterozygous for a disrupted LRP gene appear grossly normal. They exhibit normal development and a normal life span. Both sexes can transmit the mutant LRP allele through the germ line. Thus, the LRP defect is recessive.

The extracellular domain of LRP recognizes a very large number—at least 30—of ligands (Table 1). It was first shown to be a receptor for activated α_2-macroglobulin. It is now known to mediate the internalization of many ligands including chylomicron remnants, such as β–very-low-density lipoprotein and lipoprotein lipase, proteinases and protease inhibitor complexes, a minor-group human rhinovirus, and PEA. Thus, LRP is considered a general scavenging receptor that is able to clear numerous metabolites from the extracellular space. In addition, all members of the low-density lipoprotein receptor family bind a small 39-kDa protein, the receptor-associated protein (RAP). RAP is considered a chaperone present in the lumen of the endoplasmic reticulum. It assists in the proper folding of LRP and prevents premature binding of ligands to the receptor. Rather than being delivered to the cell surface in RAP-deficient cells, LRP is accumulated in the rough endoplasmic reticulum and degraded.

However, it is not just the extracellular domain of LRP that can bind a multitude of ligands. There is recent evidence that a number of cytoplasmic proteins can interact with the tail of LRP. Some of these interactions are with the endocytosis machinery whereas other interactions are with proteins that are regulators of numerous intracellular signaling pathways.

Why can LRP bind such an extraordinary number of ligands? In contrast to the well-conserved, complement-type ligand-binding repeats found in the receptor, the different ligands do not share significant sequence homology. Although mutagenesis experiments have indicated that positively charged amino acids are critical for receptor binding, no common consensus sequence has been derived that would constitute a receptor-binding domain. Some investigators propose that a secondary or tertiary structural motif, which is rich in positive electrostatic potential, could constitute a receptor-binding domain for the ligands. The binding domain of PEA contains 12 lysine residues; there are none in the translocation domain and three in the ADP-ribosylating domain. Site-directed mutagenesis

Table 1   Examples of ligands that bind the extracellular domain of LRP

Ligand	Function
α_2-Macroglobulin	Proteinase and inhibitor complexes
Plasminogen activator	
Plasminogen activator inhibitor-1	
ApoE	Lipoprotein transport and metabolism
Lipoprotein lipase	
Factors IXa, VIIIa	Blood coagulation
Lactoferrin	Antibacterial
Rhinovirus	Infection
Pseudomonas exotoxin A	
RAP	Chaperone

was used to convert the lysine residues in the binding domain of PEA to glutamate residues. These studies revealed that modification of Lys-57 greatly decreases the toxicity of PEA for cultured cells. This agrees with studies on physiologically relevant LRP ligands.

LRP has four putative ligand-binding domains or clusters (Fig. 1). LRP proteolytic fragments and LRP minireceptors have been used to determine the domains to which ligands bind. RAP binds with high affinity to domains II and IV, and with low affinity to domain III, but does not bind to domain I. RAP is the only ligand to date that has been found to bind domain III. Other studies have shown that the majority of LRP ligands can bind to both domain II and domain IV. In addition, high-affinity binding with LRP appears to require that ligands interact with multiple binding repeats on the receptor. However, in contrast to many of the other ligands studied, PEA uses only domain IV to bind to LRP. The reasons for these differences in binding remain to be elucidated.

LRP Is the PEA Receptor

Evidence presented by Kounnas et al. (3) that LRP serves as the receptor for PEA is severalfold. First, both the toxin-binding protein purified from LM cells or mouse liver and LRP have similar mobility on SDS-PAGE and are indistinguishable immunologically. Second, native PEA, but not a mutant toxin defective in its ability to bind to LM cells, binds to purified LRP that is immobilized on polystyrene or on nitrocellulose; the toxin interacts with the 515-kDa heavy chain of LRP on ligand blots, not with the 85-kDa light chain. Third, RAP both blocks binding of PEA to mouse LM cells and abolishes toxicity (Fig. 2). RAP has been shown to prevent binding of the majority of ligands that are internalized via LRP. When mouse LM cells are preincubated with a RAP-glutathione *S*-transferase (GST) fusion protein, PEA binding is blocked in a dose-dependent manner. PEA binding is reduced by 50% at a RAP-GST concentration of approximately 14 nM; this value is virtually identical to the K_d measured for the interaction of RAP with purified LRP. GST itself has no effect on the binding of PEA to cultured cells. To establish that LRP is important in the internalization of the putative toxin molecules involved in cell killing, it is necessary to establish that RAP—besides blocking toxin binding to LM cells—can also block toxicity. Preincubation of LM cells with RAP-GST effectively ablates PEA-induced inhibition of protein synthesis. In the absence of RAP, incubation with PEA reduces protein synthesis by over 80%, whereas in the presence of RAP—at concentrations that block binding—protein synthesis is normal. The fact that RAP has no effect on the toxicity of a chimeric toxin, transferrin-PE40—a chimera internalized by the transferrin receptor—suggests this effect is specific to PEA internalization via LRP.

Since LRP represents the only portal of entry for PEA, it would be expected that the level of functional LRP that is expressed on the cell surface is critical in determining cellular sensitivity. A variety of resistant cell lines have been screened, and it is established that they are resistant to toxin, either because they lack LRP or because functional receptor is maintained intracellularly and not trafficked to the cell surface. This intracellular accumulation is presumed to be due to the fact that LRP remains complexed with RAP and cannot exit the en-

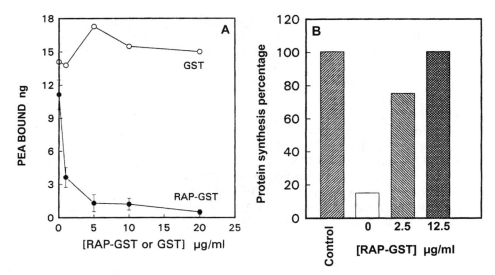

Figure 2 RAP blocks binding and toxicity of PEA for mouse LM fibroblasts. (A) Inhibition by RAP-GST of PEA binding to mouse LM cells. LM cell monolayers were incubated with medium containing RAP-GST (●) or GST (○) for 18 h at 4°C, followed by incubation with 2 μg/ml of PEA for 5 h at 4°C. Cells were then washed, harvested, and homogenized, and the concentration of cell-associated toxin assayed by enzyme-linked immunosorbent assay. (B) RAP protects LM cells from PEA-induced toxicity. Various concentrations of RAP were preincubated with LM cells for 90 min at 37°C. Cells were cooled and 40 ng of PEA added. Protein synthesis was assessed by measuring incorporation of ³H-leucine into trichloroacetic acid-precipitable material. From reference 3 with permission.

doplasmic reticulum-Golgi complex. In addition, the liver is the primary site of injury in animals following systemic challenge with PEA; this can be attributed to the high levels of LRP found in hepatocytes. Similarly, the susceptibility of several hepatic cell lines to PEA has been examined. One transformed line exhibits increased sensitivity to PEA, increased expression of surface LRP, and increased levels of LRP mRNA, compared to its normal parental cell line.

Genetic evidence also supports a role for LRP in internalization of PEA. LRP-deficient murine embryonic fibroblasts are unable to internalize activated α_2-macroglobulin or plasminogen activator/inhibitor complexes and are resistant to PEA. In addition, mutant CHO cell lines have been generated using PEA-mediated selection that display little or no surface LRP, and are unable to take up chymotrypsin–α_2-macroglobulin; the same cells are sensitive to a PEA chimeric toxin that is internalized via the transferrin receptor. These data confirm that LRP mediates the internalization of PEA.

In summary, LRP meets all the requirements for a receptor for PEA. Furthermore, it is an excellent example of a normal mammalian cell process that has been usurped by a bacterial virulence product for its own use.

DIPHTHERIA TOXIN RECEPTOR

Like PEA, diphtheria toxin (DT) is synthesized as a single polypeptide chain that is readily cleaved by proteolytic enzymes into two fragments, A and B, that are covalently linked by a disulfide bridge. The A fragment is the catalytic domain (C domain) and possesses the ADP-ribosylation activity. The B fragment consists of the transmembrane (T domain) and the receptor-binding (R domain) domains. The initial step in DT intoxication is binding of DT to a receptor on the surface of DT-susceptible cells. The toxin is internalized by receptor-mediated endocytosis. Fragment A penetrates into the endosomal membrane, aided by the T domain of fragment B. The A fragment then translocates from the endosome into the cytosol and inactivates elongation factor 2, stopping protein synthesis. Although the targets of PEA and DT are the same, the two toxins have minimal structural similarity and act on a different spectrum of cells. The latter is attributed to different receptors on the mammalian cell surface for PEA and DT.

The DT receptor has been identified, isolated, and cloned and is identical to the cell surface-expressed precursor form of the heparin-binding epidermal growth factor (EGF)-like growth factor (proHB-EGF). The mature or soluble heparin-binding EGF-like growth factor (sHB-EGF) belongs to a family of growth factors that includes EGF, transforming growth factor α, and amphiregulin. The mature molecule is synthesized as a membrane-anchored form—the proHB-EGF, which is cleaved proteolytically at the extracellular domain to yield the sHB-EGF. The shedding of the extracellular domain to generate the soluble form is regulated by multiple signaling pathways. ProHB-EGF has properties that are distinct from the soluble form; it is not only the precursor for sHB-EGF but is itself a biologically active molecule. Membrane-anchored growth factors are able to transmit signals only to neighboring cells in a juxtacrine manner compared to the autocrine and paracrine manner for the soluble form; they also can form complexes with other membrane molecules (see below). In vivo these molecules have been implicated in several physiological processes including wound healing and blastocyst implantation.

The road to identification of the DT receptor was long and arduous. Most early studies utilized monkey kidney cell lines (Vero and BSC-1) of high toxin sensitivity. The first studies suggesting the existence of a specific receptor for DT on the surface of mammalian cells were competition experiments using CRM197, a nontoxic mutant form of DT; preincubation with CRM197 protects cells from the action of native toxin. Middlebrook and colleagues (5) reported that most toxin-sensitive cell lines tested (monkey kidney derived; e.g., Vero cell) demonstrated the highest level of specific binding of ^{125}I-labeled DT, whereas cells of intermediate sensitivity (human HeLa) and resistant cells (mouse-derived cells) demonstrated no specific binding. It was estimated that intact sensitive cells have 100,000 receptors on their surface.

Numerous attempts were made to isolate the specific receptor from cell fractions, but problems were encountered because the toxin binds to numerous different proteins in solubilized cell extracts. It took nearly 10 years to identify the DT receptor in detergent-treated cell extracts at the protein level. When identified, the putative receptor was shown to be a 14.5-kDa protein—very small compared to the typical mammalian cell receptor.

In a defining series of experiments, Eidels and colleagues (6) cloned the cDNA encoding a DT receptor from highly toxin-susceptible monkey Vero cells. This receptor then was expressed in normally toxin-resistant mouse LM (TK⁻) cells, which resulted in a DT-sensitive cell line (DT^S-II) having a level of toxin sensitivity comparable to that of Vero cells, and able to specifically bind DT. Homology search analysis showed this product to be identical to proHB-EGF. Binding of DT to immobilized HB-EGF showed that DT binds with an affinity similar to that for intact Vero cells. These results clearly demonstrate that the receptor for DT is proHB-EGF.

DT Receptor Structure

The proHB-EGF cDNA encodes a protein of 208 amino acids and consists of a signal sequence, an extracellular, a transmembrane, and a carboxyl-terminal cytoplasmic domain (Fig. 3). The cell surface-expressed form (after cleavage of the signal sequence) is composed of 185 amino acids and has a molecular weight of 20,652. The extracellular domain comprises a propeptide and heparin-binding, EGF-like, and juxtamembrane domains.

As already stated, mice, rats, and tissues derived from these animals are exceedingly resistant to the action of DT, whereas humans are quite sensitive. Nevertheless, HB-EGF is expressed in multiple tissues in these resistant species. Therefore, the question can be asked as to why cells from mice and rats are

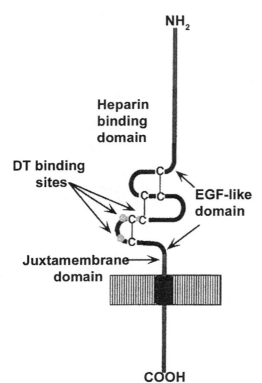

Figure 3 Structure of the cell surface-expressed precursor form of the heparin-binding EGF-like growth factor, the DT receptor.

resistant to DT. Transfection of human but not mouse HB-EGF cDNA into mouse L cells renders them sensitive to DT. These results suggest that mouse HB-EGF cannot serve as a functional receptor for DT. This deficiency is due to nonconserved amino acid substitutions in the EGF-like domain of the receptor. Three amino acid residues (Ile-133, His-135, and Glu-141) are critical for DT binding to the receptor. Conversion of these residues to the corresponding mouse proHB-EGF residues results in a complete loss of receptor activity. Conversely, substitution of the corresponding mouse residues with these three monkey residues results in a mouse proHB-EGF that is fully functional as a DT receptor. The crystal structure of a complex of DT with part of the HB-EGF molecule confirmed that these three residues of the receptor are in contact with the toxin.

DT binds to its cell surface receptor through the R domain of the B fragment. Alanine replacement studies of this region of the toxin molecule identified the positively charged Lys-516 as the most critical amino acid residue involved in binding. Similarly, the negatively charged Glu-141 of the DT receptor is the most critical residue for binding of DT. Reciprocal substitution of residues was undertaken to determine if the key interaction between toxin and receptor occurs between these two oppositely charged amino acids. These studies confirmed a specific interaction between these two key residues and, in addition, interactions between other positively charged residues of toxin and Glu-141 of the receptor.

Molecules Associated with the DT Receptor

One of the characteristics of proHB-EGF is that it forms a complex with several other membrane proteins. The first of these molecules to be identified was DRAP27, the monkey homolog of the human CD9 antigen. CD9 is a member of the tetraspanin superfamily and is characterized by four highly conserved hydrophobic transmembrane domains that form two extracellular domains. The role of CD9 on DT receptor interaction has been examined by transfection of CD9 cDNA and DT receptor cDNA into L cells. These transiently transfected cells bind 10 times more DT and are 20 times more sensitive to DT than are cells transfected with receptor alone. More recent studies provided evidence that CD9 also increases the receptor affinity of cells for DT. Studies indicate that monkey CD9 directly interacts with proHB-EGF to form a complex. It is suggested that this contact may alter the conformation of the receptor and increase the binding affinity for DT and thus increase toxin sensitivity. On the basis of these observations, it has been proposed that CD9 is a coreceptor for DT.

There also is evidence that proHB-EGF has a high affinity for heparin. Heparin-like molecules bind to the heparin-binding domain of proHB-EGF and enhance its ability to bind toxin. Vero cells have a larger concentration of surface heparan-sulfate than do L cells or CHO cells. CHO cells that are deficient in heparan-sulfate proteoglycans are 15 times less sensitive to DT than are wild-type CHO cells. Addition of heparan-sulfate or free heparin to culture medium restores the DT sensitivity of the mutant cells. This increase in sensitivity is due to increased binding of toxin to the cells. Scatchard plot analysis showed that heparin increases the affinity of the receptor for toxin but does not change the number of binding sites. Heparin addition to Vero cells has no effect on DT binding or sensitivity, probably because these cells have an abundance of heparan-sulfate on their surface. CD9 and heparin are thought to enhance DT binding by different mechanisms. One model is presented in Fig. 4.

Role of Receptor in the Toxin Entry Process

DT molecules bound to receptor on the cell surface are internalized by receptor-mediated endocytosis. Interestingly, there is a tyrosine residue (Tyr-192) in a region of the cytoplasmic domain that resembles the signal to direct receptors to clathrin-coated pits; however, deletion studies demonstrated that the cytoplasmic domain is not required for DT endocytosis.

Besides clathrin-mediated internalization, there is also evidence that the receptor might affect the translocation of DT across the endosomal membrane. The juxtamembrane domain of proHB-EGF is the linker between the transmembrane and the EGF-like domains. It has been hypothesized that the juxtamembrane domain might influence the interaction of the T domain of toxin with the endosomal membrane. Cells expressing receptors with different-length juxtamembrane domains bind DT normally; however, they exhibit reduced sensitivity to DT when compared to wild-type cells. Data suggest that this difference is due to a reduction in translocation of the A fragment from endosome into the cell cytosol. Thus, the toxin receptor serves multiple functions in DT intoxication, including recognition of toxin, involvement in internalization, and translocation to the cytosol.

SHIGA TOXINS

Shiga toxins are a family of structurally and functionally related toxins produced by several enteric bacteria including *Shigella dysenteriae* serotype 1 and entero-

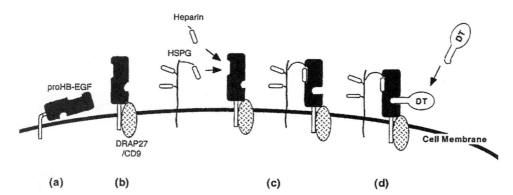

Figure 4   Proposed model for role of CD9 and heparin-like molecules in binding of DT to proHB-EGF. (a) ProHB-EGF alone in the plasma membrane cannot bind DT. (b) CD9 binds to and orients proHB-EGF so that it is accessible to DT. (c) Cell surface heparan-sulfate proteoglycans (HSPG) or free heparin binds to proHB-EGF at the heparin-binding domain and induces a conformational change; this change results in increased affinity of the receptor for DT. (d) The proHB-EGF/CD9-HSPG-DT complex is formed. From reference 8 with permission.

hemorrhagic *Escherichia coli*. These agents are associated with gastroenteritis and are spread via food and water or person to person. The Shiga toxins produced by these microorganisms have a role in disease pathogenesis and are responsible for some of the complications associated with the infections, including hemorrhagic colitis and hemolytic uremic syndrome (HUS). These toxins also are called Shiga-like toxins or verotoxins. There are at least six members of this family with a similar structure; each toxin has a single enzyme active A subunit and five B subunits involved in binding to a receptor on the target cell. The A subunit cleaves a purine residue from 28S ribosomal RNA, resulting in blockage of protein synthesis and ultimate death of the target cell. The toxins can induce apoptosis in epithelial cells and in certain endothelial cells. The mammalian cell receptor for most of the Shiga toxin family members is Gb3. The exception is the receptor for Stx2e (pig edema disease toxin) that binds preferentially to the longer glycolipid globotetraosylceramide (Gb4). Gb3 (see Fig. 5) is the Pk antigen in the P blood group system. It is also a cell differentiation marker (CD77) and a tumor marker for a B-cell lymphoma. There are differences in affinity for receptor and a range in toxicity for tissue culture cells and for mice among the various family

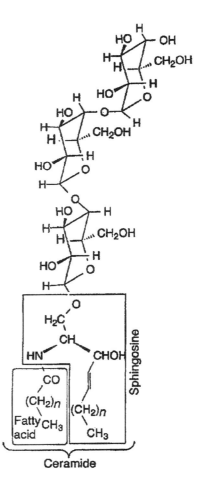

Figure 5 Structure of Gb3. Gb3 consists of a ceramide long-chain fatty acid that is embedded in the plasma membrane and a short extracellular trisaccharide chain ending in a digalactose residue. From reference 4 with permission.

members. However, this discussion will speak in general terms about receptors, focusing on the Gb3 receptor.

Identification of Shiga Toxin Receptor

The search for receptors for Shiga toxin on eukaryotic cells began in 1977 when it was reported that cells that were sensitive to Shiga toxin could remove toxin bioactivity from tissue culture medium whereas toxin-resistant cells did not have this capacity. However, it was not until 1986 that unequivocal evidence for a receptor was provided (1). Those investigators demonstrated specific binding of ^{125}I-labeled Shiga toxin to isolated brush border membrane vesicles from rabbit ileum. This is an excellent source of material as it has been recognized for many years that the toxin has an enterotoxic effect in rabbit ileum. Soon thereafter, several groups identified membrane glycolipids that contain the carbohydrate sequence Galα(1-4)Galβ(1-4)Glu-ceramide as the putative receptor. Through thin-layer chromatography, Shiga toxin was shown to bind to several membrane glycolipids including Gb3. Several lines of evidence established this glycolipid as a receptor for Shiga toxin. For example, addition of Gb3 to culture medium inhibits the biological activity of Shiga toxin. If receptors are destroyed by digestion of membrane glycolipids with a galactosidase or if the synthesis of glycolipids is blocked, cytotoxicity is lost. In addition, incorporation of Gb3 into the membrane of toxin-resistant cells was shown to restore toxin sensitivity to these cells. It also has been shown that stimulation of the biosynthetic pathways involved in Gb3 synthesis, using butyrate, increases toxin sensitivity. Last, Gb3 is developmentally regulated. The suckling rabbit is resistant to the action of Shiga toxin; after weaning, receptors are expressed on villus epithelial cells. This corresponds to the appearance of sensitivity to the action of Shiga toxin.

Mere expression of Gb3 does not always correlate with Shiga toxin sensitivity. This may be explained by the heterogeneity of the lipid moiety of Gb3, which exhibits differences in the length of the fatty acid chain. These differences have been shown to affect toxin sensitivity and binding. For example, Shiga toxin was shown to bind preferentially to Gb3 molecules that have fatty acid chain lengths greater than 20 carbons. Cells treated with sodium butyrate (a known regulator of gene transcription) exhibit increased sensitivity to toxin, in part because of an increase in the length of the fatty acid chain. Significantly, members of the normal enteric microbial flora produce butyrate and its concentration in the colon may be sufficient to sensitize cells to toxin during the infection process. In the butyrate studies it also was shown that the fatty acids have an important role in the intracellular trafficking of the toxin, suggesting that the receptor is involved in more than just toxin binding. In fact, different fatty acid isoforms of Gb3 target the toxin to different intracellular sites. These differences correlate with significant differences in cellular sensitivity to toxin.

The contribution of Gb3 expression to Shiga-like toxin-induced renal disease has also been studied. HUS is considered to be a disease primarily of the renal glomeruli. It occurs in children but only rarely in adults, except for the elderly. When sections of human pediatric kidney are stained, significant toxin binding to cells in the renal glomeruli is seen. In contrast, there is little reaction with toxin in the adult glomerulus, suggesting the absence of functional receptor. This correlates with the age-related incidence of HUS and suggests that toxin receptor distribution plays an important role in induction of pathology.

Several cell types are considered generally to be sensitive to Shiga toxins; these include endothelial cells, epithelial cells, macrophages and monocytes, and certain B lymphocytes. Shiga toxins kill epithelial, endothelial, and some B cells but do not kill macrophages and monocytes. Shiga toxins directly damage intestinal epithelium in certain animal species such as rabbit. Challenge of rabbit ligated ileal loops with purified toxin results in mucosal damage and fluid accumulation. Work with isolated enterocytes from rabbit ileum suggests that villus enterocytes, not crypt cells, are the target of the toxins, as they possess Gb3 receptor. In contrast, gastrointestinal epithelial cells from humans and mice do not express Gb3 and are resistant to the lethal effects of Shiga toxins. It should be noted that cultured epithelial cells are able to transcytose Shiga toxins without succumbing to their cytotoxic effect; this may explain how the toxin enters the systemic circulation and thus may be important in understanding the importance of the pathogenesis of these microorganisms.

Endothelial cells are considered the primary in vivo target of Shiga toxins (7). Endothelial cells growing in culture express Gb3, bind Shiga toxins, and are susceptible to the cytotoxic effects of these toxins. Endothelial cells derived from different tissues exhibit different susceptibility to toxin. Thus, human intestinal and kidney endothelial cells are highly sensitive to toxin, whereas cells derived from human umbilical vein are very resistant. In addition, it has been well documented that lipopolysaccharide, tumor necrosis factor-α, interleukin 1-β, and sphingomyelinase enzyme added to endothelial cells in culture increase receptor expression on diverse cell types and increase the toxin sensitivity of these cells. The end result of the interaction of Shiga toxins with endothelial cells is the impairment of their function, leading to decreased blood flow in small vessels in targeted organs, altered leukocyte adhesion, and an increase in thrombosis formation. It is generally accepted that Shiga toxins are involved in vascular changes including hemorrhagic colitis and HUS but have relatively little importance in the diarrheal phase of these infections.

Toxin-Receptor Interactions

The core glycolipid structures for members of the globo-series of glycolipids are unique because they have an α-linked galactose residue. This residue results in the formation of a "kink" in the conformation of the oligosaccharide. X-ray crystallography studies and site-directed mutagenesis studies have been carried out on the B-subunit pentamer of Shiga toxin and have shown that there are two to three globotriose-binding sites per B-subunit monomer and that these sites are biologically significant.

The Shiga-like toxins bind with high affinity to receptor. For example, verotoxin binds to Gb3 in a cholesterol-lecithin bilayer structure with an affinity in the range of 10^{-8} M and to the plasma membrane of Gb3-containing cells in the range of 10^{-10} to 10^{-11} M. This affinity for target cells makes Shiga toxins exceedingly active biological molecules.

It should be mentioned that certain cell lines (A341 human epidermoid carcinoma and T84 human colon cancer cell lines) do not express glycolipid receptors but still are able to internalize and traffic toxin. In addition, mutated Vero cells, which are deficient in Gb3, bind Shiga toxin. Cross-linking studies using these cells and ^{125}I-labeled Shiga toxin B subunit reveal binding to a 50-kDa protein. The role of this membrane protein in induction of toxicity remains to be determined. However, all data suggest that the interaction of these toxins with target cells is complex and that there is still much to be learned.

CONCLUSIONS

The three receptors discussed have functions essential to the normal physiologic properties of mammalian cells. Nevertheless, they represent molecules usurped by different bacterial toxins as the first step in the intoxication process. Cells lacking functional cell surface receptor are resistant to the toxin, expression of receptor correlates with the tissue specificity of toxin damage, and this in turn correlates with the disease symptoms seen in animal models and in patients.

Acknowledgments
I thank Jeong-Heon Cha, Leon Eidels, David FitzGerald, and Tom Obrig for careful reading of the manuscript and helpful discussions.

REFERENCES
1. **Fuchs, G., M. Mobassaleh, A. Donohue-Rolfe, R. K. Montgomery, R. J. Grand, and G. T. Keusch.** 1986. Pathogenesis of shigella diarrhea: rabbit intestinal cell microvillus membrane binding site for *Shigella* toxin. *Infect. Immun.* **53:**372–377.
2. **Herz, J., and D. K. Strickland.** 2001. LRP: a multifunctional scavenger and signaling receptor. *J. Clin. Invest.* **108:**779–784.
3. **Kounnas, M. Z., R. E. Morris, M. R. Thompson, D. J. FitzGerald, D. K. Strickland, and C. B. Saelinger.** 1992. The $\alpha_2$macroglobulin receptor/LDL receptor related protein binds and internalizes *Pseudomonas* exotoxin A. *J. Biol. Chem.* **267:**12420–12423.
4. **Lingwood, C. A.** 1996. Role of verotoxin receptors in pathogenesis. *Trends Microbiol.* **4:**147–152.
5. **Middlebrook, J. L., R. B. Dorland, and S. H. Leppla.** 1978. Association of diphtheria toxin with Vero cells. *J. Biol. Chem.* **253:**7325–7330.
6. **Naglich, J. G., J. E. Metherall, D. W. Russell, and L. Eidels.** 1992. Expression cloning of a diphtheria toxin receptor: identity with a heparin-binding EGF-like growth factor precursor. *Cell* **69:**1051–1061.
7. **Obrig, T. G., P. J. Del Vecchio, J. E. Brown, T. P. Moran, B. M. Rowland, T. K. Judge, and S. W. Rothman.** 1988. Direct cytotoxic action of Shiga toxin on human vascular endothelial cells. *Infect. Immun.* **56:**2372–2378.
8. **Umata, T., K. D. Sharma, and E. Mekada.** 1999. Diphtheria toxin and the diphtheria toxin receptor, p. 45–66. *In* K. Aktories and I. Just (ed.), *Bacterial Protein Toxins.* Springer-Verlag, Berlin, Germany.

SUGGESTED READING
1. **O'Loughlin, E. V., and R. M. Robins-Browne.** 2001. Effect of Shiga toxin and Shiga-like toxins on eukaryotic cells. *Microbes Infect.* **3:**493–507.

Bacterial Protein Toxins
D. Burns et al., Editors
©2003 ASM Press, Washington, DC 20036

Chapter 10

Direct Penetration of Bacterial Toxins across the Plasma Membrane

Franca R. Zaretzky, Mary C. Gray, and Erik L. Hewlett

Bacterial toxins have been categorized by their sites and mechanisms of action into the following groups: (i) toxins that act by binding to or enzymatically modifying molecules at the target cell surface (toxic shock syndrome toxin, cytolytic phospholipases); (ii) pore-forming toxins that insert into the cytoplasmic membrane of the host cell to create ion-conducting pathways (staphylococcal α-toxin, *Escherichia coli* α-hemolysin, perfringolysin); and (iii) toxins with enzymatic activity involving intracellular targets, necessitating the entry of some or all of their catalytic domain into the cytoplasm (ADP-ribosylating toxins, such as pertussis toxin and cholera toxin; deamidating toxins, such as cytotoxic necrotizing factor; and toxins with glycosidase activity, such as Shiga toxin). This latter group also includes enzymes now known to be toxins by virtue of their introduction into cells via the type III secretion system (YpkA, YopH, and exoenzyme S).

Although many of the toxins that deliver their catalytic domains to the target cell interior have been shown to possess some pore-forming activity in vitro, most bind to specific cell surface receptors and employ receptor-mediated endocytosis (RME) to enter the cytoplasm directly from the endosome or more distally via the Golgi apparatus or endoplasmic reticulum. This group includes diphtheria toxin, anthrax protective antigen (PA) plus either edema factor (EF) or lethal factor, cholera toxin, and the clostridial neurotoxins, to name a few. These important virulence factors are discussed in other chapters in this volume.

One fascinating toxin does not fit clearly into any of the categories noted above. This apparent hybrid toxin is related to and presumably derived from a family of pore-forming toxins and creates a transmembrane pore, but also contains a catalytic domain that can be delivered from the target cell surface. Adenylate (or adenylyl) cyclase (AC) toxin, produced by *Bordetella pertussis* and related organisms (*Bordetella parapertussis* and *Bordetella bronchiseptica*), is a novel

Franca R. Zaretzky and Mary C. Gray • Department of Medicine, Room 6828, Old Medical School, University of Virginia School of Medicine, Charlottesville, VA 22908. *Erik L. Hewlett* • Departments of Medicine and Pharmacology, Room 6832, Old Medical School, University of Virginia School of Medicine, Charlottesville, VA 22908.

molecule possessing these features. It is a member of the RTX (repeat-in-toxin) family of hemolysins and leukotoxins and is hypothesized to have originated by virtue of a gene fusion event between an RTX ancestral gene and an AC gene from some eukaryotic organism. AC toxin is a known virulence factor and protective antigen for *Bordetella* species. Because AC toxin is the only molecule known to deliver its catalytic domain from the target cell surface, it will be used as the model for discussion of (i) the nature of its interaction with and insertion into the membrane of target cells, (ii) the biophysical basis for delivery of its catalytic domain, and (iii) the process by which this toxin, which is primarily cell associated and located on the surface of *B. pertussis*, is delivered when whole bacteria are interacting with target cells. The aspects of AC toxin function that are novel will be noted, as will the similarities with other toxins.

BACKGROUND

AC toxin was discovered in the 1970s as a cell-associated enzymatic activity on *B. pertussis* but was not demonstrated to be a toxin until nearly a decade later. Confer and Eaton in 1982 demonstrated that urea extracts of *B. pertussis* containing AC toxin inhibited phagocytosis and oxidative burst by human neutrophils, suggesting a role for this toxin in protecting the infecting bacteria from clearance by inflammatory cells (1). AC toxin is an essential virulence factor for *B. pertussis*, as mutants lacking this molecule are virtually avirulent in an animal infection model. AC toxin is expressed as a single 1,706-amino-acid (aa) polypeptide encoded by *cyaA*, the structural gene that exists in an operon with the genes (*cyaB*, *cyaD*, and *cyaE*) whose products are involved in toxin secretion. The gene for posttranslational acylation of AC toxin, *cyaC*, is divergently transcribed from the structural and secretion genes. Acylation of AC toxin in *B. pertussis* occurs by covalent attachment of predominantly palmitate at Lys-983, whereas recombinant AC toxin produced in *E. coli* may possess acylation at Lys-983 and an additional acyl group at Lys-860. Acylation is required for binding, insertion, and all of the activities of the toxin involving target cells.

AC toxin possesses a catalytic domain (aa 1 to 400) that is responsible for conversion of ATP to cyclic AMP (cAMP), a hydrophobic domain (aa 500 to 700) believed to be involved in insertion of the toxin into the target cell membrane, and a glycine-aspartic acid repeat region (aa 1,000 to 1,600), which binds calcium and is characteristic of toxins in the RTX family (Fig. 1). Three cellular events take place when AC toxin interacts with target cells: intoxication, hemolysis, and K^+ efflux. Delivery of the catalytic domain of AC toxin to the target cell interior results in supraphysiologic levels of intracellular cAMP. This function is known as "intoxication" or "toxin/cell invasive" activity and is that which distinguishes AC toxin from other RTX molecules. The C-terminal 1,300-aa segment of AC toxin is responsible for binding to the target cell and delivery of the catalytic domain. Like other members of the RTX family, AC toxin is hemolytic for sheep erythrocytes; however, compared to other RTX toxins, AC toxin is a weak hemolysin. It is this hemolytic activity that is responsible for the zone of hemolysis around colonies of *Bordetella* cells on agar plates containing blood. The N-terminal catalytic domain (aa 1 to 373) possesses AC enzymatic activity, but nei-

AC Toxin Domain	Properties
1	• catalytic domain • binds ATP for conversion to cAMP • possesses adenylate cyclase enzymatic activity • binds eukaryotic calmodulin which activates toxin • calmodulin binding prior to toxin addition blocks intoxication and enhances hemolysis
2	• presence of this domain is antagonistic to hemolysis
3	• hydrophobic domain with 4-6 putative transmembrane segments • required for insertion into target cell membrane
4	• function unknown • contains site for acylation
5	• repeat region • binds calcium
6	• secretion signal required for toxin export from cell cytoplasm • required for intoxication and hemolysis

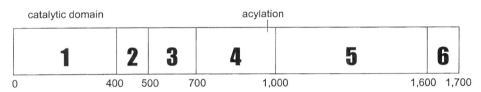

Figure 1   Domain structure of AC toxin and description of domain-associated activities.

ther the physical presence of this portion of the molecule nor cAMP production is required for hemolysis.

Intoxication occurs with virtually immediate onset after AC toxin addition to cells and appears to reflect the action of a single toxin molecule (3). In contrast, hemolysis has a lag of at least an hour and is believed to require oligomerization of three to four toxin monomers to create a pore that leads to colloid osmotic lysis (3). Insertion of AC toxin into a variety of cell types, including red blood cells, can also cause sufficient disruption of the cell membrane to allow nonlytic leakage of K^+ into the medium (3). This third toxin activity is immediate in onset and appears to be the function of toxin monomers, similar to intoxication.

All three of the activities of AC toxin are dependent on posttranslational acylation and the presence of calcium ions, some 40 of which are bound to fully occupy the RTX repeat sites on the toxin molecule. Enzymatic activity is regulated by calmodulin in the host cell cytoplasm, increasing the basal catalytic rate 100- to 1,000-fold. The fact that calmodulin is found in eukaryotic but not prokaryotic cells is one basis for the suggestion that the catalytic domain may have its origin in a eukaryotic genome.

INITIAL INTERACTION BETWEEN AC TOXIN AND TARGET CELLS

Binding to a cell surface receptor, followed by RME, is the process used by many toxins to gain access to the cytoplasm of their target cells. RME is characterized by several features that can be used to determine whether a particular toxin is entering by that pathway. First, it is generally true that ligand binding can occur at 4 and 37°C, but internalization only occurs at 37°C. Second, binding is saturable and limited by the number of receptor molecules on the target cell. Toxins generally bind with high specificity to target cells producing the required receptor, which may be limited to a small subset of cell types. Third, toxins that enter target cells by RME exhibit a lag phase between toxin addition and biological effects, due to the time necessary for uptake into an endosome (blocked by cytochalasin D and dansylcadaverine), acidification of the endosome (blocked by NH_4Cl, methylamine, chloroquine, and monensin), and resultant translocation of the catalytic portion into the cytoplasm.

Based on these criteria, several lines of evidence indicate that AC toxin does not require a protein receptor to bind and insert into target cells. The process of binding and insertion of AC toxin does not appear to be saturable and the toxin does not discriminate among cell types. AC toxin can elicit cAMP accumulation in a wide variety of cells, virtually all that have been tested, and in cells that have been exposed to trypsin or cycloheximide (a potent protein synthesis inhibitor). AC toxin does not require RME for its catalytic domain to gain access to the target cell interior. Significantly reduced binding of AC toxin to target cells is observed at 4°C compared with 37°C, and intoxication is virtually immediate upon AC toxin addition. Intoxication of target cells by AC toxin is not blocked by inhibitors of either endocytosis or endosome acidification. Furthermore, AC toxin is active in cells that demonstrate little membrane trafficking, such as erythrocytes.

These observations have been made in a number of laboratories and include a study in which direct comparison of toxin delivery was made between AC toxin and anthrax toxin, consisting of PA and EF, another calmodulin-activated adenylate cyclase (2). AC toxin from *B. pertussis* and EF/PA from *Bacillus anthracis* both produce supraphysiologic levels of cAMP in target cells. Despite these functional similarities, EF/PA utilizes RME for intoxication of target cells. Whereas cytochalasin D, chloroquine, and NH_4Cl completely blocked intoxication of CHO cells by EF/PA, intoxication by AC toxin was unaffected, indicating that different mechanisms are involved and that intoxication of target cells by AC toxin does not require RME.

Although most studies on AC toxin delivery have established that the toxin can translocate its catalytic domain across lipid membranes irrespective of the presence of a receptor, the possibility that a receptor for AC toxin exists cannot be ruled out. One recent report demonstrated that cells producing the α_M/β_2-integrin, CD11b/CD18, were more efficiently intoxicated by AC toxin than were cells without the receptor; monoclonal antibodies against CD11b inhibited AC toxin binding and cAMP accumulation in these cells (4). Transfection of CHO cells with CD11b/CD18 enhanced the sensitivity of this cell line to AC toxin action, promoting increased production of cAMP and AC toxin-mediated apoptosis compared with wild-type CHO cells that do not produce CD11b/CD18. It is clear, however, that CHO cells and other cell types without this integrin can be intoxicated by AC toxin (2), indicating that multiple pathways for entry may be operative. Because CD11b is present on neutrophils, macrophages, dendritic cells, and natural killer cells, these findings are consistent with the hypothesis that a primary role for AC toxin is to inactivate inflammatory cells involved in front-line defense against bacterial pathogens.

BIOPHYSICAL BASIS FOR TRANSLOCATION OF THE CATALYTIC DOMAIN

All toxins that act on intracellular targets must have a mechanism by which at least a part of their structure can traverse the plasma membrane of the host cell. In the case of toxins that enter by RME, a major stimulus for this translocation event appears to be the conformational change elicited by acidification of the endosome. This principle is illustrated by the ability of such molecules to enter cells when endocytosis is blocked, while the extracellular environment is made acidic to mimic the conditions in the endosome (2). The absolute requirement for acidification does not apply to AC toxin, leading to the question of what are the signals and driving forces that initiate the process that propels the catalytic domain into the cell interior.

Intoxication by AC toxin is dependent on the electrical potential across the plasma membrane (7). This voltage-dependent process is downstream from toxin binding to the cell surface and appears to be a driving force for translocation of the catalytic domain. In single-cell patch-clamp experiments, after the toxin has undergone its initial association with the cell, delivery of the catalytic domain does not occur when the cell is depolarized. The magnitude of intoxication exhibits a steep dependence on the membrane potential.

The importance of the electrostatic charge of the catalytic domain to translocation is illustrated in studies whereby the catalytic domain is used to introduce inserted T-cell epitopes into target cells for processing by the major histocompatibility complex class I pathway. Delivery of the inserted epitope, as reflected by development of a cell-mediated immune response in vivo, is prevented when the insert has a net negative charge of -4 (5).

Pore formation by AC toxin, as with other pore-forming toxins, is also voltage dependent, which may reflect an influence on either the structure of the membrane-inserted form or the stability of the resultant oligomer. Although controversial, binding of intracellular calmodulin to the catalytic domain of AC toxin has also been proposed as a mechanism to provide a driving force for rapid entry of AC toxin into target cells.

Characterization of the requirements for translocation of the catalytic domain and K^+ efflux and hemolysis has led to the realization that these events are not only dissociable but are antagonistic as well. Hemolysis and K^+ efflux can occur at 0°C, albeit at a reduced rate relative to that at 37°C (3). In contrast, delivery of the catalytic domain to the target cell cytoplasm does not occur at 0°C (Fig. 2). Similarly, the calcium requirement for translocation of the catalytic domain is different from that for K^+ efflux and hemolysis. For example, preparation of AC toxin in the presence of calcium results in K^+ efflux and hemolysis, even when toxin is added to target cells in the presence of excess calcium-

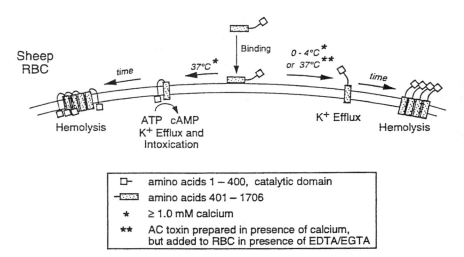

Figure 2   Characteristics of the functional activities of AC toxin.

chelating agents. In contrast, delivery of the catalytic domain requires an extra-cellular calcium concentration of greater than 300 μM, indicating differential roles for calcium binding in these activities.

A number of reports have provided evidence that intoxication and hemolysis are antagonistic. Preventing translocation of the catalytic domain by exogenously added calmodulin or with an antibody to the distal catalytic domain, deletion of the catalytic domain plus a small portion distal to the catalytic domain, or sub-stitutions in the hydrophobic domain result in enhancement of hemolysis by three- to fourfold. Thus, prevention of delivery or removal of the catalytic do-main and a segment distal to it elicits a conformation of the toxin molecule that is more favorable for hemolysis. Thus, the catalytic domain appears to act as a constraint to hemolysis. These findings support the concept that this novel toxin resulted from a fusion between a eukaryotic AC gene and a gene encoding an RTX ancestor. The result is a molecule able to deliver this newly acquired cata-lytic domain, but at the cost of hemolytic potency.

Divalent metal binding to AC toxin and the membrane potential of the target cell are clearly critical factors in the processes of insertion and translocation of the toxin, but the signals that initiate this sequence of events from the cell surface remain unknown. Pursuit of the identification of these signals has led to consid-eration of how AC toxin comes into contact with target cells during natural *B. pertussis* infection.

Can AC Toxin Bind, Insert, and Deliver from Its Location on the Bacterial Outer Membrane?

Most of the understanding of the structure and function of AC toxin comes from work utilizing soluble material obtained by urea extraction of *B. pertussis* organ-isms or recombinant *E. coli* expressing AC toxin. Unlike other toxins in the RTX family that are released in abundance into the culture medium, AC toxin remains primarily associated with the *B. pertussis* surface. It has become clear from a number of studies that although AC toxin released into the culture supernate possesses AC enzymatic activity, this material does not function as a toxin (1). Whole bacteria, however, can deliver AC toxin to target cells.

The mechanism by which AC toxin is delivered from *B. pertussis* to target cells remains a mystery, but a number of observations have been made. *B. pertussis* organisms are capable of adhering to and invading a variety of nonphagocytic cells such as HeLa cells, human tracheal epithelial cells, CHO cells, and T lymphocytes, and professional phagocytes including neutrophils and macrophages (Fig. 3A and B). The ability of *B. pertussis* to invade these cell types, in many cases, depends on whether AC toxin is expressed. For example, expression of AC toxin inhibits invasion of HeLa cells by *B. pertussis*. Mutants deficient for AC toxin production enter in significantly higher numbers than wild-type organisms, but the increased level of invasion can be restored to wild-type levels if intracellular cAMP levels are artificially increased, suggesting a direct role for AC toxin in inhibiting internalization. It has been proposed that *B. pertussis* organisms may enter nonprofessional phagocytic cells of the respiratory tract and remain within these cells for a brief time during infection to "hide out" from detrimental inflammatory cells.

Studies examining the role of AC toxin in the interaction of *B. pertussis* with professional phagocytes have led to conflicting results. Whereas *B. pertussis* organisms are killed by neutrophils (6), the bacteria appear to survive within human macrophages, in a process that requires the presence of AC toxin. Furthermore, AC toxin inhibits phagocytosis of opsonized *B. pertussis* by human neutrophils whereas AC toxin-deficient mutants are efficiently internalized; inhibition can be overcome by anti-AC toxin monoclonal antibodies (8). *B. pertussis* organisms producing AC toxin induce apoptosis in a macrophage cell line, a

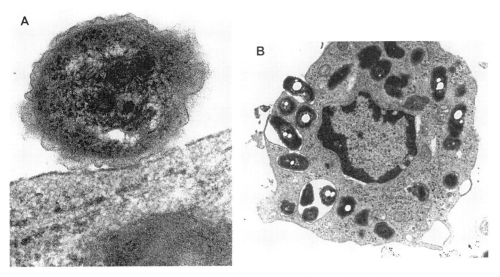

Figure 3 AC toxin is delivered by *B. pertussis* adherent (A) to human monocytes and following phagocytosis (B) by these cells.

phenomenon that does not require invasion by the bacteria. Although it is clear that *B. pertussis* cell-associated AC toxin is required for these activities, a debate still exists as to whether mere adherence of *B. pertussis* to target cells can induce delivery of AC toxin to the target cell interior or if internalization of the bacteria is required. If internalization of *B. pertussis* is required, how does AC toxin escape from the endosome? Do *B. pertussis* organisms within the endosome have access to ATP or do organisms escape from the confines of the endosome and reside within the cytoplasm of the target cell, allowing access to ATP and subsequent cAMP production? These questions have yet to be answered.

Infection with *B. pertussis* is not systemic; organisms remain localized to the respiratory mucosa. Increased mucus production, ciliostasis, and extrusion of ciliated cells from the respiratory epithelium are characteristic of pertussis. One or more of these pathological features of pertussis may be explained, in part, by increased intracellular cAMP within ciliated respiratory epithelial cells to which adherent *B. pertussis* organisms are delivering AC toxin. The key question that remains to be answered is whether purified, soluble AC toxin is delivered to target cells by the same mechanism as that delivered from the surface of whole bacteria.

REFERENCES

1. **Confer, D. L., and J. W. Eaton.** 1982. Phagocyte impotence caused by an invasive bacterial adenylate cyclase. *Science* **217:**948–950.
2. **Gordon, V. M., S. H. Leppla, and E. L. Hewlett.** 1988. Inhibitors of receptor-mediated endocytosis block the entry of *Bacillus anthracis* adenylate cyclase toxin but not that of *Bordetella pertussis* adenylate cyclase toxin. *Infect. Immun.* **56:**1066–1069.
3. **Gray, M., G. Szabo, A. S. Otero, L. Gray, and E. Hewlett.** 1998. Distinct mechanisms for K^+ efflux, intoxication, and hemolysis by *Bordetella pertussis* AC toxin. *J. Biol. Chem.* **273:**18260–18267.
4. **Guermonprez, P., N. Khelef, E. Blouin, P. Rieu, P. Ricciardi-Castagnoli, N. Guiso, D. Ladant, and C. Leclerc.** 2001. The adenylate cyclase toxin of *Bordetella pertussis* binds to target cells via the $a_M b_2$ integrin (CD11b/CD18). *J. Exp. Med.* **193:**1035–1044.
5. **Karimova, G., C. Fayolle, S. Gmira, A. Ullmann, C. Leclerc, and D. Ladant.** 1998. Charge-dependent translocation of *Bordetella pertussis* adenylate cyclase toxin into eukaryotic cells: implication for the in vivo delivery of $CD8^+$ T cell epitopes into antigen-presenting cells. *Proc. Natl. Acad. Sci. USA* **95:**12532–12537.
6. **Lenz, D. H., C. L. Weingart, and A. A. Weiss.** 2000. Phagocytosed *Bordetella pertussis* fails to survive in human neutrophils. *Infect. Immun.* **68:**956–959.
7. **Otero, A. S., X. B. Yi, M. C. Gray, G. Szabo, and E. L. Hewlett.** 1995. Membrane depolarization prevents cell invasion by *Bordetella pertussis* adenylate cyclase toxin. *J. Biol. Chem.* **270:**9695–9697.
8. **Weingart, C. L., P. S. Mobberley-Schuman, E. L. Hewlett, M. C. Gray, and A. A. Weiss.** 2000. Neutralizing antibodies to adenylate cyclase toxin promote phagocytosis of *Bordetella pertussis* by human neutrophils. *Infect. Immun.* **68:**7152–7155.

SELECTED READINGS

1. **Hewlett, E. L., and M. C. Gray.** 2000. Adenylyl-cyclase toxin from *Bordetella pertussis*, p. 473–488. *In* K. Aktories and I. Just (ed.), *Bacterial Protein Toxins.* Springer-Verlag, Berlin, Germany.
2. **Ladant, D., and A. Ullmann.** 1999. *Bordetella pertussis* adenylate cyclase: a toxin with multiple talents. *Trends Microbiol.* **7:**172–176.
3. **Rappuoli, R., and M. Pizza.** 2000. Bacterial toxins, p. 193–220. *In* P. Cossart, P. Boquet, S. Normark, and R. Rappuoli (ed.), *Cellular Microbiology.* ASM Press, Washington, D.C.

Bacterial Protein Toxins
D. Burns et al., Editors
©2003 ASM Press, Washington, DC 20036

Chapter 11

Transport of Toxins across Intracellular Membranes

Kirsten Sandvig

A number of protein toxins with a cytosolic target act on cells by first binding to cell surface receptors, then the toxins are endocytosed, and subsequently they are transported to the organelle from where they are translocated to the cytosol. These toxins include a variety of bacterial toxins with two functionally different domains, one domain (B) that binds to cell surface receptors, and another domain (A) that enters the cytosol and acts enzymatically at a cytosolic target (Fig. 1). As indicated in the figure, the binding domain may consist of several subunits. Furthermore, the A and B subunits can be noncovalently or covalently attached to each other. In some cases the toxins (for instance, diphtheria toxin, Shiga toxin, *Pseudomonas* exotoxin A) can be cleaved and activated by enzymes of the target cells, whereas in other cases the toxin molecule (cholera toxin, *Escherichia coli* heat-labile toxin) is cleaved by the bacteria that produce the toxin. In spite of the structural similarities between these toxins, they use, as described in this chapter, different strategies to enter the cytosol.

ENTRY INTO THE CYTOSOL REQUIRES ENDOCYTOSIS OF THE TOXINS

Most protein toxins have to be endocytosed before being translocated to the cytosol. Only few exceptions are known; one is the adenylate cyclase from *Bordetella pertussis* (not included in Fig. 1), which enters the cytosol directly through the plasma membrane. A number of protein toxins enter the cytosol either from acidic endosomes or from the endoplasmic reticulum (ER) after retrograde transport from endosomes to the Golgi apparatus and to the ER (Fig. 2). Why is endocytic uptake of protein toxins required? Why don't they penetrate directly through the plasma membrane? The toxins turn out to use several different strategies to enter the cytosol of target cells, but in all cases toxin internalization is

---

Kirsten Sandvig • Department of Biochemistry, Institute for Cancer Research, The Norwegian Radium Hospital, Montebello, 0310 Oslo, Norway.

Toxin		Enzymatic activity	Cellular targets

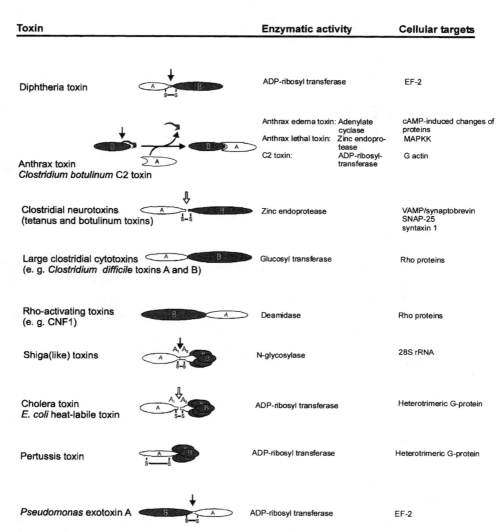

Diphtheria toxin — ADP-ribosyl transferase — EF-2

Anthrax toxin
Clostridium botulinum C2 toxin

Anthrax edema toxin: Adenylate cyclase — cAMP-induced changes of proteins
Anthrax lethal toxin: Zinc endoprotease — MAPKK
C2 toxin: ADP-ribosyltransferase — G actin

Clostridial neurotoxins
(tetanus and botulinum toxins) — Zinc endoprotease — VAMP/synaptobrevin SNAP-25 syntaxin 1

Large clostridial cytotoxins
(e. g. *Clostridium difficile* toxins A and B) — Glucosyl transferase — Rho proteins

Rho-activating toxins
(e. g. CNF1) — Deamidase — Rho proteins

Shiga(like) toxins — N-glycosylase — 28S rRNA

Cholera toxin
E. coli heat-labile toxin — ADP-ribosyl transferase — Heterotrimeric G-protein

Pertussis toxin — ADP-ribosyl transferase — Heterotrimeric G-protein

***Pseudomonas* exotoxin A** — ADP-ribosyl transferase — EF-2

Figure 1 Schematic structure and effects of A-B toxins. The enzymatically active part (A) and the binding moiety (B) can either be covalently or non-covalently attached to each other. In some toxins the binding moiety consists of several subunits. In several cases there is a disulfide bond that keeps the toxin moieties together after a proteolytic cleavage, which can occur before (white arrow) or after (black arrow) the toxin is presented to the target cell. When the toxins are cleaved by the target cells, it is usually the enzyme furin that is responsible for the activation. In the case of anthrax toxin, the furin-induced cleavage of the B moiety at the cell surface allows subsequent binding of the A moiety. The enzymatic effects and cellular targets are indicated.

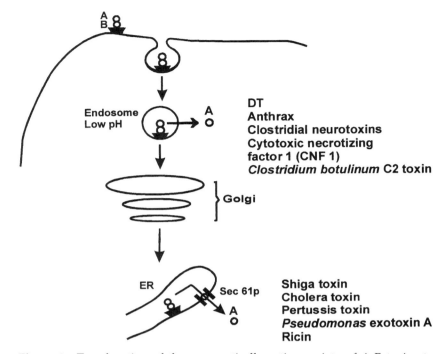

Figure 2 Translocation of the enzymatically active moiety of A-B toxins to the cytosol occurs from different intracellular organelles. After endocytosis some toxins like diphtheria toxin (DT) (see details in Fig. 4), anthrax toxin, CNF1, *C. botulinum* C2 toxin, and clostridial neurotoxins are transported from endosomes by a low pH-induced translocation process that may require the toxin receptor and in some cases requires the proteolytically cleaved toxin. In this figure no distinction has been made between early and late endosomes. Toxin entry may occur from one or the other, and this seems to be dependent on the type of toxin. Other toxins are transported to the Golgi apparatus and retrogradely to the ER before translocation of the A moiety to the cytosol. This group of toxins include the bacterial toxins *Pseudomonas* exotoxin A, cholera toxin, Shiga toxin, and pertussis toxin as well as the plant toxin ricin. Only entry of the enzymatically active part is indicated in the figure. However, a larger part of the toxin molecule might enter the cytosol.

required to obtain exposure of the toxin to conditions or components found in intracellular compartments (1–3, 5).

In the case of diphtheria toxin, clostridial neurotoxins (tetanus and botulinum toxins), *Clostridium botulinum* C2 toxin, and anthrax toxin, the low endosomal pH is crucial and sufficient for the cytosolic entry of the proteolytically processed toxin molecule. Also, cytotoxic necrotizing factor 1 (CNF1) (3) and *Clostridium difficile* toxin B (2) seem to enter the cytosol in response to low pH. For some of the other toxins, such as Shiga toxin, cholera toxin, and *Pseudomonas* exotoxin A, the toxins have to gain access to the ER before translocation to the cytosol. Several of the toxins are cleaved and activated by the proteolytic enzyme furin, which is found at the cell surface, in endosomes, and in the Golgi apparatus. This enzyme cleaves some toxins most efficiently at slightly low pH. In cells without furin other intracellular enzymes may, although less efficiently, activate the toxins. There is now evidence that some of the bacterial toxins and the structurally related plant toxin ricin (which consists of two polypeptide chains joined by a disulfide bond) use the Sec61p transmembrane protein complex to gain access to the cytosol (see below) (10). This ER-located protein complex is normally used for translocation of newly synthesized proteins into the ER and also for translocation of misfolded proteins back into the cytosol, where they are then degraded. Thus, the toxins are exploiting the ER translocation machinery to gain access to the cytosol.

It should be noted that although a toxin enters the cytosol from the ER, low pH may be required. This is the case for *Pseudomonas* exotoxin A. Compounds that increase endosomal and Golgi pH protect against even the proteolytically cleaved form of *Pseudomonas* exotoxin A, and the requirement for low pH is currently not understood.

That a toxin or toxin-receptor complex carries its own translocation machinery and that low endosomal pH is sufficient to induce translocation can be demonstrated by investigating whether cell surface-bound toxin immediately penetrates the plasma membrane upon exposure to low pH, thus mimicking the conditions in the endosome. Such direct toxin transport was first demonstrated for diphtheria toxin around 20 years ago (4, 9). The current model for diphtheria toxin translocation is described below under "Translocation of Toxins from Endosomes to the Cytosol."

ENDOCYTIC MECHANISMS INVOLVED IN TOXIN ENTRY

Ways To Interfere with Endocytosis and Toxin Uptake

Endocytosis of protein toxins can occur by several mechanisms (Fig. 3). It is now clear that there is not only endocytosis from clathrin-coated pits, the structures

Figure 3 Endocytic mechanisms proposed to be involved in toxin uptake. Clathrin-dependent endocytosis has previously been blocked by hypotonic shock and potassium depletion or by acidification of the cytosol. In more recent studies this pathway was inhibited by expression of a dominant negative mutant of dynamin 1 (dynamin K44A), by Eps15 mutants, and by expression of antisense to clathrin heavy chain. Formation of invaginated clathrin-dependent structures seems to be dependent on cholesterol since treatment of cells with methyl-β-cyclodextrin (mβCD) inhibits their formation. Only flat coated pits are seen after such treatment. Clathrin-independent endocytosis comprises more than one mechanism and can be independent from a possible uptake from caveolae. There are both dynamin-dependent and -independent forms of the clathrin- and caveolae-independent endocytosis, and uptake of a ligand by such a mechanism can in some cases be dependent on membrane cholesterol whereas in other cases it is not. Uptake by caveolae has been reported to be dependent on dynamin and can be inhibited by extraction of cholesterol or complex formation of cholesterol with filipin or nystatin. Interfering with the different endocytic pathways has revealed that diphtheria toxin (DT), *Pseudomonas* exotoxin A, Shiga toxin, and also to some extent cholera toxin are internalized by clathrin-dependent endocytosis. Both CNF1 and to some extent cholera toxin are internalized independently of caveolae.

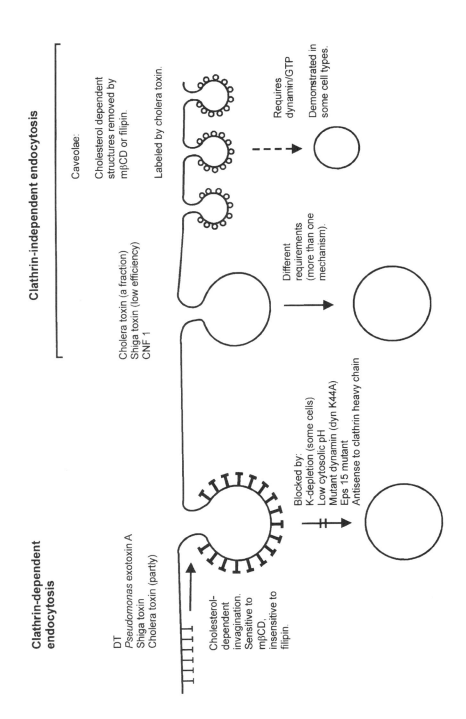

involved in uptake of a number of hormones and growth factors, but there are also clathrin-independent endocytic mechanisms. In endothelial cells and possibly also in other cell types there is evidence for endocytic uptake via the small cholesterol-rich structures called caveolae, membrane invaginations coated with the protein caveolin at the cytosolic side. Furthermore, there are endocytic mechanisms that operate independently of both clathrin-coated pits and caveolae.

The uptake from clathrin-coated pits has been studied in most detail, and a number of cellular proteins involved in this process have been identified. This knowledge has provided us with new tools to investigate whether clathrin-dependent endocytosis is involved in uptake of a given toxin. For instance, a dominant negative mutant of the GTP-binding protein dynamin blocks uptake from clathrin-coated pits, whereas some forms of clathrin-independent endocytosis continue to operate. Stably transfected cells with inducible expression of mutant dynamin or cells that transiently express this protein have been used to interfere with the clathrin-dependent pathway and to study the effect on uptake of several toxins. It should, however, be noted that also uptake from caveolae is blocked by mutant dynamin.

A number of other molecules involved in clathrin-dependent endocytosis, such as Eps15 mutants, parts of the clathrin heavy chain, or antisense to clathrin heavy chain, can also be used to interfere with clathrin-dependent endocytosis. However, even before the discovery of the different molecules that are associated with clathrin-dependent endocytosis, one was able to inhibit this pathway. A number of studies of toxin uptake have been carried out by using acidification of the cytosol to block pinching off of clathrin-coated vesicles. This method seems to work in all cell types tested so far. Another method previously employed was to remove the clathrin coats from the cell surface by using hypotonic shock and potassium depletion. However, this last method works only in some cell types. It should be noted that the effect of these last two methods on a possible vesicle formation from caveolae is unknown.

In most cell types it has not been investigated whether caveolae pinch off or whether they might be stable structures involved in, for instance, transmembrane signaling. However, a possible endocytic uptake from caveolae can be inhibited by removing membrane cholesterol. This can be done by addition of methyl-β-cyclodextrin, a polysugar that extracts cholesterol from the membrane. The uptake from caveolae can also be inhibited by addition of filipin or nystatin, drugs that bind to cholesterol in the membrane. These types of treatment will destroy the caveolar structures. Surprisingly, clathrin-dependent endocytosis also is sensitive to removal of cholesterol by methyl-β-cyclodextrin. Thus, several methods should be employed to investigate the endocytic mechanisms used by a given toxin.

The strategies outlined above have been used to demonstrate that diphtheria toxin, *Pseudomonas* exotoxin A, and Shiga toxin most efficiently enter by clathrin-dependent endocytosis in different cell types (Fig. 3). However, when diphtheria toxin is bound to a genetically modified, glycosylphosphatidyl-linked version of the normal receptor (the uncleaved precursor of the heparin-binding epidermal growth factor-like precursor), the toxin-receptor complex is internalized by clathrin-independent endocytosis, and the toxin still enters the cytosol, although less efficiently than when bound to its normal receptor. Shiga toxin can, although it binds to a glycolipid receptor (Gb3), be internalized from clathrin-coated pits by a toxin-induced process. It is not known how this occurs, since the lipid receptor does not have a sorting signal for clathrin-coated pits. One may speculate that the toxin has affinity for a transmembrane protein that enters clathrin-coated pits. However, when clathrin-dependent endocytosis is blocked, for instance, by expression of antisense to clathrin heavy chain, there is still some toxic effect of Shiga toxin. This seems to reflect a common feature of several toxins, which may use more than one pathway to enter the cell.

The endocytic pathway, which is most important for the uptake of a toxin, might vary from one cell type to another. In the case of Shiga toxin it is known that the lipid composition of the receptor, a cell type-dependent property, is important for the intracellular sorting. Similarly, this might also affect the type of endocytosis used by the toxin in a given cell. Cholera toxin is another example of a toxin that can use different endocytic mechanisms (7, 13, 15). This toxin has been used as a marker for caveolae since it has a tendency to accumulate in these structures. However, cholera toxin is able to enter cells that do not have caveolae, and although it can be visualized in caveolae, this does not imply that it enters from this location. Although caveolae have been reported to pinch off at a reasonable rate by a dynamin-dependent process in some cell types such as endothelial cells, this might be a very slow process in other cells. Interestingly, cholera toxin, which like Shiga toxin binds to a glycolipid receptor (GM1 in the case of cholera toxin), seems to be internalized partly by clathrin-dependent endocytosis in neurons and in several other cell types. In those cells it turns out that the uptake is only partially inhibited by filipin. On the other hand, the endocytic uptake can be inhibited by expression of mutant Eps15, by expression of antisense to clathrin heavy chain, or by other methods interfering with clathrin-dependent endocytosis. As indicated in Fig. 3, CNF1 also enters by a clathrin- and filipin-independent mechanism. Thus, the toxins use a number of different strategies to enter a compartment from where they can get access to the cytosol.

How To Follow Toxin Internalization

There are a number of methods that can be used to follow toxin uptake. To study endocytosis biochemically or by microscopy, the toxins are commonly labeled with radioactive iodine (^{125}I), biotin (often via a cleavable disulfide bond to differentiate between surface-bound toxin that can be reached by a membrane-nonpermeable reducing agent and truly endocytosed toxin where the disulfide bond will remain intact), horseradish peroxidase (HRP), fluorescent labels, or gold particles. In all cases it is of course essential to test that the toxin still binds specifically, and one has to be aware of the possibility that the stability of the toxin might be reduced after modification. Importantly, labeling a toxin, especially when creating multivalent complexes, might change the routing of a toxin inside the cell. Such multivalent complexes could also use endocytic mechanisms for entry other than those used by the native toxin. It is, for instance, known that multivalent complexes of the plant toxin ricin (multivalent HRP complexes or gold complexes) are sorted directly to lysosomes and never seen in the Golgi apparatus, as is the case for the native toxin or monovalent complexes with HRP.

When an invagination at the cell surface is widely open to the surrounding medium, a toxin inside this invagination is available to antibodies that might be used to quantify the amount of surface-bound toxin. However, it has been shown that in the case of clathrin-dependent endocytosis there is an almost closed version of the structure (still surface connected) where the content cannot be reached by antibodies but where small molecules (for instance, small reducing substances) still can enter through the narrow neck and act on a disulfide bond connecting biotin to the ligand. A similar situation might exist in the case of caveolae. Thus, the use of antibodies to differentiate between surface-bound toxin and internalized toxin should be avoided.

TRANSLOCATION OF TOXINS FROM ENDOSOMES TO THE CYTOSOL

It is an old observation that amines that increase endosomal and lysosomal pH also protect against diphtheria toxin. It was, therefore, suggested and later demonstrated that low pH is required for translocation of the toxin into the cytosol. Not only amines that increase endosomal pH but also ionophores that abolish the low pH and, more specifically, the proton pump inhibitor bafilomycin protect against diphtheria toxin and other toxins that require low pH for entry. Of the toxins that enter the cytosol from endosomes, most details are known about diphtheria toxin, and the emphasis will, therefore, be on this toxin. The finding in 1980 that exposure of cells with surface-bound, proteolytically processed toxin (the A-B form only connected by a disulfide bond) to low pH could mimic the endosomal conditions and induce translocation of the A chain to the cytosol greatly facilitated the studies of diphtheria toxin entry into cells.

When diphtheria toxin is exposed to low pH, there is a conformational change in the molecule leading to exposure of hydrophobic domains in the B chain. The toxin can then bind Triton X-100, partition into Triton X-114, and form cation-selective channels in lipid bilayers as well as in intact cells when bound to the toxin receptors. When the toxin is prebound to the cell surface and exposed to low pH, the A chain and part of the B chain become protected against externally added pronase, reflecting translocation of the A chain into the cytosol and insertion of part of the B chain into the membrane (Fig. 4). That the A chain is free in the cytosol can be demonstrated by showing that it is able to diffuse out of permeabilized cells. The B chain of diphtheria toxin consists of a C-terminal receptor-binding domain with a β-sheet structure and an amino-terminal, α-helical transmembrane or translocation (T) domain containing the two helices, TH8 and TH9 (Fig. 4). Upon exposure to low pH there is protonation of two amino acids, Asp-352 and Glu-349, facilitating transfer of the B chain into or across the membrane. The two helices called TH8 and TH9 and the loop between them are sufficient to form a cation-selective channel in lipid bilayers. However, the efficiency of channel formation by these helices is low. It should be noted that the possible role of the ion channel in diphtheria toxin entry is still not known. This channel is open only when there is a pH gradient across the membrane. To get diphtheria toxin translocation into cells, permeable anions also are required. The explanation for this requirement is not known.

Although studies of the interactions of diphtheria toxin with lipid membranes provide knowledge about the positioning of the different domains of the

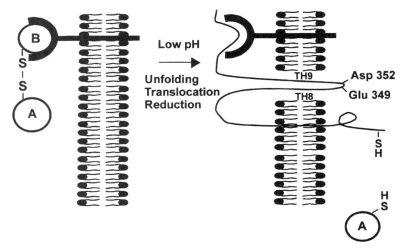

Figure 4 Membrane translocation of diphtheria toxin. Proposed mechanism of low pH-induced diphtheria toxin entry across the membrane. Low endosomal pH will induce a conformational change in the molecule as well as protonation of Asp-352 and Glu-349 in the B chain, insertion of the toxin into the membrane, and translocation and release of the A chain. Insertion of the B chain into the membrane is associated with formation of cation-selective channels.

B chain, the toxin-membrane interactions are not necessarily identical to those found in the cell membrane, where the receptor seems to be essential for toxin entry. When diphtheria toxin is bound via a receptor where the transmembrane or cytosolic domain of the normal receptor has been substituted with a glycosylphosphatidyl anchor, the translocation is less efficient. Also, other changes of the diphtheria toxin receptor are found to change the efficiency of the translocation process (14).

Importantly, at low concentrations diphtheria toxin can enter the cytosol both from endosomes and from the cell surface (upon exposure of surface-bound toxin to low pH) in such low amounts that prolonged action of the toxin in the cytosol is required for optimal protein synthesis inhibition. This is in contrast to some early suggestions about quantal release of this toxin from ruptured endosomes and an immediate and complete inhibition of protein synthesis. The lack of ruptured endosomes upon incubation with diphtheria toxin is also consistent with the finding that inhibition of unfolding of the A fragment by introduction of additional disulfides in the molecule inhibits translocation of the A fragment to the cytosol.

As indicated in Fig. 2, other toxins also are translocated from endosomes after exposure of the toxin-receptor complex to low pH. Anthrax toxin, clostridial neurotoxins, CNF1 (3), *C. difficile* toxin B (2), and *C. botulinum* C2 toxin (1) are all reported to enter from endosomes in response to low pH exposure. As earlier described for diphtheria toxin, the method involving exposure of cell with surface-bound toxin to low pH has been used for several of these toxins to clarify the importance of low pH for their entry. As for diphtheria toxin, the translocation can be associated with channel formation, but in no case is the role of the channel completely understood. There are, however, in the case of anthrax toxin, data demonstrating that an inhibition of channel formation also inhibits toxin entry. When the B chain of anthrax toxin (called protective antigen or PA) is receptor bound and cleaved by furin, there is formation of a heptamer of B chains that can bind up to seven A chains (this can either be edema factor, which is an adenylate cyclase, or lethal factor, which is a metalloprotease). Upon exposure to low pH there is formation of a channel, and one or the other of the two A chains can be translocated. However, if the heptamer consists partially of B chains with mutations in the area normally facing the lumen of the channel, both channel formation and translocation of the A chain are inhibited (12). This argues for a role for the channel but does not necessarily mean that the A chain is translocated through the channel itself. Importantly, interfering with translocation in this manner might be used in therapy.

The finding that increased endosomal pH protects against a given toxin is not sufficient to conclude that low pH is directly involved in toxin translocation through the membrane. Increased endosomal pH has been shown to affect recycling of some receptors to the cell surface, so in case of protection, it has to be tested that the toxin still binds to the cell surface. Also, as mentioned earlier in this chapter, toxins may require low pH during entry into the cell although they enter from an organelle other than the endosome. In those cases, for instance in the case of *Pseudomonas* exotoxin A and the plant toxin modeccin, one cannot obtain toxin entry from the cell surface by exposing cells with surface-bound toxin to low pH.

SORTING OF TOXINS TO THE GOLGI APPARATUS

Involvement of the Golgi Apparatus in Toxin Transport to the Cytosol

Cholera toxin was the first bacterial toxin visualized in the Golgi apparatus (6). Later, toxin transport to the Golgi apparatus has been demonstrated for several other protein toxins as well. Both bacterial toxins and plant toxins such as ricin are transported from endosomes to the Golgi apparatus. That a given toxin actually is transported to the Golgi apparatus can be shown by electron microscopy, by fluorescence microscopy, and by subcellular fractionation. Transport of toxin to the Golgi apparatus can also be quantified by using genetically modified toxin molecules containing a sulfation site that can be modified by the sulfotransferase found in the Golgi apparatus. By using radioactively labeled sulfate, one can label and quantify the amount of toxin molecules entering the Golgi apparatus.

To demonstrate that the Golgi apparatus is of importance for intoxication, several approaches have been used. Evidence for the importance of toxin transport to the Golgi apparatus can be obtained by using the drug brefeldin A, which disrupts the Golgi apparatus in a number of cell types. Treatment with this drug leads to retrograde transport of Golgi cisterna to the ER. When there is a correlation between brefeldin A-induced Golgi disruption and protection of the cells against a toxin, this can be used as evidence for involvement of the Golgi apparatus provided (i) that the treatment does not reduce the number of toxin receptors at the cell surface during the time of the experiment, (ii) that a cell surface or endosomally located proteolytic toxin cleavage is not inhibited, and (iii) that cells where the Golgi apparatus is resistant to brefeldin A, but where endosomes undergo the same structural changes as in cells with a brefeldin A-sensitive Golgi apparatus, stay sensitive to the toxin. That toxicity can be blocked at a temperature of approximately 20°C is also an indication of transport to an organelle different from early endosomes for translocation to occur. It has been known for many years that there is a temperature-sensitive step between endosomes and the Golgi apparatus.

Pathways from Endosomes to the Golgi Apparatus

As already described, a number of protein toxins are sorted to the Golgi apparatus on their way to the cytosol. A well-characterized pathway leading from endosomes to the Golgi apparatus is the Rab9-dependent pathway responsible for sorting of mannose-6-phosphate receptors (M6PRs) from late endosomes to the trans-Golgi network (TGN) of the Golgi apparatus (Fig. 5). This is the pathway also used by furin, the proteolytically active enzyme that recycles between the cell surface, endosomes, and the Golgi apparatus and that is involved in cleavage and activation of a number of protein toxins (such as Shiga toxin, anthrax toxin, and diphtheria toxin). However, transport from endosomes to the Golgi apparatus can occur by at least one other pathway, and there is now evidence that the Golgi marker TGN38 and Shiga toxin B chain, Shiga toxin, and the plant toxin ricin enter the Golgi apparatus via a different pathway (10). It has been suggested that Shiga toxin B chain enters the Golgi from endosomes via the perinuclear recycling compartment by a clathrin-dependent pathway that is also dependent on Rab11. Rab11 is known to be present both on the recycling endosome and in the Golgi apparatus. However, the plant toxin ricin can enter the Golgi apparatus by a Rab9- and Rab11-independent pathway that is also independent of functional clathrin. Since the plant toxin ricin binds to a variety

Figure 5 Endosome to Golgi transport. Endocytosed furin and M6PRs are transported to the Golgi apparatus through late endosomes by a Rab9-dependent pathway, whereas Shiga toxin, the plant toxin ricin, and the cellular protein TGN38 can enter the Golgi apparatus from an earlier endosomal compartment. In the case of ricin, transport to the Golgi apparatus occurs by a Rab9- and clathrin-independent process.

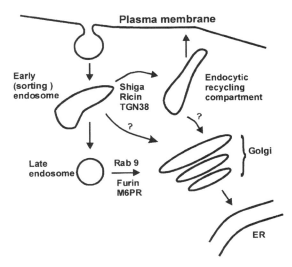

of glycolipids and glycoproteins with terminal galactose, it may be able to use more pathways than Shiga toxin, which only binds to Gb3. Thus, there may be parallel pathways circumventing late endosomes. As indicated in Fig. 5, the Rab9-independent transport could go directly from early endosomes to the Golgi apparatus, or transport could occur via the perinuclear recycling endosome.

Regulation of Endosome to Golgi Transport of Toxins

The fraction of endocytosed toxin transported to the Golgi apparatus in a certain cell type and the subsequent intoxication are under regulation and might vary from one cell type to another. Endosome-to-Golgi transport can be regulated by protein kinase A and C, and in the case of Shiga toxin, it has been found that the lipid composition of the receptor, Gb3, is important for sorting the toxin from endosomes to the Golgi apparatus. To obtain maximal toxin transport to the Golgi apparatus, the length of the fatty acid chain has to be within certain limits. The composition of Gb3 can be changed by incubation of cells with cAMP or with butyric acid, which sensitizes a number of cell types to Shiga toxin. This was first shown for A431 cells, which without butyric acid treatment bind Shiga toxin but are completely resistant to the toxin. After treatment with butyric acid the cells become sensitive, and Shiga toxin is transported to the Golgi apparatus and the ER in sufficient amounts to visualize the toxin in all the Golgi cisterns and in the ER by electron microscopy. This was actually the first demonstration of toxin transport all the way from the cell surface through the Golgi apparatus and to the ER (10).

That in polarized cells endosome-to-Golgi transport is regulated in different ways, depending on the pole from which a ligand is internalized, is in agreement with the idea that the transport in the direction of the Golgi apparatus occurs from different endosomes, depending on the pole from where the toxin was internalized. However, the details of endosome-to-Golgi transport of protein toxins are not yet known, and the subject needs further investigation.

RETROGRADE TOXIN TRANSPORT TO THE ER

A protein can be retrieved in the ER or brought in a retrograde manner from the Golgi to the ER if this protein has a KDEL sequence (or a KDEL-like sequence). Such a motif can bind to so-called KDEL receptors, which are found throughout the Golgi apparatus and which are transported in a retrograde direction in COPI-coated (coatomer I-coated) vesicles (Fig. 6). Several protein toxins fulfill this requirement: In cholera toxin the A chain with its C-terminal KDEL sequence goes through the ring of B chains (pentamer) that binds to the receptor GM1, and the KDEL sequence thereby becomes exposed to the KDEL receptors. Similarly, in *E. coli* heat-labile toxin, there is an RDEL sequence that can function in the same way. In agreement with this theory, there is a delay in the action of these toxins after introduction of mutations in these sequences. However, even the cholera toxin B chain, which is without any KDEL sequence, can move in a retrograde direction into the various Golgi cisternae. The mechanism behind this transport is unknown. Also, *Pseudomonas* exotoxin A has a KDEL-like sequence; it has the sequence RDELK, which in itself does not bind the KDEL receptor. However,

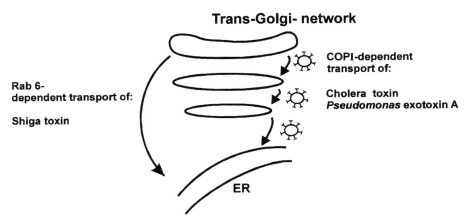

Trans-Golgi- network

COPI-dependent
transport of:

Rab 6-
dependent transport of:

Cholera toxin
Pseudomonas exotoxin A

Shiga toxin

ER

Figure 6 Retrograde transport through the Golgi apparatus. Toxins with a KDEL sequence or a KDEL-like sequence might be transported retrogradely by a COPI-dependent mechanism, whereas other toxins, such as Shiga toxin, can enter the ER by a COPI-independent, Rab6-dependent pathway. It is not clear whether this pathway goes via the different cisternae of the Golgi apparatus or whether Shiga toxin observed in the different cisternae ends up in these locations via a different mechanism.

there are proteases that can remove the K, providing the toxin with the sequence RDEL, which does bind the KDEL receptor. There is good evidence for the importance of this sequence for intoxication with *Pseudomonas* exotoxin A. Cells are sensitized to the toxin after expression of additional KDEL receptors; the toxicity is lost when the KDEL sequence is removed and when lysozyme-KDEL is expressed in cells, thereby saturating the KDEL receptors. Thus, *Pseudomonas* exotoxin A seems to be transported retrogradely mainly by binding to KDEL receptors. Further support for this notion was obtained by microinjection of antibodies to the cytosolic part of the KDEL receptor, which also protected the cells. *Pseudomonas* exotoxin A is therefore presumably transported retrogradely by COPI-coated vesicles.

In contrast to *Pseudomonas* exotoxin A and cholera toxin, Shiga toxin, Shiga-like toxins, and the plant toxin ricin do not have a KDEL or a KDEL-like sequence, but they are still transported retrogradely. How can this occur? Clearly there are mechanisms that act independently of the KDEL receptor. The conditions, described above, protecting against *Pseudomonas* exotoxin A—overexpression of lysozyme-KDEL or antibodies to the KDEL receptor—had no effect on the toxic action of Shiga-like toxin 1, which differs from intact Shiga toxin by only one amino acid in the A chain. Actually, retrograde transport of Shiga toxin B chain was recently shown to occur by a Rab6-dependent, COPI-independent pathway that might be responsible for retrograde transport of other toxins without a KDEL sequence as well. Furthermore, since mutation of the KDEL sequence in cholera toxin only caused a delay in toxin action, and cholera toxin B is transported retrogradely as well, a given toxin might be transported from the Golgi apparatus to the ER by more than one mechanism.

A very limited number of toxin molecules seem to enter the ER. In most cases the toxin cannot be localized in the ER by electron microscopy. However, also in this case, genetically modified toxin molecules have turned out to be useful. By addition of glycosylation sites to Shiga B chain or to the A chain of the plant toxin ricin one can demonstrate that they are glycosylated and therefore transported to the ER.

It was recently shown that simian virus 40 (SV40) enters the ER after uptake from caveolae and transport through a neutral compartment called a caveosome by the authors (8). Could this pathway be used by protein toxins? First of all, the uptake of SV40 might be virus induced; second, it was shown to be very slow, taking hours to lead to the ER. From the kinetics of toxin entry this is, therefore, not a very likely pathway for toxin transport to the ER.

TRANSPORT OF PROTEIN TOXINS ACROSS THE ER MEMBRANE

There is a lively transport of newly synthesized proteins from the cytosol and into the ER lumen through the Sec61p complex. It is also known that this protein complex is used for transport of misfolded protein in the other direction, back into the cytosol, where these proteins can be ubiquitinylated, deglycosylated, and degraded by proteasomes. The Sec61p protein complex was, therefore, a candidate also as a toxin translocator, and evidence has been accumulating suggesting that Sec61p is involved in toxin transport into the cytosol.

In the case of cholera toxin, it has been shown that the A1 subunit can be released from the rest of the toxin inside the ER lumen, and coimmunoprecipitation studies reveal that the A1 subunit of the toxin can be associated with the Sec61p translocator (11), suggesting involvement of this protein complex in toxin transport to the cytosol. Furthermore, there is evidence that the ER enzyme, protein disulfide isomerase, acts as a chaperone involved in unfolding of cholera toxin (13). Importantly, this enzyme binds and releases the A chain depending on whether it is in the reduced or oxidized state. Also, *Pseudomonas* exotoxin A and the plant toxin ricin seem to interact with the Sec61p channel. However, the details of the translocation step are still unknown.

Another question in connection with toxin translocation from the ER is whether toxin chains other than the enzymatically active part can enter the cytosol. In some cases epitopes bound to toxin B chains can be presented by major histocompatibility complex (MHC) class I, which is usually supplied with epitopes produced in the cytosol. However, it has not yet been demonstrated whether the B chain with the epitope actually enters the cytosol (see below). Thus, there are a number of unanswered questions concerning entry of toxin from the ER to the cytosol.

TRANSLOCATION OF IMMUNOTOXINS INTO CELLS

Bacterial and plant toxins act rather nonspecifically on different cell types. A number of attempts to target the toxins to specific cell types, such as cancer cells, have been made, and promising clinical trials have been performed. To produce such specific toxins, either the enzymatically active part of the toxin or a larger portion of the toxin can be coupled, for instance, to monoclonal antibodies or growth factors. This can be done either biochemically or at the genetic level. However, little is known about the entry mechanisms of most of these molecules, and construction of efficient hybrid toxins might be facilitated by increased knowledge about the role of the different toxin domains. With the elucidation of toxin entry mechanisms one might be able to construct new molecules with higher efficiency of penetration into the cells. It will also be useful to know from which compartment the toxin normally enters and the mechanistic details of toxin entry through the membrane.

TOXINS AS VECTORS TO BRING PEPTIDES OR PROTEINS INTO THE CYTOSOL

The ability of toxins to pass membranes and enter the cytosol can in itself be exploited for medical purposes. Several bacterial toxins are able to bring with them epitopes that can become presented by the MHC antigen I. Normally this protein complex presents antigens that are produced in the cytosol, transported into the ER by a separate transporter (the TAP transporter), and then the antigen is bound to the MHC antigen I and presented at the cell surface to cytotoxic T cells. It turns out that anthrax toxin, diphtheria toxin, Shiga-like toxin 1, Shiga toxin B fragment, *E. coli* heat-labile toxin, and *Pseudomonas* exotoxin A can be used to bring additional amino acid sequences to the cytosol and/or to deliver epitopes for presentation by MHC class I. However, it is not yet clear whether the epitopes in all these cases were actually first delivered to the cytosol or whether, for instance, the sequences added to Shiga B chain could be cleaved off in the ER and added there to the MHC molecules. Also, proteins with biological activity on their own might be added to a given toxin and brought in as passengers. Toxins or catalytically inactive derivatives of toxins exploited in this manner might be useful both in therapy and in cell biological studies.

REFERENCES

1. **Barth, H., D. Blöcker, J. Behlke, W. Bergsma-Schutter, A. Brisson, R. Benz, and K. Aktories.** 2000. Cellular uptake of *Clostridium botulinum* C2 toxin requires oligomerization and acidification. *J. Biol. Chem.* **275**:18704–18711.
2. **Barth, H., G. Pfeifer, F. Hofmann, E. Maier, R. Benz, and K. Aktories.** 2001. Low pH-induced formation of ion channels by *Clostridium difficile* toxin B in target cells. *J. Biol. Chem.* **276**:10670–10676.
3. **Contamin, S., A. Galmiche, A. Doye, G. Flatau, A. Benmerah, and P. Boquet.** 2000. The p21 Rho-activating toxin cytotoxic necrotizing factor 1 is endocytosed by a clathrin-independent mechanism and enters the cytosol by an acidic-dependent membrane translocation step. *Mol. Biol. Cell* **11**:1775–1787.
4. **Draper, R. K., and M. I. Simon.** 1980. The entry of diphtheria toxin into the mammalian cell cytoplasm: evidence for lysosomal involvement. *J. Cell Biol.* **87**:849–854.
5. **Falnes, P. O., and K. Sandvig.** 2000. Penetration of protein toxins into cells. *Curr. Opin. Cell Biol.* **12**:407–413.
6. **Joseph, K. C., S. U. Kim, A. Stieber, and N. K. Gonatas.** 1978. Endocytosis of cholera toxin into neuronal GERL. *Proc. Natl. Acad. Sci. USA* **75**:2815–2819.
7. **Nichols, B. J., A. K. Kenworthy, R. S. Polishchuk, R. Lodge, T. H. Roberts, K. Hirschberg, R. D. Phair, and J. Lippincott-Schwartz.** 2001. Rapid cycling of lipid raft markers between the cell surface and golgi complex. *J. Cell Biol.* **153**:529–542.

8. **Pelkmans, L., J. Kartenbeck, and A. Helenius.** 2001. Caveolar endocytosis of simian virus 40 reveals a new two-step vesicular-transport pathway to the ER. *Nat. Cell Biol.* **3:**473–483.

9. **Sandvig, K., and S. Olsnes.** 1980. Diphtheria toxin entry into cells is facilitated by low pH. *J. Cell Biol.* **87:**828–832.

10. **Sandvig, K., and B. van Deurs.** 2000. Entry of ricin and Shiga toxin into cells: molecular mechanisms and medical perspectives. *EMBO J.* **19:**5943–5950.

11. **Schmitz, A., H. Herrgen, A. Winkeler, and V. Herzog.** 2000. Cholera toxin is exported from microsomes by the Sec61p complex. *J. Cell Biol.* **148:**1203–1212.

12. **Sellman, B. R., M. Mourez, and R. J. Collier.** 2001. Dominant-negative mutants of a toxin subunit: an approach to therapy of anthrax. *Science* **292:**695–697.

13. **Shogomori, H., and A. H. Futerman.** 2001. Cholera toxin is found in detergent-insoluble rafts/domains at the cell surface of hippocampal neurons but is internalized via a raft-independent mechanism. *J. Biol. Chem.* **276:**9182–9188.

14. **Takahashi, T., T. Umata, and E. Mekada.** 2001. Extension of juxtamembrane domain of diphtheria toxin receptor arrests translocation of diphtheria toxin fragment a into cytosol. *Biochem. Biophys. Res. Commun.* **281:**690–696.

15. **Torgersen, M. L., G. Skretting, B. van Deurs, and K. Sandvig.** 2001. Internalization of cholera toxin by different endocytic mechanisms. *J. Cell Sci.* **114**(Pt. 20):3737–3747.

SUGGESTED READING

1. **Alouf, J. E., and J. R. Freer (ed.).** 1999. *The Comprehensive Sourcebook of Bacterial Toxins.* Academic Press, Ltd., London, United Kingdom.

2. **Iversen, T.-G., G. Skretting, A. Llorente, P. Nicoziani, B. van Deurs, and K. Sandvig.** 2001. Endosome to Golgi transport of ricin is independent of clathrin and of the Rab9- and Rab11-GTPases. *Mol. Biol. Cell* **12:**2099–2107.

3. **Sandvig, K., and B. van Deurs.** 2002. Membrane traffic exploited by protein toxins. *Annu. Rev. Cell Dev. Biol.* **18:**1–14.

4. **Schiavo, G., and F. G. van der Goot.** 2001. The bacterial toxin toolkit. *Nature* **2:**530–537.

Bacterial Protein Toxins
D. Burns et al., Editors
©2003 ASM Press, Washington, DC 20036

Chapter 12

Transcytosis of Bacterial Toxins across Mucosal Barriers

Bonny L. Dickinson and Wayne I. Lencer

A fundamental principle of mammalian physiology is the formation and maintenance of epithelial barriers. Such barriers line the tissues of organ systems that interface with the environment and are found at all mucosal surfaces, including the gastrointestinal, respiratory, and genitourinary systems. Structurally, these barriers are created by a continuous monolayer of polarized epithelial cells with distinct apical and basolateral membrane domains, each containing unique protein and lipid components. This polarity in cell structure defines the opposing lumenal (apical) and serosal (basolateral) functions of mucosal surfaces and accounts for the vectorial transport of specific solutes, gases, and water across epithelial barriers such as that found in the intestine, lung, and genitourinary tract.

To generate barrier function, the individual epithelial cells lining these tissues must also assemble circumferential intercellular tight junctions that seal one cell to another. The tight junction represents the rate-limiting barrier that restricts passive diffusion (and convection) of solutes, molecules, and water between cells. In this way, the single-cell-thick monolayer that defines the mucosal surface establishes and maintains biological homeostasis between the outside and inside environments. Ultimately, the epithelial barrier protects the host from microbial invasion and penetration of other noxious agents.

Some bacterial toxins, such as cholera toxin secreted by *Vibrio cholerae*, Shiga-like toxin secreted by enterohemorrhagic *Escherichia coli*, and botulinum toxin secreted by *Clostridium botulinum*, must breach the intestinal epithelial barrier to cause disease. These microbes do not invade the intestinal mucosa or directly assist the delivery of toxin into the host cell cytoplasm by type III secretion systems or by other mechanisms. Rather, the molecular determinants that drive entry of these toxins into and across the intestinal cell are encoded entirely within the structure of the toxin itself. As described below, these toxins are able to cross

Bonny L. Dickinson and Wayne I. Lencer • GI Cell Biology, Combined Program in Pediatric Gastroenterology and Nutrition, Children's Hospital, the Harvard Digestive Diseases Center, and the Department of Pediatrics, Harvard Medical School, Boston, MA 02115.

the mucosal barrier as fully folded and functional proteins by exploiting the normal cellular pathways endogenous to the host cells lining these surfaces.

Transporting epithelial cells, such as those found in the intestine, kidney, and lung, exhibit two fundamentally different absorptive mechanisms (Fig. 1). First, small solutes may cross the epithelial barrier and enter the subepithelial space by moving around cells through the intercellular tight junctions. This process, termed paracellular transport, is both charge and size selective. Normally, in health, only small molecules penetrate tight junctions. Macromolecules and all bacterial toxins are excluded from this pathway. Certain small solutes, such as glucose and Na^+, may efficiently cross epithelial barriers utilizing a second pathway, by moving into and across the epithelial cell. This is termed transcellular transport and depends on the action of specific solute transporters or channels expressed in the polarized epithelial cell that facilitate movement of solutes across apical or basolateral plasma membranes. Again, macromolecules and all bacterial toxins are excluded from this pathway.

Fully folded proteins do, however, move across epithelial barriers. For example, immunoglobulins (Ig) G and M and dimeric IgA (dIgA) are all transported across mucosal surfaces by binding specific membrane receptors that direct the movement of these proteins into and then across the epithelial cell, from one membrane domain to the other, by vesicular traffic. This process is termed transcytosis. Transcytosis of such immunoglobulins typically displays strict dependence on the biology of the membrane receptor of the membrane that binds the immunoglobulin and dictates trafficking of the immunoglobulin-receptor complex into the transcytotic pathway. Transcytosis has no effect on

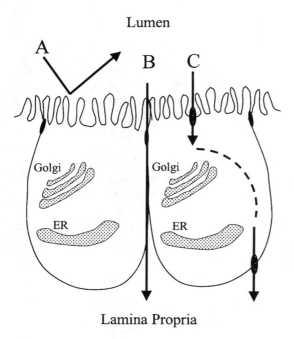

Figure 1 Absorptive mechanisms of transporting epithelia. (A) Polarized epithelial cells lining mucosal surfaces form a barrier that prevents penetration of macromolecules and bacterial toxins beyond the epithelium. (B) Some molecules move between cells and pass through the tight junctions by the process of paracellular transport. (C) In contrast, small solutes such as sodium and glucose cross the epithelial barrier by the action of specific transporters and channels that facilitate their movement in a process termed transcellular transport.

intercellular tight junctions and thus no effect on the integrity of the epithelial barrier to other macromolecules or microbes. To date, the transcytotic pathways of only two receptors, the polymeric immunoglobulin receptor (pIgR) and the neonatal Fcγ receptor (FcRn), have been studied at the molecular level and thus define our current understanding of the mechanism of transcytosis.

Some microbial toxins can move across epithelial barriers by transcytosis. However, the exact pathways and the mechanisms used by these toxins to cross the epithelial cell differ fundamentally from the "classical pathways" used by the immunoglobulins. In this chapter, we will first focus on the biology and mechanisms of transcytosis for dIgA and IgG. We will then detail the mechanism of transcytosis for cholera toxin. Finally, we will make reference to Shiga and Shiga-like enterotoxins that likely move across epithelial barriers by opportunistically exploiting similar mechanisms of vesicular transport.

TRANSCYTOSIS OF IgA AND IgG: THE CLASSICAL PATHWAYS

The pIgR

The pIgR mediates the unidirectional transport of dimeric IgA and IgM across polarized epithelial barriers (10). The pIgR is a 116-kDa transmembrane protein located on the basolateral surface and in the endosomal compartments of polarized epithelial cells lining mucosal surfaces. The receptor is expressed as a single polypeptide chain that exhibits five extracellular immunoglobulin-like domains, a short transmembrane region, and a 103-amino-acid cytoplasmic tail. pIgR-dependent transcytosis of dIgA has been effectively modeled in vitro by using MDCK cells stably expressing the pIgR. This model represents the standard by which all other systems of transcytosis are measured.

Transcytosis is initiated by binding of dIgA to the pIgR at the basolateral surface of the epithelial cell (Fig. 2, left side). Ligand binding stimulates endocytosis of the immunoglobulin-pIgR complex into clathrin-coated pits. In the early endosome, the dIgA-pIgR complex moves away from fluid-phase cargo and other membrane components targeted to the late endosome or lysosome for degradation. The dIgA-pIgR complex then traffics into the common endosome and then again to the apical recycling endosome. In doing so, the dIgA-pIgR complex traffics away from other membrane components (such as the transferrin receptor) that are rapidly recycled back to the basolateral plasma membrane (3). Thus, trafficking of the dIgA-pIgR complex out of the common endosome and into the apical recycling endosome represents the step or steps that define polarized sorting of the pIgR in the transcytotic pathway.

After budding from the apical recycling endosome, vesicles containing the dIgA-pIgR complex fuse with the apical plasma membrane, where the pIgR undergoes proteolytic cleavage by a serine protease localized to the apical membrane. The domains of the pIgR that bind dIgA are left intact, and thus cleavage at this site releases "secretory" IgA (sIgA; composed of dIgA bound to the extracellular domain of the pIgR) from the mucosal surface into the lumen.

Trafficking of the dIgA-pIgR complex through the transcytotic pathway depends on information encoded in the structure of the extracellular and cytoplasmic tail domains of the pIgR itself. Sorting motifs contained in the pIgR

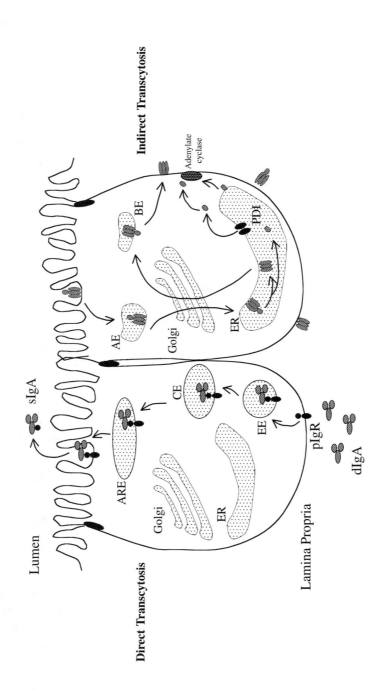

cytoplasmic tail play a dominant role in biogenesis of the receptor and in driving the pIgR into the apically directed transcytotic pathway. At least four mechanisms are thought to regulate trafficking of this receptor through the transcytotic pathway: (i) phosphorylation of conserved serine residues within the cytoplasmic tail, (ii) calmodulin binding to the cytoplasmic tail, (iii) ligand binding to the extracellular domain (stimulating dimerization), and (iv) regulated interactions between vesicles containing the pIgR and components of the actin cytoskeleton.

Several features of this system should be emphasized to distinguish the pIgR pathway of transcytosis from the transcytotic pathway utilized by cholera and possibly Shiga and Shiga-like toxins. First, the itinerary and mechanisms of transcytosis for dIgA are defined by the biology of the pIgR, a protein receptor that moves in only one direction across the cell, from basolateral to apical membrane. Thus, the pIgR directs secretion of sIgA onto mucosal surfaces, where this antibody functions in mucosal immunity. The pIgR cannot perform in the opposite direction to absorb sIgA from the mucosal surface. Finally, trafficking of the dIgA-pIgR complex across the cell does not require movement of the pIgR through the Golgi apparatus or endoplasmic reticulum (ER). Cholera toxin, on the other hand, moves in the opposite direction, from the apical to the basolateral membrane, and thus enters the host by this mechanism. Furthermore, as described below, transcytosis of cholera toxin may depend critically on movement of toxin into the Golgi and possibly ER (5).

The Neonatal Receptor for IgG, FcRn

The neonatal FcRn also functions to transport immunoglobulins across mucosal barriers. The FcRn is expressed at both the apical and basolateral membrane domains of polarized cells as a heterodimer composed of a glycosylated heavy

Figure 2 Absorptive mechanisms of transporting epithelia. Direct transcytosis of dIgA-pIgR complexes (left) is initiated by binding of dIgA to the pIgR at the basolateral surface of the epithelial cell, which stimulates endocytosis of the complex into an early endosome (EE). Here, the dIgA-pIgR complex moves away from fluid-phase cargo and other membrane components targeted to the late endosome or lysosome for degradation. The dIgA-pIgR complex then traffics into the common endosome (CE) and then again to the apical recycling endosome (ARE). Transport from the common endosome into the apical recycling endosome is a hallmark of polarized sorting of the pIgR in the transcytotic pathway. In indirect transcytosis (right panel), CT enters the polarized epithelial cell by binding GM1 at the apical membrane and then moves via an early endosomal compartment (AE) into the Golgi complex. In the Golgi, the C-terminal KDEL motif on the A-subunit facilitates retrograde movement of the CT-GM1 complex through the Golgi stack and into the ER. In the ER, the A-subunit separates from the B-subunit-GM1 complex, unfolds, and translocates across from the ER into the cytosol in a process catalyzed by the ER lumenal chaperone protein disulfide isomerase (PDI). Alternatively, the A-subunit may remain attached to the B-subunit after translocation and moves to the basolateral membrane by entering the anterograde transport vesicles (BE) and moving back out the secretory pathway.

(α) chain associated noncovalently with β_2-microglobulin (11). FcRn binding to IgG exhibits a strong pH dependence, with high-affinity binding at acidic pH (pH $\leq$ 6.0) and weak or no binding at neutral pH (pH $\geq$ 7.0).

The FcRn may best be characterized for its role in the absorption of IgG across the intestinal epithelium of suckling mice and rats. In the neonatal rodent, the FcRn is abundantly expressed at the enterocyte apical surface, where it binds IgG from breast milk and then transports the immunoglobulin across the cell into the systemic circulation by transcytosis (1). In humans, the passive transfer of IgG from mother to fetus occurs across the placental barrier and this likely depends on the FcRn.

In addition to the function of the FcRn in the acquisition of humoral immunity during early life, recent data indicate that the FcRn may function critically in adult life to regulate IgG metabolism. For example, adult endothelial tissues express the FcRn where the receptor binds IgG after endocytosis and recycles the immunoglobulin away from the late endosome and lysosome and back into the bloodstream. In doing so, the FcRn rescues IgG from lysosomal degradation, and this extends the half-life of IgG relative to other serum proteins.

FcRn is also expressed in the polarized epithelial cells lining the intestine of adult humans (Fig. 3) (see Color Plates following p. 256) (4). These data suggest that IgG, like IgA, may also be transported across epithelial barriers by transcytosis in adult life. To test this idea, we developed the polarized human intestinal T84 cell line as a model for studying the FcRn-dependent transport of IgG across mucosal surfaces. T84 cells form well-differentiated, polarized monolayers and express the FcRn in a pattern analogous to that found in crypt enterocytes of normal human jejunum (Fig. 3a–c) (4). The FcRn specifically transported IgG across polarized T84 cell monolayers in both the lumenal (apically directed) and ablumenal (basolaterally directed) directions. Moreover, the bidirectional IgG transport in T84 cells was shown to depend on endosomal acidification, a characteristic of the FcRn binding to IgG.

The exact itinerary of the FcRn trafficking across epithelial cells, however, remains to be defined. Clearly, the pathway of transcytosis for the FcRn must differ from that of the pIgR. First of all, the FcRn moves in both directions across the cell and binds its ligand, IgG, in the acidic endosome rather than at the cell surface. Nonetheless, early studies on FcRn-mediated absorption of IgG in the neonatal rodent indicate that, just like the pIgR, FcRn-IgG complexes probably do not traffic through the Golgi apparatus. This "direct" pathway of transcytosis for the FcRn and the pIgR differs fundamentally from that exploited by bacterial toxins, as described below.

TRANSCYTOSIS OF BACTERIAL PROTEIN TOXINS: THE "INDIRECT PATHWAY"

Cholera Toxin

Colonization of the small intestine by *V. cholerae* results in diarrhea due to massive salt and water secretion without epithelial damage. The secretory diarrhea is induced by the direct action of cholera toxin (CT) on the polarized epithelial cells that line the intestine. To cause disease, CT must first enter the intestinal

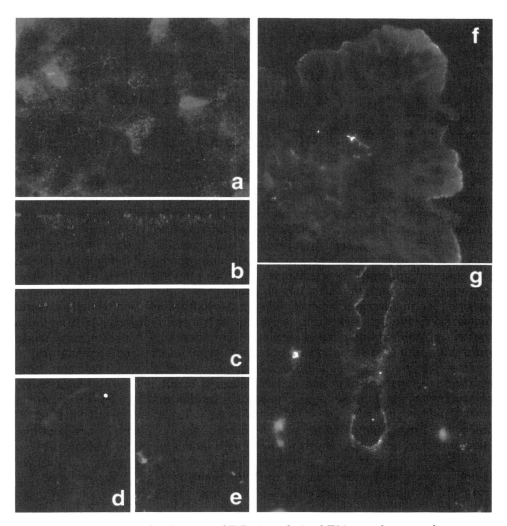

Figure 3 Immunolocalization of FcRn in polarized T84 monolayers and normal adult human small intestinal mucosa. (a) Whole-mount T84 monolayers visualized en face show a diffuse, punctate staining pattern. The Z0-1 image was captured slightly above the focal plane of FcRn. (b) FcRn staining of whole-mount T84 monolayers visualized as confocal vertical sections. (c) FcRn staining was absent in the presence of an isotype-matched, irrelevant antiserum. (d and e) FcRn staining was absent in the presence of an irrelevant antiserum or with secondary antibody alone. (f) Villous enterocytes of normal adult human small intestine show delicate linear staining in the region of the apical cytoplasmic membrane. (g) Crypt enterocytes show an apical and punctate staining pattern visible not only at the level of the apical cytoplasmic membrane, but also in the apical cytoplasm below the level of the apical membrane. (See Color Plates following p. 256.)

epithelial cell from the apical membrane where it encounters the mucosal barrier after being secreted by *V. cholerae* in the intestinal lumen.

CT binds a ganglioside receptor on the apical cell surface, enters the intestinal cell by non-clathrin-mediated endocytosis, and then moves retrograde into Golgi cisternae and ER to enter the cytoplasm and induce disease (see reference 5 for review). CT can also breach the epithelial barrier, from lumen to serosa, as an intact protein by moving across the cell by transcytosis (Fig. 2, right) (7). Our recent studies indicate that such trafficking of CT into the Golgi apparatus and possibly into the transcytotic pathway may depend on lipid-based sorting motifs defined by the toxin's cell surface receptor ganglioside GM1 (13). We also find that the itinerary of toxin trafficking across the cell by transcytosis likely involves movement of CT through the Golgi apparatus before arrival at the contralateral membrane. Thus, transcytosis of CT, unlike that defined by trafficking of the pIgR, depends on the biology of a membrane lipid and on a pathway that intersects the Golgi apparatus and possibly the ER. To distinguish this pathway from that utilized by the pIgR and the FcRn, we have termed this process indirect transcytosis. We propose that such lipid-based trafficking may account for transcytosis of other bacterial toxins with similar structure and function, such as that observed for Shiga and Shiga-like toxins, and *E. coli* heat-labile toxins.

CT Structure and Function

In CT, five identical polypeptides (≈11 kDa) assemble into a highly stable and ordered pentameric ring termed the B subunit (≈55 kDa) that binds with high affinity and specificity to ganglioside GM1 (Fig. 4). Binding to GM1 tethers the toxin to the plasma membrane of host cells and distributes the CT-GM1 complex into specialized apical plasma membrane microdomains termed lipid rafts. CT-B binding to GM1 is stoichiometric, with one B-subunit pentamer cross-linking five GM1 gangliosides at the cell surface. The single A subunit is composed of two major structural domains termed the A_1- and A_2-peptides linked together by an exposed loop containing a serine-protease-sensitive "nick" site subtended by a single disulfide bond. The A_2-peptide (≈5 kDa) noncovalently tethers the

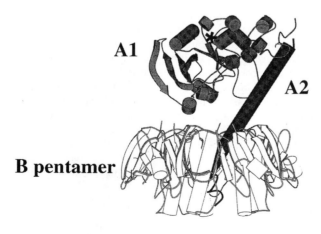

A1

A2

B pentamer

Figure 4 Crystal structure of CT. A ribbon diagram based on the crystal structure of the CT holotoxin is shown viewed from the side. Each subunit is labeled. This figure was kindly provided by Ethan Merritt and Wim Hol, University of Washington, Seattle.

A_1-peptide to the B subunit and contains a C-terminal KDEL motif (9). The KDEL motif is known to be a sorting signal that allows endogenous lumenal ER proteins of the eukaryotic cell to be retrieved efficiently from post-ER compartments. The A_1-peptide ($\approx$22 kDa) is the enzymatically active subunit that must eventually dissociate from the B subunit, translocate across a cellular membrane (presumably the ER membrane), and act inside the cell to activate adenyl cyclase by catalyzing the ADP-ribosylation of the heterotrimeric GTPase Gsα.

Toxin Action Depends on Trafficking of CT into the ER of Host

To induce disease, CT must move from the cell surface into the endosome and then retrograde into the ER. After arriving in the ER, the A subunit separates from the B-subunit-GM1 complex, unfolds, and translocates across the ER limiting membrane to reach the cytosol (5, 12). Toxin unfolding is catalyzed by the ER lumenal chaperone protein disulfide isomerase. Movement of toxin into the ER is required for bioactivity. To enter this pathway efficiently, CT has evolved a mechanism that exploits the trafficking of GM1, a membrane lipid (see below). Retrograde trafficking into the ER is also required for Shiga toxin, ricin, and *Pseudomonas* exotoxin A. It is our view that such trafficking into the Golgi apparatus (and perhaps ER) sets the stage for toxin transcytosis across epithelial barriers.

Transcytosis of CT

Much of our current understanding of the cellular mechanisms exploited by CT to enter host cells has been elucidated by modeling CT action on the polarized human epithelial cell line T84. This system is particularly relevant because the model requires that CT enter the cell and transduce a signal from the apical membrane as must occur in vivo. In T84 cells, the physiologic site of toxin binding to the apical membrane is separated spatially from adenyl cyclase on the cytoplasmic surface of the basolateral membrane by circumferential tight junctions. Neither CT nor any other protein toxin can move through tight junctions to reach the basolateral membrane.

  Several years ago, we tested the idea that CT may arrive at the basolateral membrane by vesicular traffic (7). To do so, we assayed for trafficking of the CT B subunit. The B subunit, unlike the A_1-peptide, does not translocate across cell membranes, and this allowed us to measure toxin transcytosis directly. These studies showed that CT B subunit moved from apical to basolateral membrane domains by transcytosis. The time course and temperature dependency of B-subunit transcytosis correlated closely with the time course of toxin action, providing some evidence that the process of toxin transcytosis may be mechanistically related to toxin action.

  We also found that the time course of transcytosis of B-subunit was delayed in CT variants that contained an inactivating mutation in the ER-targeting motif KDEL (Fig. 5) (6). The KDEL receptor is localized to the Golgi apparatus and functions as a retrieval mechanism for ER lumenal proteins that inadvertently move out of the ER in budding vesicles containing nascent secretory or membrane proteins. The KDEL mutation also caused a delay in toxin action, and

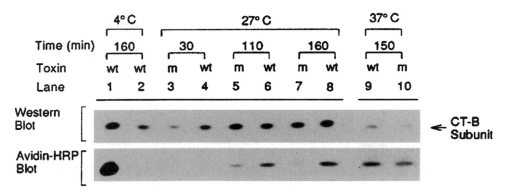

Figure 5 Transcytosis of CT B-subunit across polarized T84 cell monolayers and effect of mutation in KDEL on CT trafficking through the transcytotic pathway. Transcytosis of KDEL-mutant (m) and wild-type (wt) CT from apical to basolateral membranes was assessed by selective cell surface biotinylation. T84 cell monolayers grown on permeable supports were exposed continuously to 1 nM of KDEL-mutant or wt CT at the apical cell surface at the indicated temperatures and times followed by cell surface biotinylation at either the apical (lane 1) or basolateral (lanes 2–10) surface. The upper panel shows a Western blot to demonstrate that each lane contains equivalent amounts of immunoprecipitated CT B subunit. The lower panel shows the avidin-horseradish peroxidase (HRP) blot of the same experiment, revealing biotinylated toxin. The first lane shows that CT B subunit can be labeled with biotin while bound to GM1 at the apical plasma membrane (positive control). At 4°C, a temperature that completely inhibits vesicular traffic, CT B does not reach the basolateral membrane, as indicated by the absence of biotinylated CT B subunit (lane 2). Lanes 7 and 8 show that in cells incubated at 27°C for 160 min, basolaterally applied biotin has now labeled a fraction of the B-subunit at the basolateral membrane, indicating that both the KDEL-mutant and wt CT reach the basolateral membrane by transcytosis. The amount of the KDEL-mutant CT reaching the basolateral membrane is approximately fivefold less than that of wt CT, indicating that transcytosis for the KDEL-mutant CT proceeds more slowly than for wt toxin. From reference 6 with permission.

this correlated closely with the delay in transcytosis. These data suggested that CT may require trafficking first into the Golgi apparatus or ER before moving across the cell by transcytosis.

   In this hypothesis, we reason that by binding GM1, CT (and other bacterial toxins that require entry into the ER for bioactivity) has evolved an efficient strategy to exploit an endogenous lipid-based retrograde trafficking pathway from cell surface to Golgi cisternae and ER, and this then leads to transcytosis as described below.

   The Golgi and ER represent the central organelles in the biosynthetic pathway for all intrinsic membrane and secretory proteins, and the Golgi apparatus ultimately feeds nascent proteins to both the apical and basolateral domains of polarized epithelial cells. Sorting of nascent apical and basolateral membrane proteins occurs in the trans-Golgi network. Thus, once CT has arrived inside the

Golgi apparatus or ER, the toxin now has the opportunity to recycle back to the plasma membrane from an organelle that normally delivers proteins to both sides of the cell.

Although it is possible that the CT-GM1 complex may define a motif that dictates sorting into a recycling pathway from Golgi to basolateral plasma membrane specifically, it is equally plausible that recycling CT out of the Golgi apparatus represents a stochastic process of membrane transport. Such a stochastic process of trafficking out of the Golgi would result in redistributing CT to both apical and basolateral membrane domains. Thus, CT that moves efficiently from apical plasma membrane into Golgi and ER could recycle (though perhaps inefficiently) back out the biosynthetic pathway to the opposite basolateral membrane domain by randomly associating with membrane domains that move anterograde in the secretory pathway. The result would be transcytosis. Once across the epithelial barrier, CT would then be in a position to interact with other cells present in the subepithelial space or to redistribute systemically, as may occur for Shiga and vero toxins (see below).

To distinguish this transcytotic pathway that involves transit through the Golgi apparatus from that defined by transcytosis of the pIgR and the FcRn, we have termed the process of CT trafficking across epithelial cells indirect transcytosis (7). The idea, however, has not yet been tested directly, and our data do not preclude the possibility that CT may also move across the cell by a direct transcytotic mechanism, such as that defined for the pIgR and possibly the FcRn.

Finally, with respect to mechanisms of toxin transcytosis across epithelial barriers, it is important to emphasize here that the lipid-based sorting motif defined by the CT-GM1 complex represents an important feature that distinguishes the indirect transcytotic pathway of CT from that defined by the immunoglobulins. Transcytosis of both dIgA and IgG depends on sorting motifs embedded within the protein structures of the pIgR and the FcRn. Trafficking of CT in the transcytotic pathway, however, depends on the lipid-based membrane anchor provided by the toxin's receptor GM1 and on the ability of GM1 to concentrate CT in detergent-insoluble glycolipid-rich apical membrane microdomains or "lipid rafts" (Box 5). These structures are present in polarized human intestinal epithelial cells and are critical to toxin action (13). Lipid rafts are distinct membrane structures rich in cholesterol and glycolipids that function as membrane organizing centers for signal transduction, protein and lipid sorting, endocytosis, and transcytosis across endothelial cells. Such lipid-based sorting may also account for Shiga toxin trafficking into the host cell, as Shiga toxin, like CT, binds pentavalently to its own glycolipid receptor GB3.

Functional Consequences of Toxin Transcytosis

It is well documented that CT and LTI represent the most potent mucosal immunogens and adjuvants recognized to date. Such efficiency in eliciting inductive immunity after application to mucosal surfaces, not seen with other ingested proteins of comparable size, implies that CT exhibits an ability to encounter and perhaps act on antigen-presenting cells of the mucosal immune system. Immunocompetent cells are abundant in both the organized lymphoid follicles of the Peyer's patch and the follicle-associated epithelium, and also in the lamina pro-

BOX 5

Lipid rafts

Membranes of nearly all eukaryotic cells contain distinct microdomains that display low density and resistance to detergent extraction by virtue of their lipid (rich in glycosphingolipids and cholesterol) and protein composition. These structures are variously termed lipid rafts, detergent-resistant membranes, or detergent-insoluble glycosphingolipid-rich membranes. These specialized membrane microdomains exist on both intracellular and plasma membranes. Noncoated 50- to 70-nm plasmalemmal vesicles (first discovered in epithelial cells of the mouse gallbladder and termed caveolae by Yamada in 1955) typify such detergent-resistant structures. Caveolae in many cell types have characteristic flask-shaped morphology and contain a unique 21-kDa membrane protein, caveolin. It is now recognized that caveolae represent only a subset of lipid rafts, whose composition and function may vary among cell types. Lipid rafts are now known to mediate key cellular functions that include ligand-induced signal transduction, protein and lipid sorting, solute transport, cholesterol metabolism, endocytosis, and (in vascular endothelium) transcytosis. Ganglioside GM1 clusters in lipid rafts and cholera toxin have been used widely to localize and examine the biology of these structures.

pria of the intestine. Given the ability of these toxins to breach the mucosal barrier by transcytosis, we propose that such trafficking may facilitate toxin access to relevant antigen-presenting cells in the lamina propria, and thus contribute significantly to the potent effects of orally delivered CT on mucosal and systemic immune responses.

Finally, as alluded to above, both Shiga and botulinum toxins have been shown to cross epithelial barriers in vitro by transcytosis (2, 8). The clinical consequences of such transepithelial transport are readily apparent. Both toxins can elicit systemic disease. Like CT, botulinum toxin must breach the epithelial barrier in the absence of bacterial invasion of the intestinal mucosa. All botulinum toxins exhibit lectin-like activity, and thus these toxins may bind a wide range of glycoprotein and glycolipid membrane components. The itinerary and mechanisms by which botulinum toxin crosses the epithelial cell, however, remain completely unknown.

Shigella dysenteriae and the Shiga-like toxin-producing *E. coli* strains are food-borne pathogens responsible for the diarrheal dysentery and hemolytic-uremic syndrome, respectively. Disease correlates with the secretion of Shiga toxin and the Shiga-like toxins. In hemolytic-uremic syndrome, Shiga toxin must breach the intestinal barrier to gain access to the systemic circulation, where the toxin eventually induces severe injury to the kidney, thrombocytopenia, and finally a hemolytic uremia. Like CT, Shiga toxin also binds pentavalently to a membrane lipid and must traffic backward in the biosynthetic pathway to reach the ER of host cells for toxicity. As outlined above, we propose that transcytosis of Shiga toxin may also occur by a lipid raft-mediated mechanism of indirect transcytosis and involve transit of toxin first through Golgi cisternae or ER before arriving at the contralateral basolateral membrane domain.

SUMMARY

In summary, certain bacterial toxins can enter and move across polarized epithelial cells that represent the fundamental barrier against invasion of microbes and noxious agents at all mucosal surfaces. CT represents the paradigm for such transport. Unlike the mechanisms of transcytosis for the pIgR and the FcRn, CT moves across the cell by opportunistically binding a membrane lipid, ganglioside GM1, and enters the Golgi complex before arrival at the contralateral membrane domain. Binding GM1 associates CT with lipid rafts and this likely drives sorting of the CT-GM1 complex into the Golgi apparatus and perhaps ER. Such trafficking is required for CT-induced toxicity and probably for transcytosis of toxin across the epithelial cell. The functional consequences of such transport likely affect inductive mechanisms of mucosal immunity and the distribution of bacterial toxins to the systemic circulation.

Acknowledgments
This work was supported by research grants DK/AI-53056 and DK-48106 from the National Institutes of Health to W. I. Lencer, DK-34854 to the Harvard Digestive Diseases Center; and 1K01DK59945-01 and 1F32DK10038-01 to B. L. Dickinson.

REFERENCES
1. **Abrahamson, D. R., A. Powers, and R. Rodewald.** 1979. Intestinal absorption of immune complexes by neonatal rats: a route of antigen transfer from mother to young. *Science* **206:**567–569.
2. **Acheson, D. W., R. Moore, S. De Breucker, L. Lincicome, M. Jacewicz, E. Skutelsky, and G. T. Keusch.** 1996. Translocation of Shiga toxin across polarized intestinal cells in tissue culture. *Infect. Immun.* **64:**3294–3300.
3. **Brown, P. S., E. Wang, B. Aroeti, S. J. Chapin, K. E. Mostov, and K. W. Dunn.** 2000. Definition of distinct compartments in polarized Madin-Darby canine kidney (MDCK) cells for membrane-volume sorting, polarized sorting and apical recycling. *Traffic* **1:**124–140.
4. **Dickinson, B. L., K. Badizadegan, Z. Wu, J. C. Ahouse, X. Zhu, N. E. Simister, R. S. Blumberg, and W. I. Lencer.** 1999. Bidirectional FcRn-dependent IgG transport in a polarized human intestinal epithelial cell line. *J. Clin. Invest.* **104:**903–911.
5. **Lencer, W. I.** 2001. Cholera: invasion of the intestinal epithelial barrier by a stably folded protein toxin. *Am. J. Physiol. Gastrointest. Liver Physiol.* **280:**G781–786.
6. **Lencer, W. I., C. Constable, S. Moe, M. Jobling, H. M. Webb, S. Ruston, J. L. Madara, T. Hirst, and R. Holmes.** 1995. Targeting of cholera toxin and *E. coli* heat labile toxin in polarized epithelia: role of C-terminal KDEL. *J. Cell Biol.* **131:**951–962.
7. **Lencer, W. I., S. Moe, P. A. Rufo, and J. L. Madara.** 1995. Transcytosis of cholera toxin subunits across model human intestinal epithelia. *Proc. Natl. Acad. Sci. USA* **92:**10094–10098.
8. **Maksymowych, A. B., and L. L. Simpson.** 1998. Binding and transcytosis of botulinum neurotoxin by polarized human colon carcinoma cells. *J. Biol. Chem.* **273:**21950–21957.

9. **Merritt, E. A., S. Sarfaty, F. van der Akker, C. L'Hoir, J. A. Martial, and W. G. J. Hol.** 1994. Crystal structure of cholera toxin B-pentamer bound to receptor GM1 pentasaccharide. *Protein Sci.* **3:**166–175.

10. **Mostov, K. E., M. Friedlander, and G. Blobel.** 1984. The receptor for transepithelial transport of IgA and IgM contains multiple immunoglobulin-like domains. *Nature* **308:**37–43.

11. **Simister, N. E., and A. R. Rees.** 1985. Isolation and characterization of an Fc receptor from neonatal rat small intestine. *Eur. J. Immunol.* **15:**733–738.

12. **Tsai, B., C. Rodighiero, W. I. Lencer, and T. Rapoport.** 2001. Protein disulfide isomerase acts as a redox-dependent chaperone to unfold cholera toxin. *Cell* **104:**937–948.

13. **Wolf, A. A., M. G. Jobling, S. Wimer-Mackin, J. L. Madara, R. K. Holmes, and W. I. Lencer.** 1998. Ganglioside structure dictates signal transduction by cholera toxin in polarized epithelia and association with caveolae-like membrane domains. *J. Cell Biol.* **141:**917–927.

SECTION IV
TOXIN ACTION

Bacterial toxins have evolved a number of mechanisms to alter critical metabolic processes within the host cell. In general, toxins act either by damaging host cell membranes or by modifying proteins that are critical to the maintenance of the physiology of the cell. These modifications can result in cell death, as observed by the inactivation of protein synthesis by diphtheria toxin, or a modification of host cell physiology, such as that observed following elevation of cyclic AMP (cAMP) levels induced by cholera toxin and heat-labile enterotoxin of *Escherichia coli*.

Some toxins exert their effects from a location on the surface of eukaryotic cells. These toxins include membrane-damaging toxins, such as pore-forming toxins, and the superantigens. The latter bind to specific proteins on immune cells and stimulate antigen-independent cytokine production.

Other bacterial toxins exert their effects on eukaryotic cells only after entering the cell. These toxins include those that catalyze covalent modification of host proteins or nucleic acids, those that synthesize the second messenger cAMP (such as edema factor of anthrax toxin, *Bordetella pertussis* adenylate cyclase toxin, and ExoY of *Pseudomonas aeruginosa*), and those that modulate small GTPases that are involved in signal transduction. Covalent modifications catalyzed by bacterial toxins include mono-ADP-ribosylation, mono-glucosylation, and deamidation of host proteins. The proteolytic properties of the clostridial neurotoxins and the DNase activity of the cytolethal distending toxins have been recently described. A family of toxins covalently modifies ribosomal RNA by deadenylation. Recent studies have identified a new class of bacterial toxins, which modify the activation state of Rho GTPases without covalent modification to the GTPase. Each of these toxins mimics the action of two host proteins, either by activating the Rho GTPase by serving as a nucleotide exchange factor (Rho GEF), or by inactivating Rho GTPases by serving as a GTPase activating protein (Rho GAP). In contrast to covalent modification, which is often irreversible or slowly reversible relative to the life cycle of the cell, modulation of the nucleotide state of the Rho GTPases by the bacterial GEFs and GAPs offers a fine-tuning of host cell physiology.

Many protein toxins are produced by the bacterial pathogen in a proenzyme form, which is then activated by limited proteolysis or disulfide bond reduction.

In addition to the need to be processed to express enzymatic activity, some toxins show another level of regulation, which involves the requirement to bind a eukaryotic protein to express their catalytic activity. Included among these toxins are *Bordetella* adenylate cyclase toxin and edema factor of anthrax toxin, which are activated by binding calmodulin. This provides the toxins with an additional level of regulation, which requires that the toxin is localized to a specific site within the host to allow activation.

Our understanding of the molecular actions of bacterial toxins and their effects on host cell physiology has paralleled advances in the physical analysis of toxin structure-function and eukaryotic cell biology. New developments in other areas of science, including neurobiology and immunology, will undoubtedly provide novel tools to enhance our perspective of how bacterial toxins modulate host physiology by yet undefined mechanisms.

Bacterial Protein Toxins
D. Burns et al., Editors
©2003 ASM Press, Washington, DC 20036

Chapter 13

Membrane-Damaging Toxins: Pore Formation

F. Gisou van der Goot

A number of pathogenic bacteria secrete exotoxins that are capable of inserting into a membrane and of forming a pore, thereby impairing the permeability of the bilayer. This family includes the translocation domain or B subunit of a number of toxins discussed in other chapters as well as bona fide pore-forming toxins (PFTs). The latter will be discussed here with the exception of RTX toxins, which are treated in a separate chapter. The role of PFTs in bacterial pathogenesis will be discussed first; then their general mode of action will be outlined and the events that lead to pore formation will be described at the structural level, using two examples, the PFTs from *Staphylococcus aureus* and the cholesterol-dependent toxins (CDTs). Next the role of mammalian surface molecules as receptors for PFTs will be discussed. Finally, some of the consequences of pore formation will be reviewed.

WHY IS IT USEFUL FOR BACTERIA TO PRODUCE PFTs?

The production of PFTs favors survival and multiplication of a variety of pathogenic bacteria. How PFTs contribute to virulence is not always clear. Four scenarios will be mentioned in which PFTs appear to aid in the survival or cell-to-cell spreading of the bacteria (Fig. 1).

Pore formation can occur at the plasma membrane, as is the case for aerolysin and α-hemolysin produced by *Aeromonas hydrophila* and *Staphylococcus aureus*, respectively, two bacteria that generally remain extracellular. Permeabilization of the plasma membrane may allow release of useful nutrients from the perforated or dying cells (Fig. 1, step 1). In addition, killing of the host cell might be useful in itself, for example, if the targets are white blood cells, because the antibacterial immune response would be attenuated and bacterial spread promoted. Alternatively, pore formation at the plasma membrane may allow injec-

F. Gisou van der Goot • Department of Genetics and Microbiology, Centre Médicale Universitaire, 1, rue Michel-Servet, CH-1211 Geneva 4, Switzerland.

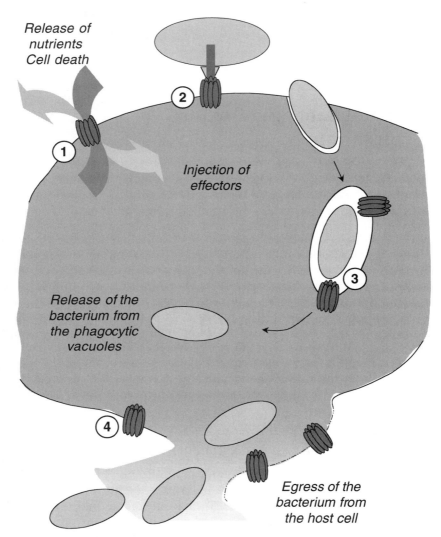

Figure 1 Roles of PFTs. A variety of nonmutually exclusive roles for PFTs have been proposed. Certain PFTs punch holes in the plasma membrane of the target cell with the possible purpose of releasing nutrients or killing the host cell (step 1 example: aerolysin). Others might serve at the tip of a type III secretion system to perforate the host cell plasma membrane and allow injection of bacterial effectors into the host cytoplasm (step 2 example: strep-tolysin O). In the case of invading bacteria, production of PFT within the phagocytic vacuole allows rupture of the vacuolar membrane and release of the bacterium in the host cytoplasm (step 3 example: listeriolysin O). Once the invading bacterium has sufficiently multiplied, egress from the host cell can require the secretion of a PFT (step 4 example: IcmS).

tion of bacterial effectors into the host cell cytoplasm, as was recently shown for *Streptococcus pyogenes*. Streptolysin O was found to form the tip of a type III-like secretion system that perforates the host cell membrane (Fig. 1, step 2).

Pore formation can also occur once the bacterium has invaded the host cell, either when the bacterium is in the phagocytic vacuole or when it is in the cytoplasm. In the first case, pore formation results in rupture of the phagocytic membrane and release of the bacterium into the cytoplasm, as shown for *Listeria monocytogenes*, which produces listeriolysin O (Fig. 1, step 3). In the latter case, production of the PFT occurs only when the bacterium has finished multiplying in the cytoplasm. Pore formation then allows egress of the bacterium from the host cell, as was observed for *Legionella pneumophila*. Production of IcmS by this organism leads to pore formation in the host plasma membrane from the cytoplasmic side, allowing local rupture and release of the microorganisms into the cell surroundings (Fig. 1, step 4).

GENERAL MODE OF ACTION OF PFTs

Most PFTs are able to form pores in artificial membranes such as liposomes. This allowed researchers to study the mechanisms that lead to pore formation in great detail using in vitro approaches. All characterized PFTs appear to have a similar overall mode of action (Fig. 2). Discussion here will be restricted to toxins rich in β-sheet structure, which constitute the vast majority of PFTs that have mammalian targets. PFTs are produced by the bacterium as soluble, generally monomeric proteins. Monomers diffuse toward the target membrane and bind via specific interactions with host surface molecules (Table 1). These can be lipids, such as cholesterol in the case of streptolysin O; sugar moieties, such as the glycan core of glycosylphosphatidyl (GPI)-anchored proteins in the case of aerolysin; or proteins, as recently shown for the protective antigen of *Bacillus anthracis* (this is not a bona fide PFT since its role is to translocate the enzymatic moieties of the anthrax toxin into the cytoplasm, but it shares many similarities with PFTs). The surface-bound monomeric PFT then diffuses in the two-dimensional space of the membrane until it encounters other monomers and oligomerizes into a ring-like structure. The number of monomers composing this ring varies from 7, in the case of staphylococcal α-hemolysin and aerolysin, to up to 50 in the case of streptolysin O and other members of the CDTs. The ring-like complex, which exposes hydrophobic surfaces as discussed later, inserts spontaneously into the lipid bilayer and forms a transmembrane pore. Whether the pores are small (15 to 35 Å, corresponding to the heptameric pores) or large (up to 300 Å for CDTs), they are generally nonselective and act rather as molecular sieves. Depending on their size, they will therefore allow either only the passage of small ions or the passage of macromolecules as large as antibodies.

PFTs: Janus-Like Proteins

The most striking feature of PFTs, which has fascinated structural biologists for decades, is that these proteins are able to adopt two generally incompatible states, hence the name of Janus-like proteins in reference to the two-faced god. These proteins can be either water soluble (in the beginning of their existence)

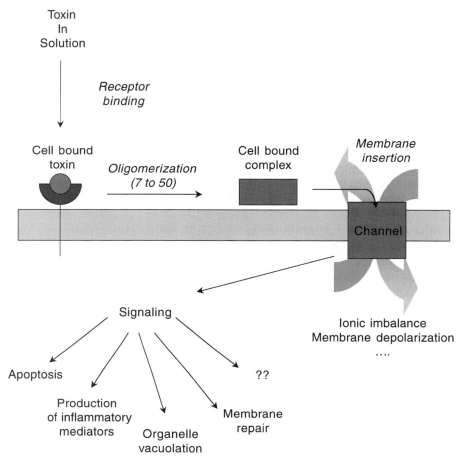

Figure 2   General mode of action of PFTs. All β-barrel PFTs have a similar overall mode of action. They are secreted as soluble proteins that bind, often with great specificity, to the host cell membranes. There, upon encounter with other toxin molecules, they undergo circular polymerization into a ring-like structure called the prepore; a subsequent conformational change leads to the exposure of hydrophobic surfaces. The complex then inserts into the membrane and forms a pore. Pore formation at the plasma membrane can have a number of consequences that might be dependent on the toxin, the toxin concentration, and the cell type.

or transmembrane (at the end of their existence), meaning that they can convert from a stable folded state in water to a different stable folded state in a membrane. Importantly, this transition does not require the assistance of other proteins such as chaperones or translocons, which are generally responsible for the insertion of membrane proteins into the endoplasmic reticulum or the bacterial plasma membrane. All the information required for this transition is therefore contained in the primary structure of the PFT.

Table 1 Examples of PFTs[a]

Toxin	Mol wt (kDa)	Producing bacterium	Roles	Subunits forming the channel	Receptor
α-Hemolysin	33	*Staphylococcus aureus*		7	–
Leukotoxins		*Staphylococcus aureus*		6	–
Aerolysin	47	*Aeromonas hydrophila*	Spreading of the bacterium	7	GPI-anchored proteins
α-Toxin	41	*Clostridium septicum*		Unknown (7?)	GPI-anchored proteins
LLO (listeriolysin O)	56	*Listeria monocytogenes*	Escape from the vacuole	Up to 50	–
SLO (streptolysin O)	61	*Streptococcus pyogenes*	Type III secretion	Up to 50	Cholesterol
PFO (perfringolysin O)	53			Up to 50	Cholesterol
VacA	88	*Helicobacter pylori*		6–7	–
IcmS	12	*Legionella pneumophila*	Egress from host cell	Unknown	Maybe none

[a] A few examples of PFTs were selected. A given PFT is likely to have more than one role to play in the infectious process. Only a few documented roles have been listed.

A brief introduction to transmembrane proteins

To understand the metamorphosis that PFTs must undergo, a brief introduction to the structure of membrane proteins is required. Since soluble proteins are surrounded by water, their entire solvent-exposed surface is hydrophilic; otherwise aggregation would occur (Fig. 3A). In marked contrast, membrane proteins

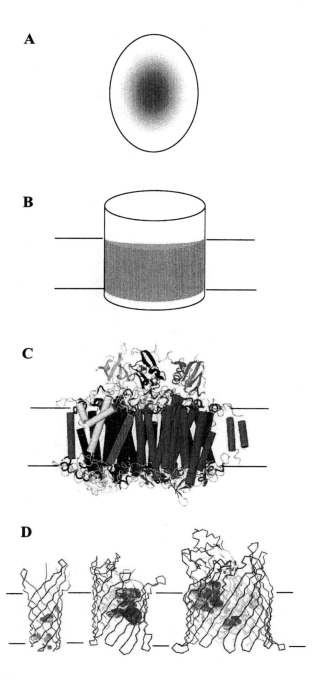

Figure 3 Structure of transmembrane proteins. (A) Soluble proteins, although they might have a hydrophobic interior (depicted in gray-black), only expose hydrophilic surfaces. (B) Transmembrane proteins have a surface of mixed hydrophobicity. Indeed, regions of the protein surface (depicted in gray) must interact with the acyl chains of the lipids while other regions of the surface must interact with the intracellular or lumenal aqueous environment (depicted in white). (C) Photosystem I, an example of a transmembrane α-helical protein complex (from reference 2 with permission). (D) Side views of bacterial outer membrane proteins OmpA, OmpF, and FhuA (from left to right) (from reference 4 with permission).

must interact, on one hand, with the lumenal or extracellular and the cytoplasmic milieus, which are aqueous and therefore hydrophilic, and, on the other hand, with the acyl chains of the lipids, which are hydrophobic (Fig. 3B). This particular property of membrane proteins makes them very difficult to handle and explains why so few structures are known despite the fact that about 30% of the proteins encoded by the human genome are membrane proteins. Despite the limited number of known structures, it clearly appears that all membrane proteins fall into one of two major structural classes.

In members of the first class, the transmembrane region is composed of individual or bundled hydrophobic helices, as illustrated in Fig. 3C for photosystem I, one of the multiprotein complexes involved in the conversion of light into useful energy forms. Most eukaryotic membrane proteins, with the possible exception of mitochondrial and chloroplast proteins, belong to this first class. Also, the translocation domains of a variety of toxins, such as diphtheria toxin, cross the membrane by this means, and it is possible that RTX toxins do as well.

In members of the second class, the transmembrane structure is a monomer, dimer, or trimer of β-barrels (Fig. 3D). This structure was first elucidated for the bacterial protein OmpF. The protein was found to span the membrane as a beautiful, regularly arranged β-barrel composed of 16 antiparallel β-strands. Since then, a variety of outer membrane bacterial proteins have been found to span the membrane as a β-barrel. The β-barrel is a highly modular structure since the number of strands can vary depending on the desired size of the pore, as illustrated in Fig. 3D. At present no mammalian protein has been found to use this motif to span the membrane, although it has been proposed that the membrane-interacting region of the acetylcholine receptor is a β-barrel. Also, mitochondrial and chloroplast proteins are likely to have porin-like structures, probably due to their bacterial origin. This latter β-barrel motif has been preferentially chosen by PFTs, and in particular by those discussed in this chapter. For a protein that must initially be soluble, this motif offers the great advantage that individual "transmembrane-to-be" segments are moderately hydrophobic, in contrast to transmembrane α-helices.

These two transmembrane folds, hydrophobic α-helix and β-barrel, are dictated by the necessity to form oriented hydrogen-bonded secondary structures. Indeed, partitioning of a peptide bond into a lipid bilayer is costly (1.2 kcal/mol). Hydrogen bond formation has been estimated to reduce the cost of peptide partitioning by about 0.5 kcal/mol. The cumulative effect of these relatively small (per residue) free energy reductions can be very large when hundreds of residues are involved, as is the case during pore formation by bacterial toxins (see below).

Two Well-Characterized PFT Families

Many PFTs form heptameric channels of 15 to 35 Å in diameter. These include *Staphylococcus aureus* α-hemolysin and *Aeromonas* aerolysin, as well as the protective antigen of the anthrax toxin, which resembles a PFT. The best-characterized members of this family, in terms of structure, are staphylococcal PFTs, the mechanistic features of which will therefore be detailed below. In contrast, CDTs make far larger pores formed of up to 50 monomers and leading to channels of up to 300 Å in diameter. The mechanism of membrane insertion is somewhat different from that of heptameric PFT, as discussed below.

Staphylococcal PFTs

Staphylococcus aureus secretes a variety of membrane-damaging toxins, including the α-hemolysin and the bicomponent leukotoxins, the active toxin of which comprises the combination of two similar subtype proteins. The elucidation of the structures of leucocidin F of Panton and Valentin in its soluble form (5) and that of the α-hemolysin in its heptameric form (8) has been a major breakthrough in our understanding of pore formation by heptameric β-barrel toxins.

The primary sequence of these toxins is very hydrophilic, with the exception of a central glycine-rich loop. Secondary structure predictions and circular dichroism studies had suggested a high degree of β-sheet structure, reminiscent of outer membrane bacterial porins, which was confirmed by the crystal structures.

Understanding the mechanism leading to pore formation is facilitated by describing the structure of the heptamer before that of the soluble monomer. The heptameric complex is a mushroom-shaped molecule of approximately 100 Å in height and in diameter (Fig. 4). A water-filled channel, 15 Å in diameter at its narrowest point, perforates the structure along the central sevenfold symmetry axis. The most striking feature of this complex is the transmembrane stem domain, which is an extremely regular 14-stranded β-barrel. Unlike bacterial porins, where all strands belong to the same polypeptide chain, the stem domain of the α-hemolysin pore is composed of seven antiparallel β-hairpins, contributed by seven different circularly arranged monomers (Fig. 4B). The soluble monomer has a very similar structure, with the exception of the loops connecting the secondary structure elements: the N terminus (in black in Fig. 4) and, most important, the prestem domain (in light gray in Fig. 4), which is folded into three short antiparallel β-strands with the hydrophobic residues positioned against the protein core.

The sequence of events resulting in pore formation is thought to be the following. The monomers bind, via specific or nonspecific interactions, to the target membrane. By lateral diffusion at the cell surface, monomers encounter and form a circular heptameric structure. At this stage, the prestem domain and the N terminus are still folded against the core of the corresponding monomer to shelter the small hydrophobic surface. This state has been called the prepore. The N terminus is then released from the core of the protein to interact with the neighboring monomer. Concomitantly, the stem domain unfolds, following an unknown trigger, and reassembles with neighboring stem domains into a β-barrel as membrane insertion occurs. A membrane partition-folding coupling mechanism is likely to take place upon contact of the α-hemolysin prepore with the bilayer, as shown upon folding and insertion of OmpA; i.e., membrane partitioning and folding are concomitant.

The barrel configuration fulfills the crucial requirement for a transmembrane protein, i.e., fully internally satisfied hydrogen bonding. In addition, alternating hydrophilic and hydrophobic residues form the β-strands of the stem domain, generating a hydrophobic exterior, in contact with the lipids, and a hydrophilic interior, the water-filled channel. It is important to note that since β-barrel formation involves the exposure of a hydrophobic surface, this event must occur as

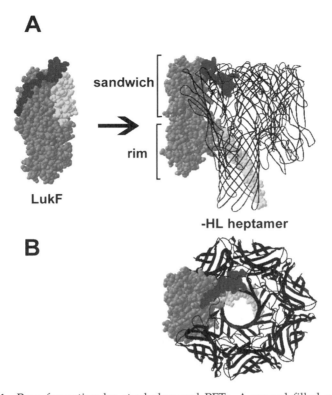

Figure 4 Pore formation by staphylococcal PFTs. A spaced filled model of the structure of leucocidin F illustrates the structure of monomeric α-hemolysin. The heptameric pore-forming complex is shown in a top (B) and a side (A) view in a ribbon diagram. One monomer is, however, shown as a spaced filled model to illustrate the conformational changes associated with the transition from water-soluble to transmembrane. Two regions of the protein undergo major and apparently concomitant conformational changes. The amino-latch (black) is initially folded against the core of the soluble monomer. In the heptamer it has folded out and interacts with the following monomer in the heptameric complex. Similarly, the stem domain (light gray) is initially folded onto the core of the protein to hide hydrophobic residues. In the heptamer this region has completely unfolded away from the rest of the protein into a curved β-hairpin, forming with the same region in neighboring monomers a closed circular β-sheet (from reference 3 with permission).

close as possible to the target membrane. Premature barrel formation would indeed otherwise lead to aggregation of the complex and thereby inactivation.

CDTs

CDTs form a large family of homologous proteins of least 23 members, the best characterized from a structural point of view being streptolysin O and perfringolysin O, although pneumolysin O has also received much attention. The channels they form differ substantially from those formed by staphylococcal PFTs since the pores can reach diameters of 300 Å and include up to 50 monomers. Although the pore has not been crystallized, it appears that, as for α-hemolysin, the final structure that perforates the target membrane is a β-barrel. The barrel is, however, far larger not only because more monomers form the oligomer but also because, in contrast to α-hemolysin, each monomer contributes two β-hairpins instead of one. The other peculiarity of CDTs, which led to their name, is that they only form pores in membranes that contain cholesterol. It has, however, never been unambiguously shown that cholesterol is the CDT receptor. It is equally possible that these proteins have a proteinaceous receptor and that cholesterol is required to trigger a required conformational change.

The soluble crystal structure of perfringolysin O shows that the molecule is L-shaped and can be divided into four domains as shown in Fig. 5 (6). At the surface of the target cell, CDTs undergo oligomerization when a threshold concentration is encountered, as for staphylococcal PFTs. In this prepore state, the transmembrane regions "to be" are still in a preinsertion state. The strategy adopted by CDTs to hide the hydrophobic surfaces of the β-hairpins is totally different from that of staphylococcal PFTs. In the water-soluble form, each pre-stem domain is folded into three short α-helices in domain 3 (Fig. 5B) (7). At the appropriate moment, each group of three helices extends into a β-hairpin. Since insertion of a single β-hairpin into the membrane is energetically unfavorable due to the presence of unsatisfied hydrogen bonding, this event probably occurs only when the ring structure is completed.

One of the mysteries that remain is what happens to the lipids that must be displaced to create a pore. This is a particularly striking problem for CDTs, as illustrated in the top panel of Fig. 5A. The surface occupied by a CDT pore of 300 Å in outer diameter is $\approx 71,000$ Å^2. In a membrane composed of a 1:1 mol: mol cholesterol-phospholipid mixture, this corresponds to approximately 800 phospholipid and 800 cholesterol molecules per leaflet, i.e., 3,200 lipid molecules in total. How these lipids are displaced is totally unknown. The same problem, although less crucial, exists for staphylococcal α-hemolysin-type PFTs, where only about 60 lipid molecules should be displaced.

The Crucial Role of the Toxin Receptor

For the two above-described toxin families, little has been mentioned about the role of the cell surface receptors in channel formation because the receptors for staphylococcal PFTs have not been identified and the role of cholesterol as a receptor for CDTs is controversial. It is, however, clear that the presence of a receptor at the target cell surface is crucial for the toxicity. This step has been best characterized for aerolysin from *A. hydrophila*. Aerolysin does not recognize a specific polypeptide chain at the surface of the mammalian host but instead recognizes a posttranslational modification, a GPI anchor. This anchor is added, in the endoplasmic reticulum, to the carboxy terminus of newly synthesized proteins that bear a GPI-anchoring signal and targets them to the plasma membrane. Cells that lack GPI-anchored proteins are over a thousand times less sensitive to aerolysin, indicating the crucial role of the receptors. They not only allow the toxin to find its target cell, but they concentrate the toxin, thereby favoring heptamer formation. They do so in two ways. First, cell surface binding leads to

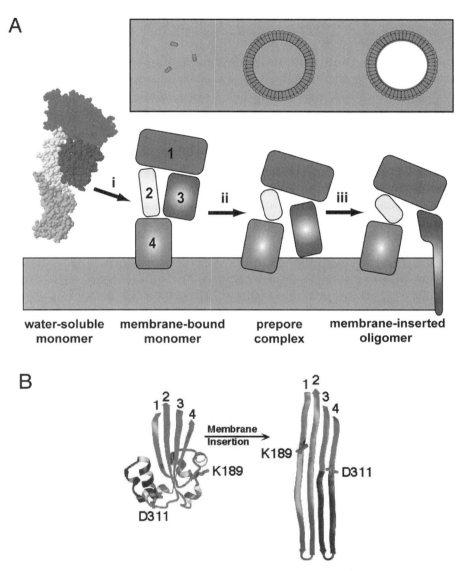

Figure 5 Pore formation by CDTs. (A) Interaction of cholesterol-dependent toxins with the lipid bilayer. CDT can be divided into four domains. Domain 4 interacts first with the membrane. This interaction triggers a conformational change in the molecule, resulting in membrane contact with domain 3. As shown in (B), unfolding of domain 3 occurs with a concomitant α-helix-to-β-sheet transition that leads to the insertion of the membrane. The inset shows a schematic top view of the channel. Monomers associate in a circular membrane. What remains a mystery is the fate of the lipids inside the ring-like structure. (B) Conformational changes in domain 3 of perfringolysin O that accompany pore formation. As shown on the left, domain 3 is composed by a four-stranded β-sheet that terminates with two bundles of three helices. It is proposed that, upon membrane insertion, each of these two bundles unfolds to prolong the β-sheet of domain 3 as illustrated on the right. This extension of the β-sheet would penetrate the membrane (from reference 3 with permission).

concentration due to the fact that the toxin moves from a three-dimensional space (the extracellular medium) to a two-dimensional space (the membrane surface). Second, once in the two-dimensional space of the membrane, the GPI-anchored aerolysin receptors further promote concentration of the toxin because of their ability to cluster in specific microdomains of the plasma membrane called lipid rafts. These are heterogeneities in the plasma membrane, highly enriched in cholesterol and glycosphingolipids as well as certain proteins. Rafts have been shown to be important in signal transduction and membrane trafficking events as well as during infection by certain bacteria and viruses or attack by certain toxins.

The role of rafts in promoting toxin oligomerization is most likely not specific to aerolysin but will probably apply to other toxins, PFTs, or other pore-forming proteins (1). A variety of toxins such as *Clostridium septicum* α-toxin, *Helicobacter pylori* VacA, and tetanus toxin indeed rely on GPI-anchored proteins for efficient intoxication. Others require cholesterol (CDTs) or sphingomyelin (*Vibrio cholerae* cytolysin), which are also major raft components. Further studies will, however, be required to confirm this raft-facilitated intoxication mechanism.

Consequences of Pore Formation in the Host Cell Plasma Membrane

Although the structural changes associated with pore formation have been extensively studied and a rather clear picture of the various mechanisms has emerged, the consequences of pore formation, in particular at the plasma membrane, have received little attention until recently. What is clear is that at low (physiological) toxin concentrations, cells, other than erythrocytes, do not rapidly swell and undergo osmotic lysis as had been supposed for a long time. Instead, they remain viable for several hours, during which a number of events occur that depend on the toxin, the toxin concentration, and the type of target cell. A number of the observations that have been made will be reviewed.

A variety of PFTs trigger the formation of large vacuoles in the cytoplasm of the host cell. In the case of VacA (Box 6), these were found to originate from late endocytic compartments because the toxin is endocytosed and impairs late endosomal membrane integrity. In contrast, vacuoles induced by aerolysin, staphylococcal α-hemolysin, *Serratia marcescens* hemolysin, and *V. cholerae* hemolysin are not acidic and still form after inhibition of the vacuolar proton pump, indicating that they do not originate from endocytic compartments. In the case of aerolysin it was unambiguously shown that vacuoles originated from the endoplasmic reticulum. The process required membrane dynamics, but there is no evidence indicating that the toxin is actually transported to this compartment. It could, therefore, be a secondary consequence of pore formation.

A variety of PFTs were also shown to trigger the release of calcium from intracellular stores (e.g., staphylococcal PFTs, aerolysin, and streptolysin O). By unknown mechanisms, these toxins lead to activation of G proteins, production of inositol(1,4,5)-triphosphate, and opening of calcium channels in the endoplasmic reticulum. Pore formation in the plasma membrane also allows entry of extracellular calcium. What the direct consequences of increased cytoplasmic calcium are is not clear at present.

BOX 6

The vacuolating toxin VacA of *H. pylori*

H. pylori produces a toxin, VacA (vacuolating toxin A), as one of its major virulence factors. Research on the mechanism of action of VacA has generated a complex and puzzling scenario. VacA was recently found to form small anion-selective, voltage-dependent channels in biological membranes at acidic pH, hence its presence in this chapter on PFTs. As other PFTs do, it forms ring-like oligomeric (hexamer and/or heptamer) structures. Its most striking property is that it triggers the formation of large cytoplasmic vacuoles that originate from late endocytic compart-

ments. The currently proposed model is the following: The toxin binds to specific receptors on the target cell via its 58-kDa carboxyl-terminal domain. The monomer inserts into the plasma membrane and then assembles into oligomeric complexes leading to the formation of small anionic channels of low conductance. Endocytosis of the toxin leads to acid activation of the channel and permeabilization of endosomes to anions. This, in turn, leads to vacuolation of the compartment. VacA has, however, also been shown to affect mitochondrial function, trigger apoptosis, and target the cytoskeleton. Therefore, VacA might be an AB-type toxin in addition to being a PFT.

One consequence could be the onset of apoptosis. PFTs such as staphylococcal α-hemolysin and aerolysin were shown to trigger programmed cell death in certain cells such as lymphocytes, but not in others, as witnessed by the degradation of genomic DNA. As shown for aerolysin, toxin-induced apoptosis of lymphocytes could be prevented by overexpression of the antiapoptotic protein Bcl-2, suggesting activation via the intrinsic, mitochondrion-dependent death pathway. Activation of the mitochondrial apoptosis machineries, independently of death receptors, has recently been demonstrated for staphylococcal α-hemolysin.

Importantly, a variety of PFTs (e.g., staphylococcal α-hemolysin, aerolysin, and pneumolysin) were shown to trigger transcriptional activation of cells and production of proinflammatory mediators. This might be important for the pathogenesis of the producing bacteria. Further studies will be required to understand the underlying mechanisms.

REFERENCES

1. **Fivaz, M., L. Abrami, and F. G. van der Goot.** 1999. Landing on lipid rafts. *Trends Cell Biol.* **9:**212–213.
2. **Fromme, P., P. Jordan, and N. Krauss.** 2001. Structure of photosystem I. *Biochim. Biophys. Acta* **1507:**5–31.
3. **Heuck, A. P., R. K. Tweten, and A. E. Johnson.** 2001. Beta-barrel pore-forming toxins: intriguing dimorphic proteins. *Biochemistry* **40:**9065–9073.
4. **Koebnik, R., K. P. Locher, and P. Van Gelder.** 2000. Structure and function of bacterial outer membrane proteins: barrels in a nutshell. *Mol. Microbiol.* **37:**239–253.
5. **Pedelacq, J. D., L. Maveyraud, G. Prevost, L. Baba-Moussa, A. Gonzalez, E. Courcelle, W. Shepard, H. Monteil, J. P. Samama, and L. Mourey.** 1999. The structure of

a *Staphylococcus aureus* leucocidin component (LukF-PV) reveals the fold of the water-soluble species of a family of transmembrane pore-forming toxins. *Struct. Fold Des.* **7:** 277–287.

6. **Rossjohn, J., S. C. Feil, W. J. McKinstry, R. K. Tweten, and M. W. Parker.** 1997. Structure of a cholesterol-binding, thiol-activated cytolysin and a model of its membrane form. *Cell* **89:**685–692.

7. **Shatursky, O., A. P. Heuck, L. A. Shepard, J. Rossjohn, M. W. Parker, A. E. Johnson, and R. K. Tweten.** 1999. The mechanism of membrane insertion for a cholesterol-dependent cytolysin: a novel paradigm for pore-forming toxins. *Cell* **99:**293–299.

8. **Song, L., M. R. Hobaugh, C. Shustak, S. Cheley, H. Bayley, and J. E. Gouaux.** 1996. Structure of staphylococcal α-hemolysin, a heptameric transmembrane pore. *Science* **274:**1859–1866.

Bacterial Protein Toxins
D. Burns et al., Editors
©2003 ASM Press, Washington, DC 20036

Chapter 14

Membrane-Damaging Toxins:
Family of RTX Toxins

Camilla Oxhamre and Agneta Richter-Dahlfors

A century ago, Kayser noted that certain strains of *Escherichia coli* caused lysis of erythrocytes from different mammalian species. Today this feature (hemolysis) is considered a hallmark for a family of bacterial exotoxins termed RTX toxins that represents a specific subset of the larger family of pore-forming toxins. The RTX toxin family remains distinct from the pore-forming toxins discussed in the previous chapter in that the members of this family display several important characteristics that are not shared with other pore-forming toxins.

This family of toxins is expressed by a wide variety of gram-negative bacteria of clinical origin. The major cellular effect of RTX toxins has been ascribed to their pore-forming capacity, resulting in plasma membrane lesions and osmotic lysis. More recently, data have been presented showing that the action of several RTX toxins is biphasic. Thus, a low dose of the toxin does not cause cell lysis but rather affects intracellular signaling pathways in nucleated host target cells. Since this effect may be more relevant for the disease process, future characterization of the consequences of toxin-membrane interaction and pore formation on intracellular signaling processes may advance our understanding of the role of RTX toxins in the pathogenesis of toxin-expressing bacteria.

GENERAL INTRODUCTION TO RTX TOXINS

The term RTX, short for "repeats in toxin," originates from the early discovery that members of this toxin family shared a number of nonapeptide repeats that bind Ca^{2+}. Moreover, they share a common gene organization as well as structural features, and they utilize the same mechanism of extracellular secretion via type I transporters, which is described in chapter 2 of this volume. Despite these genetic and structural similarities, RTX toxins differ significantly in their target cell specificity, which forms the basis for subdivision of the RTX family (Table

Camilla Oxhamre and Agneta Richter-Dahlfors • Microbiology and Tumor Biology Center, Karolinska Institutet, S-171 77 Stockholm, Sweden.

1). One group of RTX toxins consists of hemolysins, such as *E. coli* α-hemolysin (HlyA) and *Actinobacillus pleuropneumoniae* (ApxIA), that are toxic for a wide range of cell types from various species including humans and ruminants. The other category embraces leukotoxins produced by *Actinobacillus actinomyce-temcomitans* (LtxA) and *Pasteurella haemolytica* (LktA), which display a more re-stricted target cell cytolytic activity.

HlyA has been implicated as one of the important virulence factors of uro-pathogenic *E. coli*, and the genes encoding this exoprotein are present in ap-proximately 50% of all pyelonephritogenic isolates of *E. coli*. Due to the large repertoire of tools for molecular and genetic analysis of *E. coli*, HlyA is by far the best-characterized member of the RTX family, and will, as such, serve as the main RTX model toxin in this chapter.

GENETIC ORGANIZATION AND EXPRESSION OF RTX

The operon structure and pattern of transcription are common for a majority of the RTX toxins. The operons normally consist of four linked genes, *hlyCABD*, where the *A* gene encodes the major structural portion of the toxin (Fig. 1). TolC is required for the unusual mechanism of toxin secretion, and this gene (*tolC*) is located distally to the Hly operon. Expression of the *E. coli* Hly operon is regu-lated by one additional gene, *hlyR*, that is located upstream of the *hlyCABD* operon. HlyR acts by stabilizing the mRNA transcript of *hlyA*, which ensures efficient translation of the toxin's structural gene.

E. coli HlyA (110 kDa) is synthesized as an immature and nontoxic protoxin (proHlyA), which must be activated by a posttranslational modification per-formed by HlyC (19.8 kDa) (Fig. 2). This occurs by HlyC-mediated direction of two covalently attached acyl chains to the internal lysine residues Lys-564 and Lys-690 in proHlyA using the cytosolic activating factor acyl-ACP as a donor of the acyl groups. Although little is known about the biochemical properties of HlyC, it is well established that the enzyme acts like an acyltransferase that is distinct from all other bacterial acyltransferases. *E. coli* strains lacking functional HlyC produce a nonhemolytic hemolysin, indicating that other acyltransferases present in *E. coli* are unable to substitute for HlyC. Two independent HlyC rec-ognition domains, FAI and FAII, have been identified in HlyA. They span ap-proximately 50 to 80 amino acids around each of the target lysine residues, which is larger than the recognition domain of many other acyltransferases. However, the intragenic location of the acylation sites in HlyA may explain HlyC's require-ment of a larger recognition topology. Correct acylation of HlyA is required to obtain full hemolytic activity of the toxin and to induce second messenger re-sponses in target epithelial cells.

Although other RTX toxins are matured by a mechanism similar to that of *E. coli* HlyA, there seems to be little primary sequence homology of FAI and FAII among the family members. Activation of the adenylate cyclase toxin (AC toxin) (Fig. 3) in *Bordetella pertussis*, the causative agent of whooping cough, requires fatty acylation performed by the HlyC homolog CyaC. Here, acylation occurs only at one site, Lys-983, which corresponds to Lys-690 in HlyA. When *cyaA* (which encodes AC toxin) and *cyaC* are coexpressed in *E. coli*, acylation of CyaA also occurs at Lys-860, which corresponds to Lys-564 of HlyA. Dual acylation of

Table 1 Members of the RTX family

Toxin	Bacterium	Cell-type specificity	Species specificity	Size (kDa)
HlyA	*Escherichia coli*	Erythrocytes and epithelial and other cells	Human, murine, bovine, and others	110
EhxA	*E. coli*	Erythrocytes and leukocytes	Bovine	107
MmxA	*Morganella morganii*	Erythrocytes and epithelial and other cells	Human, murine, bovine, and others	110[a]
ApxIA	*Actinobacillus pleuropneumoniae*	Erythrocytes and epithelial and other cells	Human, murine, bovine, and others	105
ApxIIA	*A. pneumoniae*	Leukocytes	Porcine	103
ApxIIIA	*A. pneumoniae*	Leukocytes	Porcine	120
LtxA	*Actinobacillus actinomycetemcomitans*	Leukocytes	Human	114
LktA	*Pasteurella haemolytica*	Leukocytes	Bovine	105
CyaA	*Bordetella pertussis*	Broad	Broad	200

[a]Sodium dodecyl sulfate-polyacrylamide gel eletrophoresis.

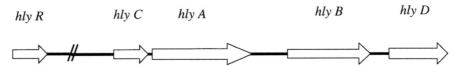

hly R   *hly C*   *hly A*   *hly B*   *hly D*

Figure 1 The Hly operon. The operon responsible for secretion of active HlyA from *E. coli* consists of four linked genes, *hlyCABD*. The gene *hlyA* (1,321 to 4,395 amino acids [aa]) is the structural gene encoding the immature and nonactive proHlyA. This immature form of the toxin undergoes posttranslational maturation by the action of *hlyC* (796 to 1,308 aa). Two genes, *hlyB* (4,468 to 6,591 aa) and *hlyD* (6,610 to 8,046 aa), encode membrane proteins that are involved in extracellular translocation of the hemolytic active HlyA toxin. Located approximately 1.5 kbp upstream of *hlyA* is *hlyR*, which acts as an enhancing sequence.

AC toxin results in reduced hemolytic activity as compared to the native, monoacylated form, which implicates that correct maturation and formation of the tertiary structure are required for the toxin to interact with the target cell membrane, thus providing its cytotoxic activity.

HlyB (80 kDa) and HlyD (54 kDa) are two membrane-bound proteins that participate in the transport of HlyA across the bacterial membranes. Expression of HlyB and HlyD is dependent on an antitermination event that is regulated by RfaH. This protein was originally described as a positive regulator of *rfa* genes that are involved in the assembly of the major bacterial outer membrane component lipopolysaccharide (LPS). Thus, RfaH may indirectly enhance the cytotoxicity of HlyA, since the presence of LPS has been reported to increase the lytic activity of HlyA.

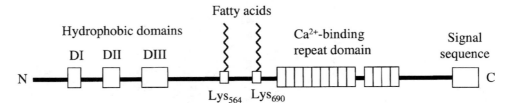

Figure 2 Linear model of HlyA. RTX toxins contain three functional domains. The hydrophobic region of HlyA is subdivided into three domains, DI to DIII (amino acids 238 to 410), which are involved in the formation of transient pores in target cell membranes. Inactive proHlyA is posttranslationally modified by covalent attachment of two fatty acyl chains to Lys-564 and Lys-690. The fatty acyl groups are required to achieve hemolytic activity. The Ca^{2+}-binding repeat domain is common in all RTX toxins. Upon binding of Ca^{2+} the repeat domain mediates conformational changes of the tertiary structure of the toxin and enables it to interact with the target cell membrane. The transport signal sequence, together with HlyB, HlyD, and the outer membrane protein TolC, is required for secretion of HlyA.

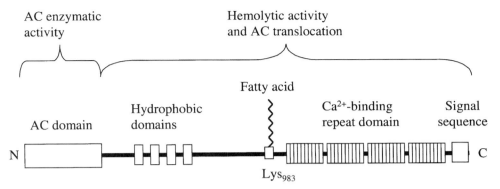

Figure 3. Linear model of AC toxin. Organization of the functional domains in CyaA resembles that of HlyA, e.g., a hydrophobic domain, a covalently attached fatty acyl chain, a Ca^{2+}-binding repeat domain, and an export signal sequence. In addition, CyaA contains an adenylate cyclase (AC) enzymatic domain located in the very N-terminal part of the protein. The hydrophobic domains of CyaA are involved in the hemolytic activity of the toxin and are required for the delivery of the AC enzymatic domain into the cytosol of the target cell.

FUNCTIONAL DOMAINS IN RTX TOXINS

Most RTX toxins contain three functional domains that are required for target cell recognition and hemolytic activity: (i) the fatty acyl groups, (ii) the hydrophobic domains, and (iii) the Ca^{2+}-binding repeat domain. The significance of each of these domains is described in the following sections.

Fatty Acyl Groups

It has been suggested that the acyl groups of the RTX toxins may function as anchoring points to the target cell membrane surface. Indeed, many membrane-associated proteins contain fatty acyl groups that increase the hydrophobicity of the protein, thereby improving their membrane-binding affinity. Both the number and the length of the fatty acyl groups differ among the RTX toxins. For example, mature HlyA contains mainly two myristic (C_{14}) fatty acyl chains, whereas AC toxin is monoacylated by one palmitoyl (C_{16}) chain. The acylation state may vary for the individual protein. Use of mass spectrometry analysis of protein isolated in vivo showed that HlyA is heterogeneously acylated with a mixture of saturated 14- (68%), 15- (26%), and 17- (6%) carbon amide-linked side chains. Originally, this posttranslational modification was used to describe the cell and species specificity of the toxin's activities. Later studies showed that the acylation state of the toxins is not sufficient to explain cell targeting, since both hemolysins and leukotoxins can be either mono- or biacylated. It has been calculated that the hydrophobic energy supplied by either myristic or palmitoyl fatty acyl chains is not sufficient to mediate a firm association between the toxin and the target cell membrane. Although the membrane affinity is strongly enhanced by incorporation of a second acyl chain, this is still not sufficient for complete membrane binding, probably due to the distance between the two groups. Thus, the acyl groups can be considered as two single chains, where each possesses too little hydrophobic force to mediate the binding alone. In the case of HlyA, the distance between the acylated leucines is 126 amino acids, and the toxin does not have clusters of hydrophobic or positively charged groups close to the sites' acylations that could improve the membrane affinity.

In an alternative model, it was proposed that the acyl chains are involved in the RTX toxin's pore-forming activities in target membranes. This model suggests that the hydrophobic force from the N-terminal domain of HlyA is sufficient to initiate the membrane interaction. Structural changes, resulting from Ca^{2+} binding to the repeat domain of the protein, aid in the protein's interaction with the hydrophobic interior of the lipid membrane. Reorientation of the fatty acyl groups causes invasion of some parts of the toxin molecule, resulting in the formation of a transient pore in the plasma membrane. Although no crystallographic data are available on the structure of mature HlyA, it appears that the most dramatic effect of the absence of the fatty acyl chains is on the toxin's ability to form membrane pores. Loss of one or both of the acyl groups in HlyA results in an almost completely abrogated hemolytic activity. However, the effect of low, sublytic concentrations of the toxin on induction of target cell signaling responses also requires the presence of the acyl chains, showing that the acyl chains provide specific features to the toxin that promote membrane interaction that does not necessarily lead to lysis.

Hydrophobic Domains

The hydrophobic domain of the otherwise hydrophilic HlyA protein can be subdivided into three continuous regions, DI to DIII, which are located between amino acids 238 and 410. This domain, which is predicted to be rich in β-turns, constitutes a very conserved region within the RTX toxin family and contributes to the pore-forming ability of these toxins. It was suggested that this hydrophobic stretch forms eight membrane-spanning α-helices, of which half are hydrophobic while the other half are amphipathic. Together, these α-helices create the membrane-spanning pore. Mutants of HlyA were created to evaluate the hemolytic function of the hydrophobic domains. Although these mutants carried small deletions in each subdomain, they were still able to bind to the plasma membrane. However, deletions in DII and DIII completely abolished the pore-forming capacity, while deletions in DI considerably reduced the activity. This experiment shows that the hydrophobic domains are of great importance for the hemolytic activity, but the exact mechanism is not fully understood. This study was recently supported by experiments showing that binding of Ca^{2+} to HlyA in solution induces conformational changes in the toxin molecule and that the hydrophobic region is inserted into the apolar core of the plasma membrane. It has been postulated that the very N-terminal region of RTX toxins forms amphiphilic α-helices that are capable of inserting into membranes; however, contrary to this idea, deletions in this area showed an improved hemolytic activity. These "improved" mutants form defined pores with higher conductivity and with a 50-fold longer lifetime compared to the wild-type HlyA. Monoclonal antibodies directed against different specific areas of the HlyA topology revealed that certain surface-exposed regions of the soluble toxin become unavailable to the antibodies in the membrane-bound phase. Only epitopes located at the far N terminus of the toxin and in the range of amino acids 594 and 640 remain accessible to the specific antibodies after HlyA had bound to the target cell. This suggests that a great proportion of the protein is embedded in the plasma membrane of the cell or that the interaction between the toxin and the target membrane involves huge alterations of the tertiary structure, thus preventing the epitopes from being recognized by the antibodies.

Calcium-Binding Domain

The characteristic feature of RTX toxins is a Ca^{2+}-binding repeat domain (GGXGXDXUX, where U represents a large hydrophobic residue and X represents any amino acid) located in the C-terminal part of the protein. The Ca^{2+}-binding domain is essential for the cytotoxicity of the proteins. Although the Ca^{2+}-binding repeat domain is present in all RTX toxins, the total number of repeats varies among the different toxins. HlyA contains 13 repeats of the consensus nonapeptide. It is most likely that the structure and function of this domain resemble those of the alkaline protease of *Pseudomonas aeruginosa*, which possesses a small number of these Asp-Gly-rich repeats. The crystal structure of *P. aeruginosa* alkaline protease reveals that the repeat domain contains β-sandwiches, where each repeat is organized in β-turns forming the two half sites for calcium binding. Proper binding is dependent on the correct radius of the ion, thus accounting for the selectivity for Ca^{2+}. Fluorescence analysis shows that binding of Ca^{2+} to this domain affects spectral shifts throughout the entire HlyA sequence, suggesting that Ca^{2+} binding mediates correct organization of the tertiary conformation, which enables the hydrophobic domains to adopt the structure of a functional pore within the membrane. This analysis contradicts some previous studies by suggesting that Ca^{2+} is not required for binding of the toxin to the plasma membrane, but only to induce the correct pore-forming structure of the toxin. The actual tertiary structure and the function of the repeat domain remain to be discovered.

The most diverse Ca^{2+}-binding domain of the RTX toxins is the one of AC toxin produced by *B. pertussis* (Fig. 3). AC toxin is a bifunctional toxin that exhibits both invasive (cytotoxic) and hemolytic (pore-forming) activities (see chapter 4). To enable these activities, AC toxin probably harbors two classes of calcium-binding repeats with different affinities for the ion. Binding of Ca^{2+} to a small number (three to five) of high-affinity binding sites appears to be critical for the membrane-binding ability and the hemolytic activity of the toxin, whereas binding to the low-affinity repeats is necessary for the cytotoxic activity of AC toxin. There are approximately 40 repeats of the low-affinity sites, which upon Ca^{2+} binding mediate major structural rearrangements of the toxin that may be involved in the translocation of the catalytic adenylate cyclase domain across the membrane of the target cell. Once intracellularly located, the toxin exerts its cytotoxic effect.

RTX INTERACTIONS WITH TARGET CELLS LEADING
TO MEMBRANE DAMAGE

It is possible that other components, bacterial or host cell encoded, act in synergism with the RTX toxin to enhance or specify its mode of action. One of the potential bacterial cofactors is LPS, whose participatory role in RTX activity for long has been discussed. The problem has turned out to be difficult to solve, since even the most rigorous purification method has until recently failed to completely remove all LPS from bacterial culture supernatants while retaining an active toxin. Studies on HlyA secreted from rough *E. coli* strains (lacking the O antigen) show a greatly reduced toxin activity compared to the hemolytic activity from clinical isolates expressing the O antigen (smooth strains). The molecular mechanism for this discrepancy is yet unknown. When LPS-free preparations of HlyA were tested for cytolytic activity, the toxin showed the same hemolytic activity against sheep red blood cells and Jurkat cells (a T-cell line) as bacterial culture supernatant containing both secreted HlyA and LPS. However, a 100-fold reduced sensitivity was observed when human renal epithelilal cells were exposed to the LPS-free form of the toxin as compared to LPS-containing HlyA preparations. This suggests that LPS enhances hemolytic activity against a subset of cells, like renal epithelial cells, while it is of less importance for lysis to occur in other cell types. This finding may be of utmost importance when considering the role of the toxin as a virulence factor and its function on the natural target cell during an in vivo infection. Indeed, uropathogenic *E. coli*, which frequently expresses HlyA, must cross the epithelial lining of the kidney before the bacterium even meets erythrocytes or lymphocytes, and it is not unlikely that the toxin plays different roles at different stages of infection.

Although the exact role of LPS in the cytotoxic mechanism of HlyA remains unclear, suggestions have been put forward that LPS forms a complex with the RTX toxin. Mutations in the LPS biosynthetic genes reduce the hemolytic activity, probably as a consequence of formation of large extracellular toxin aggregates. The reason for this is not evident, but LPS possesses calcium-binding sites that may act as extracellular reservoirs of calcium. An alteration in the LPS structure could reduce the availability of Ca^{2+} for the calcium-dependent toxin, resulting in an inactive and unstable HlyA that is prone to aggregate. An alternative model suggests that the direct contact between wild-type LPS and HlyA may prevent toxin aggregation, which thereby inactivates the toxin.

The restricted species and host cell specificity of RTX toxins suggests that some, or perhaps all, RTX toxins exert their cytotoxic effect by binding to target cell membrane receptors. By use of monoclonal antibodies directed against a specific cell surface receptor on the cancer cell line HL-60 (human leukocyte-60), the receptor for the leukotoxins LtxA produced by *A. actinomycetemcomitans* and LktA produced by *P. haemolytica* was identified and isolated. It was shown that these toxins bind to CD11a and CD18, which are subunits of the β-integrin leukocyte function-associated antigen LFA-1. This β-integrin is found in most circulating leukocytes but not in cells of nonhematopoietic origin; thus, the restricted cytotoxicity of the RTX leukotoxins LktA and LtxA can be explained by their ability to bind specifically to LFA-1-expressing cells. The host and cell-type specificity may further be restricted by the sequence divergence between the toxins and the different LFA-1 receptors. Alternatively, the toxins may recognize LFA-1-positive cells in combination with other host cell receptor molecules.

When attempting to identify receptors on target cells, it is important to design the model systems to properly mimic the in vivo conditions. The microbe's target cell type may differ at different stages of disease, as exemplified by uropathogenic *E. coli*. Initially, when bacteria ascend through the urinary tract, they interact with epithelial cells in the renal tubules. Once the mucosal lining is penetrated, bacteria may be exposed to various leukocytic cells in the interstitium, while erythrocytes are encountered once bacteria have reached the bloodstream. This is also reflected in the toxicity of HlyA, which, in contrast to the RTX leukotoxins, is toxic to a wide range of cell types including erythrocytes, leukocytes, and epithelial cells. The leukotoxin-receptor LFA-1 is also reported to act as a receptor for *E. coli* HlyA. This could explain the cytolytic activity that HlyA exerts on Jurkat cells in vitro but fails to do so for epithelial cells since they are of nonhematopoietic origin. One of the main surface-exposed proteins in erythrocyte membranes is glycophorin. Recently, it was published that erythrocytes pretreated with antiglycophorin antibodies were protected against the hemolytic activity of *E. coli* HlyA, indicating that glycophorin acts as an erythrocyte receptor for HlyA. It is quite probable that a different receptor is utilized on epithelial cells that awaits to be identified.

Little is known about the concentration of HlyA that is present during in vivo conditions. Whether expression of the toxin is differently regulated in vivo or whether cells are differently exposed at different sites in the urinary tract is unknown. It can be envisaged that in close proximity of bacterial microcolonies located on the luminal side of renal proximal tubule cells, the concentration of HlyA may be higher as compared to epithelial cells located distal to the microcolony. This will certainly affect the toxin's interaction with putative receptors, since low-affinity receptors may be utilized on cells where the toxin concentration is high.

LOW DOSES OF RTX TOXINS AFFECT INTRACELLULAR SIGNAL TRANSDUCTION PATHWAYS

Over the years, the action mechanism of RTX hemolysins on target cells has been considered as just punching holes in erythrocyte membranes. Despite massive research, we still do not fully understand the mechanism of pore formation. More recently, data have been presented that suggest that RTX toxins also affect intracellular signaling pathways without causing lysis of the cells. Thus, it appears now that the physiological effect of sublytic concentrations of the toxins is far more relevant to pathogenesis of diseases than originally believed.

E. coli HlyA Induces Intracellular Calcium Oscillations in Epithelial Cells

A novel feature of sublytic concentrations of HlyA on target cells is its ability to cause a continuous low-frequency oscillation of intracellular Ca^{2+} in primary renal epithelial cells from rat. Oscillations occurred without leading to any signs of cell lysis, whereas a 10-fold increase of the toxin rapidly lysed the cells. Both culture supernatants from *hlyA*-positive pyelonephritogenic isolates of *E. coli* and supernatants from strains where HlyA was expressed in a rough *E. coli* K-12 strain induced oscillations, suggesting a minor role of the LPS structure for the induction of oscillations. However, full acylation of HlyA was required, again pointing to the importance of the two acyl chains for a productive interaction with the target cell membrane. Although the exact mechanisms of the toxin-membrane interaction and the signaling pathways involved are not yet clear, the use of specific inhibitors showed that both L-type voltage-gated Ca^{2+} channels and inositol 1,4,5-triphosphate (IP_3) receptor-gated intracellular Ca^{2+} stores are involved to achieve the oscillatory pattern rather than influx of Ca^{2+} through a putative pore formed by HlyA. A suggested role of a particular receptor is further strengthened by the fact that phospholipase C, which normally is activated upon ligand binding to a G-protein-coupled receptor, is involved in the triggering of Ca^{2+} oscillations.

Spectral analysis revealed that HlyA-induced Ca^{2+} oscillations occur at a periodicity of 12 ± 0.7 min. Artificial induction of oscillations in the minute range have been shown to direct activation of different transcription factors as well as their downstream target genes. Thus, NF-κB is activated at different frequencies as compared to NF-AT and OCT/OAP. Since NF-κB plays a central role in regulation of inflammatory genes, the effect of HlyA-induced Ca^{2+} oscillations was tested for their role in activation of the proinflammatory cytokine interleukin (IL) 6 and chemokine IL-8, both of which were found to be activated. This result correlates with the clinical finding that women who suffer from acute pyelonephritis frequently show elevated urine concentrations of these two proinflammatory cytokines.

P. haemolytica Leukotoxin LktA Induces Apoptosis

Another RTX toxin that exhibits dual physiological functions is the leukotoxin LktA produced by *P. haemolytica*. This toxin is an important virulence factor for the establishment of bovine pulmonary infection. HL-60 cells exposed to high doses of the toxin cannot maintain osmotic equilibrium, resulting in influx of water that causes swelling and eventually bursting the cell. Today we also know that cells exposed to low doses of the toxin show features consistent with cells undergoing apoptosis. This involves typical mechanisms such as increased intracellular Ca^{2+} concentration, translocation of phosphatidylserine from the inner to the outer leaflet of the cell membrane, and internucleosomal fragmentation of DNA. The apoptotic process is inhibited if cells are pretreated with a monoclonal antibody directed against the β_2-integrin, showing that binding of LktA to the specific target cell receptor is required for induction of target cell apoptosis.

B. pertussis AC Toxin Induces Apoptosis

A final example of a bifunctional RTX toxin given in this chapter is AC toxin, which is secreted from *B. pertussis*. As mentioned previously, AC toxin exhibits both extracellular pore-forming ability as well as invasive adenylate cyclase activity. The hemolytic part of the toxin, which is similar to the pore-forming domain of *E. coli* HlyA, is responsible for cell targeting and pore formation of the plasma membrane. Similar to the LktA toxin, AC toxin exhibits a lytic activity when present in high concentrations, while lower toxin doses exert the cytotoxic effect via induction of apoptosis. Once the enzymatic domain of the toxin has entered eukaryotic cells, adenylate cyclase is activated by endogenous calmodulin to catalyze high-level synthesis of cAMP. Uncontrolled synthesis of this second messenger disrupts the physiological functions of the cell, thus leading to the cytotoxic phenotype of AC toxin and apoptosis of the host cell.

These findings collectively indicate that RTX toxins exhibit a dual, dose-dependent activity on the target cell. High doses of RTX toxins result in oligomerization and formation of large transmembrane pores that could be responsible for an extremely rapid destruction of the target cell membrane, providing no time for the host cell to defend itself by induction of the inflammatory response or apoptosis. Instead, disintegration of the target cell plasma membrane rapidly results in necrosis. However, at sublytic doses the RTX toxin may act via receptor-mediated mechanisms to induce less dramatic effects on the target cell. Increased life span of the target cell ensures time for induction of proinflammatory responses leading to neutrophil recruitment, as well as induction of apoptosis to avoid severe tissue damage as observed in necrotic lesions. Whether this is advantageous or disadvantageous for the bacteria remains to be determined.

ROLE OF RTX TOXINS IN DISEASE

RTX toxins are expressed from a variety of bacteria that cause diseases in the urinary tract, the respiratory tract, and the bloodstream. In addition, novel family members include toxins expressed by intestinal pathogens. Thus, RTX toxins may provide a selective advantage for pathogens that interact with target cells present in all mucosal surfaces of the human and animal body and in the bloodstream. Here, we review the role of two RTX toxins in the urinary and respiratory tract, respectively. Future research using improved techniques for in vivo analysis will enlighten us about their exact role in disease.

E. coli HlyA in Urinary Tract Infections

Uropathogenic *E. coli* (UPEC) strains are responsible for the vast majority of urinary tract infections in humans. Approximately 50% of the isolated UPEC strains have been found to express and secrete HlyA. Due to the combination of expressed bacterial virulence factors and host factors, bacteria may ascend to the upper urinary tract where they can establish pyelonephritis. If this infection is not properly treated, it may lead to irreversible renal failure and life long functional disorders. Previously, it was believed that the major virulence feature of HlyA was to lyse immune cells that were recruited to the site of infection, thereby promoting bacterial survival and multiplication. This may be true when bacteria have penetrated the epithelial lining, reaching the interstitium. Here, bacteria are no longer exposed to the flushing effect of urine, and bacterial multiplication can probably produce high enough concentration of the toxin for it to act cytolytic on recruited immune cells.

At the initial phase of infection, bacteria are located in the proximal tubules where they continuously are exposed to the flushing effect of urine. The concentration of secreted HlyA becomes diluted, probably leading to a low degree of toxin exposure to the epithelial lining. In this early phase of infection, HlyA-induced second messenger responses in epithelial cells may cause mobilization of host defenses toward the attacking bacteria by production of proinflammatory mediators. Thus, each of the dual activities of the toxin may be of importance at different stages of UPEC's disease process.

B. pertussis AC Toxin in Whooping Cough

Whooping cough is one of the most common causes of death from infectious diseases worldwide. The World Health Organization has reported approximately 50 million cases of whooping cough per year, among which unvaccinated individuals from developing countries are overrepresented. The role of individual virulence factors involved in the pathogenesis of whooping cough has been evaluated. Strains of B. pertussis that are unable to produce AC toxin exhibit drastically reduced ability to cause infection in animal models. At low toxin concentrations, AC toxin appears to be essential for bacterial colonization of the respiratory tract. Although the hemolytic domain and the enzymatic domain of AC toxin are equally required for virulence, it does not seem that the ability to lyse target cells is the major part of the infectious process. Instead, the invasive property of the toxin seems more relevant for target cell intoxication. AC toxin can invade phagocytic cells, in which the toxin causes uncontrolled intracellular production of cAMP. This leads to paralysis of the cell's killing functions. Also, AC toxin promotes apoptosis in phagocytic cells, perhaps as a mechanism to prolong bacterial persistence and to favor multiplication in the host. This suggests an early pathogenic role of AC toxin in the infectious process, using its activities to inactivate resident alveolar phagocytic cells.

SUGGESTED READING
1. **Andersen, C., C. Hughes, and V. Koronakis.** 2000. Chunnel vision. Export and efflux through bacterial channel-tunnels. *EMBO Rep.* 1(4):313–318.
2. **Ladant, D., and A. Ullmann.** 1999. *Bordetella pertussis* adenylate cyclase: a toxin with multiple talents. *Trends Microbiol.* 7(4):172–176.
3. **Ludwig, A., A. Schmid, R. Benz, and W. Goebel.** 1991. Mutations affecting pore formation by haemolysin from *Escherichia coli*. *Mol. Gen. Genet.* 226:198–208.
4. **Schindler, S., A. Zitzer, B. Schulte, A. Gerhards, P. Stanley, C. Hughes, V. Koronakis, S. Bhakdi, and M. Palmer.** 2001. Interaction of *Escherichia coli* hemolysin with biological membranes. A study using cysteine scanning mutagenesis. *Eur. J. Biochem.* 268:800–808.
5. **Stanley, P., V. Koronakis, and C. Hughes.** 1998. Acylation of *Escherichia coli* hemolysin: aunique protein lipidation mechanism underlying toxin function. *Microbiol. Mol. Biol. Rev.* 62(2):309–333.
6. **Uhlén, P., Å. Laestadius, T. Jahnukainen, T. Söderblom, F. Bäckhed, G. Celsi, H. Brismar, S. Normark, A. Aperia, and A. Richter-Dahlfors.** 2000. Alpha-haemolysin of uropathogenic *E. coli* induces Ca^{2+} oscillations in renal epithelial cells. *Nature* 405:694–697.
7. **Welch, R. A.** 2001. RTX toxin structure and function: a story of numerous anomalies and few analogies in toxin biology. *Curr. Top. Microbiol. Immunol.* 257:85–111.

Bacterial Protein Toxins
D. Burns et al., Editors
©2003 ASM Press, Washington, DC 20036

Chapter 15

Bacterial Toxins that Covalently Modify Eukaryotic Proteins by ADP-Ribosylation

Joseph T. Barbieri and Drusilla L. Burns

The discovery that diphtheria toxin inhibited eukaryotic protein synthesis by catalyzing the covalent transfer of the ADP-ribose portion of NAD to elongation factor 2 defined a new era in bacteriology, which focuses on the molecular basis for disease production. Since the discovery of ADP-ribosyltransferase activity of diphtheria toxin, a number of other bacterial toxins were shown to transfer the ADP-ribose moiety of NAD to eukaryotic proteins (see below).

$$NAD^+ + \text{target protein} \rightarrow \text{ADP-ribose-target protein} + \text{nicotinamide} + H^+$$

Each ADP-ribosylating toxin possesses a unique structure-function organization and ADP-ribosylates specific host proteins, eliciting a pathology that results from either the inactivation or activation of the target protein. The study of ADP-ribosylating toxins has advanced our understanding of bacterial pathogenesis and provided insight into the molecular basis of eukaryotic physiology, especially G-protein-coupled signal transduction.

BIOLOGICAL ACTIVITIES OF SEVERAL WELL-CHARACTERIZED ADP-RIBOSYLATING TOXINS

Diphtheria toxin of *Corynebacterium diphtheriae* and exotoxin A of *Pseudomonas aeruginosa* ADP-ribosylate elongation factor 2 and are the most cytotoxic ADP-ribosylating toxins. Introduction of a single molecule of the catalytic domain of diphtheria toxin into a eukaryotic cell is sufficient to inhibit protein synthesis and to promote cell death. The gene encoding diphtheria toxin is located within the genome of a bacteriophage that lysogenizes *Corynebacterium diphtheriae*. While nonlysogenic strains of *Corynebacterium diphtheriae* (dt^-) colonize the nasopharynx of susceptible humans, the infection remains localized and pathology is limited to the formation of a pseudomembrane composed of bacteria and host cellu-

Joseph T. Barbieri • Department of Microbiology, Medical College of Wisconsin, 8701 Watertown Plank Rd., Milwaukee, WI 53226-0509. *Drusilla L. Burns* • Food and Drug Administration, 8800 Rockville Pike, Bethesda, MD 20892.

lar material. *Corynebacterium diphtheriae* is not invasive and does not enter the peripheral tissue or the circulatory system. Likewise, lysogenic strains of *Corynebacterium diphtheriae* (dt^+) colonize the nasopharynx and produce the characteristic pseudomembrane, but they also secrete diphtheria toxin, which is transported throughout the body to produce a systemic pathology. Diphtheria toxin damages most organs, and fatality is often due to cardiac failure. Thus, diphtheria toxin does not contribute to colonization but is responsible for the pathology associated with the disease, a characteristic common to many bacterial pathogens that cause disease through the action of toxins.

ADP-ribosylating toxins can modulate host cell physiology without eliciting a cytotoxic pathology. Cholera toxin and pertussis toxin are the best-characterized ADP-ribosylating toxins that modulate eukaryotic cell physiology. Cholera toxin ADP-ribosylates the α-subunit of G_s and inhibits the intrinsic GTPase activity of this heterotrimeric G protein. ADP-ribosylated $G_{s\alpha}$ activates eukaryotic adenylate cyclase to produce elevated amounts of cyclic AMP (cAMP), which stimulates the secretion of electrolytes into the lumen of the intestine accompanied by the secretion of H_2O to maintain osmolarity. Ingestion of purified cholera toxin is sufficient to elicit the purging of liters of fluid. Cholera-mediated diarrhea does not contain blood cells, indicating that *Vibrio cholerae* does not invade and damage the intestinal tract. This watery diarrhea is the clinical hallmark of cholera.

Pertussis is a complex disease that results from the action of several virulence factors produced by *Bordetella pertussis*. The pathology associated with pertussis follows a classical toxin profile. *B. pertussis* colonizes ciliated epithelial cells of the lung and remains at the site of infection. During colonization, *B. pertussis* produces a number of biologically active agents that modulate host cell physiology, including pertussis toxin. Pertussis toxin ADP-ribosylates the α subunit of the heterotrimeric G protein, G_i or G_o. ADP-ribosylated $G_{i/o}$ does not respond to stimulation by its G-protein-coupled receptor, uncoupling signal transduction. Thus, the inhibitory component of the adenylate cyclase response is eliminated, which results in an attenuation of cAMP response to inhibitory agents and a potentiation of cAMP response to stimulatory agents of the adenylate cyclase system. The physiological outcomes of intoxication by pertussis toxin include histamine sensitization, modulation of insulin secretion, and stimulation of lymphocytosis.

A growing number of gram-negative and gram-positive bacteria have been shown to produce ADP-ribosylating toxins (Table 1). These ADP-ribosylating toxins possess conserved active-site organization and active-site residues, but differ with respect to the host proteins that are targeted for ADP-ribosylation and consequences of the ADP-ribosylation of the target protein. Each bacterial ADP-ribosylating exotoxin elicits a specific pathology for the intoxicated host.

PHYSICAL ORGANIZATION OF BACTERIAL ADP-RIBOSYLATING EXOTOXINS

Bacterial ADP-ribosylating exotoxins possess up to three independent domains (Fig. 1). The catalytic domain is referred to as the A domain. The receptor-binding domain, R domain, binds to receptors on the surface of eukaryotic cells, whereas

Table 1 Members of the family of bacterial ADP-ribosylating toxins

Toxin	AB	Target protein	Amino acid	Physiological effect
Diphtheria toxin and exotoxin A	AB	Elongation factor 2	Diphthamide	Inhibition of protein synthesis, cell death
Cholera toxin and heat-labile enterotoxin	AB_5	$G_{s\alpha}$	Arginine	Stimulate adenylate cyclase, diarrhea
Pertussis toxin	AB_5	G_i	Cysteine	Attenuate adenylate cyclase, multiple effects
C2 toxin	A-B bipartite	Actin	Arginine	Actin reorganization
C3 toxin	A	Rho	Asparagine	Inhibition of Rho GTPase, actin reorganization
ExoS	A	Ras GTPases and other proteins	Arginine	Uncouple Ras signal transduction
EDIN	A	Unknown	Unknown	Unknown

Figure 1 The three-dimensional structure of bacterial toxins provides insight into the organization and function of the AB domains. (Top) Diphtheria toxin. Although biochemical studies predicted that bacterial toxins had a discrete functional organization, solving the three-dimensional structure of several bacterial toxins verified that these domains were distinct but at the same time interactive. Examination of each toxin's crystal structure provided the framework to predict how the domains interacted and how each domain performed its biological function. Assessment of the crystal structure of diphtheria toxin provided insight into the organization and function of the translocation domain and how this domain interacted with the membranes of the endosome to transport the A domain into the cell cytosol. Adapted from reference 3. (Bottom) Heat-labile enterotoxin of *E. coli*. Although it was apparent that the A domains of cholera toxin and the heat-labile enterotoxin of *E. coli* were cleaved by proteases into A_1 and A_2 peptides, the physical relationships between A_1 and A_2 and the B_5 domain were not resolved solely by biochemical characterization. The three-dimensional structure of the heat-labile enterotoxin showed that the A_2 peptide interacted with the A_1 peptide and the B_5 domain and that the C-terminal residues of the A_2 peptide entered the central hole of the B_5 domain. This structure provided a framework to predict how the AB_5 toxins bind and enter the eukaryotic cell. Structures were obtained from the Protein Data Bank: 1SGK (from reference 1) and 1LTS (from reference 5), and edited with RasTool.

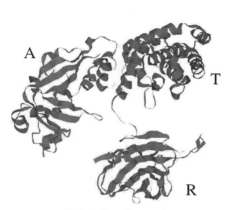

Diphtheria toxin

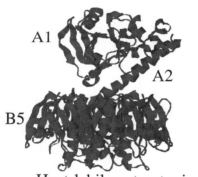

Heat-labile enterotoxin

the translocation domain, T domain, is responsible for the delivery of the A domain across a eukaryotic cell membrane into the cytosol of eukaryotic cells. Depending on the toxin, the point of translocation can be the plasma membrane or an intracellular vesicle membrane. Together, the receptor-binding and translocation domains are referred to as the B domain. Bacterial exotoxins exist in one of several physical organizations with respect to their A and B domains.

The A and B domains may be associated covalently within a protein, termed AB, as observed with diphtheria toxin. In diphtheria toxin, these functional domains are physically organized as sequential domains. The N terminus comprises the ADP-ribosyltransferase A domain, the internal domain contains the translocation T domain, and the C terminus includes the receptor-binding R domain.

Each domain has a unique physical organization that reflects its role in toxin action.

AB domains may also be associated noncovalently within a complex composed of a single A polypeptide and five B polypeptides termed AB_5, as observed with cholera toxin, the heat-labile enterotoxin of *Escherichia coli*, and pertussis toxin. In AB_5 toxins, the A polypeptide is divided into two regions, the N-terminal catalytic domain, which comprises ~190 amino acids, termed A_1, and a contiguous C terminus, which comprises ~25 amino acids, termed A_2. A_2 provides a noncovalent link between A_1 and B_5. B_5 polymerizes to form a donut-shaped structure that may be symmetrical (cholera toxin) or asymmetrical (pertussis toxin). In either case, B_5 makes multiple contacts with cell surface receptors. The significance of the hole in the center of B_5 remains speculative but appears to contribute to transport of the toxin during intracellular movement within the eukaryotic cell. In several AB_5 toxins, A_2 extends through the hole in the center of B_5 to potentially contact the eukaryotic cell membrane and intracellular trafficking molecules. The C-terminal sequence of A_2 domains can include a Lys-Asp-Glu-Leu (KDEL)-like sequence, which is an endoplasmic retention sequence for eukaryotic proteins.

In certain cases, the A and B domains of a toxin are synthesized as separate proteins that do not associate in solution. This class of toxin has been termed bipartite toxin or binary toxin; an example is the C2 toxin of *Clostridium botulinum*. During intoxication, C2B binds to the cell surface and is cleaved by a eukaryotic protease. C2A then binds to the processed surface-bound C2B, and this complex enters cells through the host's endocytic pathway.

Certain ADP-ribosylating toxins lack a B domain. C3, a toxin produced by *Clostridium botulinum*, was serendipitously identified in screens for strains of *Clostridium botulinum* that overproduced C2 toxin. C3 toxin is a small-molecular-mass A domain of approximately 26 kDa. Lacking a B domain, C3 toxin does not efficiently intoxicate cultured cells, but when introduced into cells by microinjection, proficiently ADP-ribosylates intracellular Rho. ADP-ribosylation of Rho results in a dramatic reorganization of the actin cytoskeleton. C3 toxin ADP-ribosylates Rho, but not Rac or Cdc42. This specificity makes C3 a useful pharmacological tool to study Rho-signaling pathways. Although the biological implication of C3 production by *Clostridium botulinum* is not obvious, *Staphylococcus aureus* and members of the genus *Bacillus* also produce C3-like toxins.

Similar to C3 toxin, ExoS of *Pseudomonas aeruginosa* is an A domain ADP-ribosyltransferase. However, ExoS is efficiently delivered into eukaryotic cells by the type III transport machinery of the bacterium. Type III cytotoxins do not possess defined B domains but include an N-terminal sequence of approximately 100 amino acids that is recognized by the type III apparatus.

TARGETS OF THE ADP-RIBOSYLATION REACTION

The pathology associated with an ADP-ribosylating toxin is due to alterations in the activity of specific eukaryotic proteins. To date, essentially all of the eukaryotic proteins that are ADP-ribosylated by bacterial toxins are nucleotide-binding proteins, most often GTP-binding proteins. The preference for ADP-ribosylating nucleotide-binding proteins may be due to the fact that these proteins share a common structural motif recognized by ADP-ribosylating toxins or that nucleotide-binding proteins play a strategic role in the regulation of host cell physiology, making each an ideal target.

Each ADP-ribosylating toxin modifies a specific subset of eukaryotic proteins, with an acceptor amino acid: arginine, cysteine, asparagine, or diphthamide. Diphthamide is a unique posttranslationally modified histidine found at a single position within elongation factor 2; no other eukaryotic protein possesses this modified amino acid. Diphthamide is targeted for ADP-ribosylation by diphtheria toxin and exotoxin A. Cholera toxin ADP-ribosylates the α-subunit of G_s at an internal arginine, Arg-176, whereas pertussis toxin ADP-ribosylates a cysteine located near the C terminus of the α-subunit of G_o and G_i. In contrast to the specific ADP-ribosylation catalyzed by these toxins, ExoS of *P. aeruginosa* ADP-ribosylates an arginine on numerous eukaryotic proteins in a eukaryotic cell lysate; however, its primary substrate appears to be members of the monomeric G protein family, including Ras. While most ADP-ribosylating toxins target a single amino acid for ADP-ribosylation, ExoS ADP-ribosylates Ras at two independent arginines, Arg-41 and Arg-128. Although ADP-ribosylation of target proteins at more than one site is unusual for bacterial ADP-ribosylating toxins, a eukaryotic ADP-ribosyltransferase has been reported to ADP-ribosylate actin at two independent arginines.

ADP-ribosylation can either activate or inhibit the physiological function of the modified protein. ADP-ribosylation of Arg-176 of $G_{s\alpha}$ by cholera toxin interferes with the expression of GTPase activity, placing $G_{s\alpha}$ in a constitutively active conformation. Arg-176 of G_s is an active-site residue, which is located within an arginine finger that coordinates the hydrolysis of the γ-phosphate from GTP. In contrast to the ADP-ribosylation of $G_{s\alpha}$ by cholera toxin, the ADP-ribosylation of the cysteine near the C terminus of $G_{i\alpha}$ by pertussis toxin prevents G_i from interacting with or recognizing its activated G-protein-coupled receptor, which prevents G_i from contributing to the development of an adenylate cyclase response by modulating ligands.

Toxin affinity for its target protein is typically within physiologic concentrations. With extended incubation, several toxins have been shown to completely ADP-ribosylate the intracellular pool of target protein, indicating that, over time, the entire pool of target protein becomes accessible for ADP-ribosylation. This time dependence factor was observed for elongation factor 2, because diphtheria toxin ADP-ribosylates elongation factor 2 that is in solution, but does not ADP-ribosylate elongation factor 2 when the target protein is complexed to the ribosome. The ability to ADP-ribosylate all available target protein implies that the modification is unique and that the host cell does not possess a mechanism to hydrolyze ADP-ribose from the modified target protein. In contrast, the eukaryotic ADP-ribosylation pathways appear to be cyclic, involving both the ADP-ribosyltransferase and an ADP-ribose hydrolyase.

In addition to modifying eukaryotic proteins, several toxins can ADP-ribosylate low-molecular-weight compounds. For example, pertussis toxin ADP-ribosylates a 20-mer peptide that corresponds to the primary amino acid of the C terminus of $G_{i\alpha}$, its natural target for ADP-ribosylation. As with native $G_{i\alpha}$, cysteine is the acceptor amino acid for ADP-ribose in this peptide. Cholera toxin can ADP-ribosylate agmatine. Agmatine is a neurotransmitter that shares structural properties with arginine. Although ADP-ribosylation of these low-molecular-mass targets does not appear to be physiologically significant with respect to intoxication, they are useful for in vitro characterization of the toxin's catalytic properties.

In the absence of a target protein, ADP-ribosylating toxins often utilize H_2O as the acceptor of ADP-ribose in an NAD glycohydrolase reaction as shown below.

$$NAD^+ + H_2O \rightarrow ADP\text{-ribose} + \text{nicotinamide} + H^+$$

Expression of NAD glycohydrolase activity does not elicit a pathological phenotype for eukaryotic cells. This shows that the pathology elicited by these toxins is the direct effect of ADP-ribosylating a host protein. Nonetheless, the NAD glycohydrolase reaction is a useful in vitro assay for analysis of kinetic properties. Both the K_m for NAD and the k_{cat} for NAD hydrolysis can be measured during the NAD glycohydrolase reaction. This allows an accurate determination of the effects of mutations that affect substrate-binding affinities and enzyme turnover rate. Kinetic studies on many ADP-ribosylating toxins have demonstrated the affinity for NAD in the ADP-ribosylation and the NAD glycohydrolase reactions to be similar, whereas the k_{cat} for ADP-ribosylation is greater than the k_{cat} for the NAD glycohydrolase reaction. One reason for the greater k_{cat} during ADP-ribosylation may be due to the nucleophilic properties and molecular alignment of the acceptor amino acids for ADP-ribosylation relative to H_2O, the ADP-ribose acceptor in the NAD glycohydrolase reaction.

CONSERVATION OF THE ACTIVE SITE OF ADP-RIBOSYLATING TOXINS

Diphtheria toxin and exotoxin A ADP-ribosylate elongation factor 2 with essentially indistinguishable enzymatic properties. Thus, it was unexpected that their deduced primary amino acid sequences contained only limited homology. Two key studies provided insight into how these toxins could catalyze identical reactions, despite having little amino acid homology. The first study identified the presence of a conserved active-site glutamic acid within diphtheria toxin and exotoxin A, whereas the second study showed that the three-dimensional structures of the A domains of bacterial ADP-ribosylating toxins were superimposable. Thus, despite having limited primary amino acid homology, ADP-ribosylating toxins possess structurally similar A domains that include the active-site glutamic acid.

ADP-Ribosylating Toxins Possess an Active-Site Glutamic Acid

The paradox that biochemical studies predicted molecular conservation of the active sites of bacterial ADP-ribosylating toxins, whereas subsequent DNA sequencing of the genes encoding these toxins failed to identify high amino acid homology, was unraveled by studies showing that several ADP-ribosylating toxins could undergo an unusual cross-linking reaction with NAD where UV light stimulated the chemical modification of NAD and one of the toxin's glutamic acids (Fig. 2). The decisive studies that identified the active-site glutamic acid

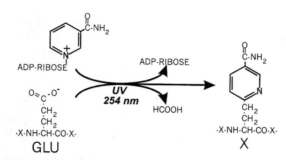

Figure 2 Photochemical cross-linking of the nicotinamide ring of NAD to a decarboxylated γ-carbonyl of the active-site glutamic acid of an ADP-ribosylating toxin. Adapted with permission from reference 2.

were performed on diphtheria toxin (2) and showed that UV light stimulated the decarboxylation of the R group of Glu-148 within the A domain of diphtheria toxin with the addition of the nicotinamide ring of NAD to the β-methylene carbon of the R group of decarboxylated Glu-148. This modification incorporated one mole of nicotinamide per mole of diphtheria toxin A domain and produced an enzymatically inactive protein. This finding implicated a role for Glu-148 in catalysis, which was confirmed during studies on recombinant forms of diphtheria toxin. These studies also showed that Glu-148 was not involved in binding NAD or target protein, but rather contributed to the catalytic rate of the ADP-ribosylation reaction. Subsequent studies showed that exotoxin A, C2 toxin, and pertussis toxin could also be covalently modified with NAD via this photochemical reaction. In each case, the chemical modification inactivated the ADP-ribosyltransferase activity of the respective toxin. This chemically modifiable, active-site glutamic acid has become a signature residue for members of the family of bacterial ADP-ribosylating toxins, as well as other proteins that catalyze the mono- or poly-ADP-ribosylation of proteins.

ADP-Ribosylating Toxins Possess Superimposable Active Sites (A Domains)

Exotoxin A and the heat-labile enterotoxin of *E. coli* were the first two ADP-ribosylating toxins to have their three-dimensional structures solved. Alignment showed that these toxins possess structurally similar ADP-ribosyltransferase domains despite having limited amino acid homology (6) (Box 7). This conservation of structure was not predicted, since the two toxins ADP-ribosylated different eukaryotic proteins and had different AB organizations. The conserved structures of the two active sites of exotoxin A and heat-labile enterotoxin are composed of seven noncontiguous peptide sequences within their A domains. In this alignment, only 3 of the 43 amino acids are conserved. However, one aligning amino acid was the signature catalytic glutamic acid that had been identified earlier by photochemical cross-linking experiments. Subsequent studies showed that the three-dimensional structures of active sites of diphtheria toxin and pertussis toxin could also be superimposed on the active sites of exotoxin A and the heat-labile enterotoxin.

Despite limited amino acid homology, ADP-ribosylating toxins have an ordered, sequential organization within three regions of the primary amino acid sequence that comprise their ADP-ribosyltransferase domain (Fig. 3) (4). Region

---

BOX 7

Discovery that the active sites of *P. aeruginosa* exotoxin A and the heat-labile enterotoxin of *E. coli* (LT) were superimposable

Solving the LT crystal structure was a lengthy process, and the A subunit took the longest. Once it was built, we wanted to find the active site. We had asked Dave McKay for the exotoxin A coordinates (this was in the days before the World Wide Web and immediate access to the Protein Data Bank), and I think they came on a tape. We looked at the models and decided that they both had two adjacent β-sheets, so I superimposed the central β-strands by eye on the graphics. This seemed amazingly good, and using the computer program Suppos, I could build up the superposition, including more and more strands, despite a virtual lack of sequence similarity. But one of the three conserved residues was the exotoxin active site glutamate 553, superimposing on glutamate 112 in LT

beautifully! This was exciting, and we concluded that it must be the active site in LT as well. When we later learned about the paper of Tsuji et al., which was out, but which we hadn't yet seen in those days before continuous MEDLINE access, we felt it merely confirmed what we had deduced from the exotoxin A comparison!

The conservation of structure between exotoxin A and LT surprised us, but the even greater structural similarity of the B subunits between LT and Shiga-like (or Vero) toxin was yet more impressive. Taken together with the verotoxin A subunits' similarity to ricin, it indicated to us the amazing evolutionary relationships in the AB toxin family, where similar subunits have been mixed and matched in an evolutionary puzzle.

1. **Tsuji, T., T. Inoue, A. Miyama, K. Okamoto, T. Honda, and T. Miwatani.** 1990. A single amino acid substitution in the A subunit of *Escherichia coli* enterotoxin results in a loss of its toxic activity. *J. Biol. Chem.* **265:**22520–22525.

Wim Hol

---

1 is located at the N terminus and includes a basic amino acid, either histidine or arginine. Region 2 is located within the central portion of the primary amino acid sequence and comprises a structural motif with a β-strand–α-helix organization. Region 3 is located in the C terminus and contains the signature catalytic glutamic acid that was identified by photochemical cross-linking experiments. The basic amino acid within region 1 is required for the expression of ADP-ribosyltransferase and is specific for each toxin. Current models predict that for some toxins this basic amino acid hydrogen bonds to the amino group on the nicotinamide ring of NAD to orient NAD for cleavage of the *N*-glycosidic linkage by the attacking nucleophile acceptor amino acid of the target protein. Since conservative mutations (arginine to lysine or histidine to arginine) produced recombinant proteins that were defective in the ADP-ribosylation reaction, spatial orientation of the respective R group also appears critical. Region 2 is located within a contiguous β-strand–α-helical structure. The primary amino acid sequence within region 2 is not completely conserved among the ADP-ribosylating toxins. Whereas region 2 for many of the ADP-ribosylating toxins and enzymes

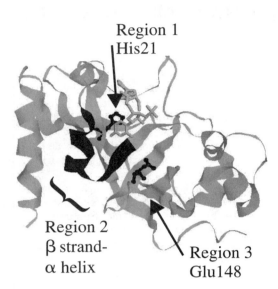

Region 1
His21

Region 2
β strand–
α helix

Region 3
Glu148

Figure 3 The primary amino acid sequences of the A domains of bacterial toxins include sequential regions of limited homology (regions 1, 2, and 3). Active site of diphtheria toxin cocrystalized with an NAD analog (shaded). Bacterial ADP-ribosylating toxins contain (in black): region 1 (a basic amino acid, either histidine or arginine at the N terminus), region 2 (a structural motif that is organized in β-strand–α-helical structure located within the central portion of the A domain), and region 3 (a catalytic glutamic acid, or diglutamic acid, located in the C terminus of the A domain). Structure (1DTP) obtained from the Protein Data Bank was deposited by M. S. Weiss and D. Eisenberg, and edited with RasTool.

includes a serine-threonine-serine sequence, region 2 of diphtheria toxin does not include this primary amino acid sequence, but contains the contiguous β-strand–α-helical structure. Region 2 contributes to the structural integrity of the NAD-binding cleft but does not directly interact with either NAD or target protein. Region 3 includes the catalytic glutamic acid that is conserved among all members of the family of ADP-ribosylating toxins. This may be represented as a single conserved glutamic acid, as observed for diphtheria toxin, or be present as a diglutamic acid organization (Glx-X-Glu), as observed for cholera toxin. Some ADP-ribosylating toxins that are included within the diglutamic acid group contain either glutamine or glutamic acid in the first position. The importance of the acidic nature of the R group and the spatial properties of this glutamic acid became apparent during the characterization of mutated forms of these toxins. Substitution of glutamic acid with either aspartic acid or serine resulted in a dramatic reduction in catalysis, and deletion of the glutamic acid produced a protein that lacked essentially all catalytic activity. This glutamic acid is proposed to make the *N*-glycosidic linkage of NAD more amenable to attack by the nucleophilic ADP-ribose-accepting amino acids of the target protein. For specific bacterial toxins, several other amino acids have been implicated in the catalytic reaction mechanism. Each residue appears to contribute to catalysis by making the active-site residues perform their functions more efficiently. Since these residues are specific for a given toxin, their function may reflect the nucleophilic amino acid that attacks the glycosidic linkage of NAD or the need to provide the correct spatial orientation to optimize the ADP-ribosylation reaction.

THE ADP-RIBOSYLATION REACTION MECHANISM

The stereochemistry of the ADP-ribosylation reaction involves the inversion of the 1′ carbon of ADP-ribose. Although the substrate, NAD, has a β-conformation at this site, the product of the ADP-ribosylation reaction possesses an α-conformation at the ribose ring. This inversion indicates that the ADP-ribosylation reaction follows an S_N2 reaction mechanism, where the attacking nucleophilic acceptor amino acid transiently associates with the leaving nicotinamide group at the 1' carbon of the ribose ring of NAD. Based on the crystal structure and biochemical characterization of native and mutated forms of ADP-ribosylating toxins, the following reaction mechanism has been proposed. Upon entry of NAD into the active site, the amino group on the nicotinamide ring hydrogen bonds to the basic amino acid in region 1 to stabilize and orient NAD. The catalytic glutamic acid extracts a hydrogen from the 2' hydroxyl on the ribose ring of the nicotinamide-linked ribose, which promotes the formation of an oxocarbonium-like intermediate. This intermediate weakens the N-glycosidic bond, making it susceptible to nucleophilic attack by the acceptor amino acid. In some toxins the attacking amino acid is made a better nucleophile by interaction with an adjacent basic amino acid, as observed in pertussis toxin.

Within the family of ADP-ribosylating toxins, the toxins exhibit a varied affinity for NAD. Some toxins, like diphtheria toxin, possess a high affinity for NAD, while others, like the heat-labile enterotoxin of *E. coli*, possess a low affinity for NAD. Currently, two models have been proposed for the observed differences in affinity between the two toxins. The first hypothesis proposes that the high affinity of diphtheria toxin for NAD is due to the presence of two tyrosines that stabilize NAD within the active site. The second hypothesis proposes that NAD has limited accessibility for the heat-labile enterotoxin of *E. coli*, since the crystal structure of this toxin identifies a short loop that obstructs accessibility of NAD to the active site.

ADP-RIBOSYLATING TOXINS ARE SYNTHESIZED AS PROENZYMES

ADP-ribosylating toxins are synthesized in an inactive form that must be modified for expression of enzyme activity. Two types of activation have been observed, covalent modification and association with eukaryotic proteins or cofactors.

Activation by Covalent Modification

The first indications that bacterial exotoxins were synthesized as proenzymes originated from studies on diphtheria toxin. Although purified preparations of diphtheria toxin were cytotoxic for cultured eukaryotic cells, they exhibited only minimal ADP-ribosyltransferase activity in vitro. Subsequent studies showed that two posttranslational modifications activated the expression of ADP-ribosyltransferase activity by diphtheria toxin in vitro, limited proteolysis and disulfide bond reduction. Limited proteolysis cleaves diphtheria toxin within an arginine-rich region (arginine-valine-arginine-arginine) that formed the junction between the A and B domains. Proteolysis produces a partially active toxin that could be further activated by addition of a reducing reagent, such as dithiothreitol. Disulfide bond reduction breaks the link between the A and B domains. Nicked and reduced diphtheria toxin efficiently ADP-ribosylates elongation factor 2 in vitro. Other bacterial toxins can be activated by limited proteolysis and disulfide bond reduction, including cholera toxin, which processes the A domain into A_1 and A_2.

Activation upon Association with Eukaryotic Proteins and Cofactors

In vitro activation of pertussis toxin, an AB_5 toxin, requires the addition of a reducing agent and a nucleotide, such as ATP. Addition of these factors reduces the affinity of the A domain for the B_5 oligomer, allowing expression of ADP-ribosyltransferase activity. It is not clear whether the activation by ATP reflects a need for ATP under in vivo conditions or whether ATP is a substitute for another agent found within the cell. In vivo, a processing event follows entry of pertussis toxin into eukaryotic cells.

Both cholera toxin and ExoS are produced as proenzymes that are activated upon interactions with the eukaryotic proteins ADP-ribosylating factor (ARF) and factor activating exoenzyme S (FAS), respectively. The activation of cholera toxin by ARF is a complex event, where ARF stimulates the ADP-ribosyltransferase activity but is not required for absolute expression of ADP-ribosyltransferase activity. In contrast, under physiological conditions, the binding of FAS to ExoS is an absolute requirement for the expression of ADP-ribosylation activity in vitro. Subsequent studies identified FAS to be a 14-3-3 protein. 14-3-3 does not modulate the affinity of ExoS for either NAD or target protein but increases the catalytic rate of the ADP-ribosyltransferase reaction. 14-3-3 proteins place ExoS in a more active conformation but do not contribute directly to the reaction mechanism. Characterization of these activation processes provided information on intoxication, the physiological role of ARF as a protein involved in the initiation of vesicle formation, and the 14-3-3 proteins, which are molecular scaffolding proteins involved in the bridging of proteins within the eukaryotic cell.

ADP-RIBOSYLATING TOXINS AS TOOLS TO PROBE HOST CELL PHYSIOLOGY

The study of bacterial toxins involves a sequential evolution of knowledge and strategies. First, the toxin is isolated and its catalytic and intoxication mechanisms determined; next, strategies are developed to determine whether the toxin is a useful vaccine candidate by empirically attempting to chemically or genetically inactivate the toxin while retaining its immunogenicity. The final stage in toxin research is to determine whether it can be used as a pharmacological reagent. Several toxins have proven to be useful pharmacological reagents, especially for the characterization of G-protein-mediated signal transduction. The key to using toxins to study eukaryotic cell biology is to define their mode of action, which ideally modifies a single or limited number of proteins within the cell. Pertussis toxin is one of the most frequently used reagents to implicate a role for G-protein-coupled signal transduction. Although pertussis toxin ADP-ribosylates hetero-trimeric G proteins of the G_i and G_o subclasses, G_i and G_o are rarely coexpressed in tissue. Thus, pertussis toxin can be used to implicate G_i or G_o signal transduction cascades. C3 toxin has also been used to implicate RhoA in eukaryotic signal transduction pathways. Rho GTPases modulate the organization of the actin cytoskeleton and include members of the Cdc42, Rac, and Rho GTPases. The utility of C3 toxin is that it specifically ADP-ribosylates the Rho subfamily of GTPases without modifying Cdc42 or Rac GTPases.

OTHER PROTEINS THAT CATALYZE THE ADP-RIBOSYLATION REACTION

Early recognition of ADP-ribosylation as an important enzyme reaction was limited to bacterial toxins. The discovery that bacterial toxins catalyzed the ADP-ribosylation reaction was facilitated by the fact that ADP-ribosylation caused dramatic changes in cell physiology and because the bacterial toxins often modified the entire pool of target proteins. Although the discovery of bacterial ADP-ribosylating toxins stimulated the search for eukaryotic proteins that catalyzed this reaction, direct evidence to support the existence of eukaryotic ADP-ribosylating enzymes has occurred recently. Indeed, eukaryotic proteins capable of catalyzing this reaction were purified and their genes cloned. The eukaryotic mono-ADP-ribosylating enzymes possess little overall homology with bacterial ADP-ribosylating toxins, but have regions 1, 2, and 3 that are conserved in ADP-ribosylating toxins. Several members of the family of viral and eukaryotic mono-ADP-ribosylating enzymes have now been identified. Some of these enzymes are ectoenzymes that are secreted into the culture medium or glycosylphosphatidyl-anchored to the cell membrane. One noteworthy example is RT6, a T-cell regulatory protein that was first predicted to be a mono-ADP-ribosylating protein through homology studies. Continued studies on the eukaryotic mono-ADP-ribosylating enzymes should provide new insight into how this posttranslational modification regulates cell physiology, as studies on bacterial ADP-ribosylating toxins have contributed to our understanding of bacterial pathogenesis.

MEASURING THE IN VIVO ADP-RIBOSYLATION OF EUKARYOTIC PROTEINS

Detection of phosphorylation as a posttranslational modification was facilitated by the ability of inorganic $^{32}P_i$ to be efficiently transported across a eukaryotic membrane. Thus, protein phosphorylation was readily detected as the incorporation of radioactivity into proteins upon the addition of $^{32}P_i$ to the culture medium. Detection of ADP-ribosylation as a posttranslational modification was impeded by the limited transport of NAD across the cell membrane, placing constraints on the ability to detect ADP-ribosylation in cultured cells. Two approaches have been successfully used to measure in vivo ADP-ribosylation of eukaryotic proteins by bacterial toxins. The first exploits the observation that, in many cases, bacterial toxins completely ADP-ribosylate the intracellular pools of target proteins and that hydrolysis of this modification is relatively slow in the eukaryotic cell. Thus, subtraction experiments can be performed on cell lysates that have been incubated alone or with toxin to measure the loss of ADP-ribosylable target proteins in lysates from cells treated with toxin relative to lysates from control cells. The second method to detect in vivo ADP-ribosylation of eukaryotic proteins by bacterial toxins follows the observation that the migration of an ADP-ribosylated protein by polyacrylamide gel electrophoresis is slower than that of the respective native protein. The utility of these techniques was illustrated during the measurement of in vivo ADP-ribosylation of the α-subunit of heterotrimeric G proteins by pertussis toxin. These techniques have been subsequently used for the measurement of the kinetics of in vivo ADP-ribosylation of eukaryotic proteins by other bacterial toxins.

CONCLUSION

Since the discovery that diphtheria toxin ADP-ribosylates elongation factor 2, the study of bacterial pathogenesis has progressed toward a more inclusive understanding of the mode of action of bacterial toxins. Progress in understanding the molecular and cellular properties of ADP-ribosylating toxin continues to provide a more complete understanding of their modes of action, with future studies addressing their interaction with the host that will provide new insights into the workings of the cell. Although ADP-ribosylation of eukaryotic proteins represents the first covalent modification attributed to a bacterial toxin, subsequent studies have identified additional covalent and noncovalent modifications catalyzed by bacterial toxins. The advances made concerning ADP-ribosylating toxins will provide a model for the characterization of these new families of bacterial toxins.

REFERENCES
1. **Bell, C. E., and D. Eisenberg.** 1997. Crystal structure of nucleotide-free diphteria toxin. *Biochemistry* **36**:481–488.
2. **Carroll, S. F., J. A. McCloskey, P. F. Crain, N. J. Oppenheimer, T. M. Marschner, and R. J. Collier.** 1985. Photoaffinity labeling of diphtheria toxin fragment A with NAD: structure of the photoproduct at position 148. *Proc. Natl. Acad. Sci. USA* **82**:7237–7241.
3. **Choe, S., M. J. Bennett, G. Fujii, P. M. Curmi, K. A. Kantardjieff, R. J. Collier, and D. Eisenberg.** 1992. The crystal structure of diphtheria toxin. *Nature* **357**:216–222.
4. **Domenighini, M., and R. Rappuoli.** 1996. Three conserved consensus sequences identify the NAD-binding site of ADP-ribosylating enzymes, expressed by eukaryotes, bacteria and T-even bacteriophages. *Mol. Microbiol.* **21**:667–674.
5. **Sixma, T. K., K. H. Kalk, B. A. van Zanten, Z. Dauter, J. Kingma, B. Witholt, and W. G. Hol.** 1993. Refined structure of *Escherichia coli* heat-labile enterotoxin, a close relative of cholera toxin. *J. Mol. Biol.* **230**:890–918.
6. **Sixma, T. K., S. E. Pronk, K. H. Kalk, E. S. Wartna, B. A. van Zanten, B. Witholt, and W. G. Hol.** 1991. Crystal structure of a cholera toxin-related heat-labile enterotoxin from *E. coli. Nature* **351**:371–377.

SUGGESTED READING
1. **Collier, R. J.** 1975. Diphtheria toxin: mode of action and structure. *Bacteriol. Rev.* **39**: 54–85.
2. **Kahn, R. A., and A. G. Gilman.** 1986. The protein cofactor necessary for ADP-ribosylation of G_s by cholera toxin is itself a GTP binding protein. *J. Biol. Chem.* **261**: 7906–7911.
3. **Tamura, M., K. Nogimori, S. Murai, M. Yajima, K. Ito, T. Katada, M. Ui, and S. Ishii.** 1982. Subunit structure of islet-activating protein, pertussis toxin, in conformity with the A-B model. *Biochemistry* **21**:5516–5522.

Bacterial Protein Toxins
D. Burns et al., Editors
©2003 ASM Press, Washington, DC 20036

Chapter 16

Glucosylating and Deamidating Bacterial Protein Toxins

Klaus Aktories

LARGE CLOSTRIDIAL CYTOTOXINS, A FAMILY OF GLUCOSYLATING TOXINS

The family of large clostridial cytotoxins comprises *Clostridium difficile* toxins A and B, the lethal and the hemorrhagic toxins from *Clostridium sordellii,* the α-toxin from *Clostridium novyi,* and various toxin isoforms mainly produced by *C. difficile.* The toxins are grouped together because they have molecular sizes exceeding 250 kDa, are glucosyl- or *N*-acetylglucosaminyltransferases, and share the same functional topology (Fig. 1). These protein toxins are single-chain toxins and are about 40 to 90% identical in their amino acid residues.

Pathological Activity of the Toxins

In the late 1970s it was found that *C. difficile* was the major pathogen responsible for antibiotic-associated diarrhea and for pseudomembranous colitis, a severe drug side effect, occurring under or subsequent to antibiotic therapy (4). Later it was recognized that *C. difficile* produced two potent toxins (toxins A and B) responsible for these complications. Toxin A and toxin B from *C. difficile* were originally termed enterotoxin and cytotoxin, respectively, because toxin A induced fluid accumulation, diarrhea, and colitis in animal models, whereas toxin B was most potent (100- to 1,000-fold more potent than toxin A) at inducing cytotoxic effects in cell cultures (5). At first, the minor enterotoxic effects of toxin B in animal models suggested that toxin A "cleared the way" for toxin B by damaging intestinal epithelial cells, allowing entry and accessibility of toxin B to sensitive cells. However, recent studies indicate that human colonic epithelial cells are sensitive to toxin B. These data support findings showing that certain *C. difficile* strains that do not produce toxin A cause diarrhea and colitis.

Klaus Aktories • Institut für Experimentelle und Klinische Pharmakologie und Toxikologie der Albert-Ludwigs-Universität Freiburg, Otto-Krayer-Haus, Albertstr. 25, D-79104 Freiburg, Germany.

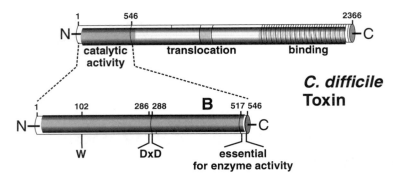

Figure 1 Scheme of the primary structure of *C. difficile* toxin B, the prototype of large clostridial cytotoxins. Toxin B consists of three domains: (i) N-terminal enzyme domain harboring glucosyltransferase activity; (ii) small hydrophobic domain in the middle of the toxin, which is most likely involved in toxin translocation; and (iii) C-terminal part consisting of polypeptide repeats suggested to be involved in receptor binding. The minimal length of the glucosyltransferase enzyme domain includes the 546 N-terminal amino acids. The conserved DXD motif is essential for transferase activity and likely to be involved in divalent cation and/or sugar nucleotide binding. The conserved Trp-102 is also important for nucleotide binding.

C. sordellii toxins are involved in diarrhea and enterotoxicemia in domestic animals and in gas gangrene in humans. Also, the α-toxin from *C. novyi* is a major cause of gas gangrene in humans and animals. Recently, the α-toxin was implicated as responsible for an increase in fatal outcomes after intravenous application of drugs by British addicts. These toxins, including toxin A, induce similar cytotoxic effects in cultured cells, stimulating retraction of the cell body and rounding up of cells (Fig. 2), where the actin cytoskeleton is drastically redistributed and the stress fibers disappear quite early (5). Most, if not all, of these cytotoxic effects appear to be caused by glucosylation and inactivation of Rho GTPases (10).

Enzyme Activities of Large Glucosylating Toxins

Large clostridial cytotoxins are potent agents and act at picomolar concentrations, which suggested that an enzyme activity was responsible for the observed pathology. However, the biochemical mechanism underlying the action of the toxins was enigmatic for years. To obtain an answer to the question of the underlying molecular mechanism, three important cell biological findings turned out to be essential. First, it was observed that Rho proteins, a subfamily of the superfamily of Ras proteins, are master regulators of the actin cytoskeleton in mammalian cells. Second, it was found that inactivation of Rho proteins (e.g., by bacterial C3 transferases or by negative mutants) induces gross changes of the cytoskeleton, also known to be elicited by the *C. difficile* toxins. Third, it was observed that treatment of cells with *C. difficile* toxins blocks the subsequent ADP-ribosylation of RhoA by C3 exoenzyme. This was most compelling to propose that Rho pro-

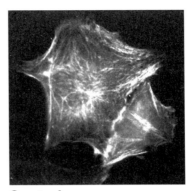

Control

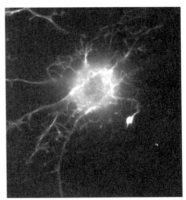

Figure 2 Cytotoxic effects of *C. difficile* toxin B on NIH3T3 fibroblasts.

Toxin B

teins are the eukaryotic targets of *C. difficile* toxins. Eventually, it was found that treatment with *C. difficile* toxins increased the size of RhoA by exactly 162 Da, indicating the attachment of a hexose residue onto the protein target (6). Further studies revealed that the toxins inactivate Rho GTPases by glucosylation of a functionally pivotal threonine residue:

$$\text{RhoA + UDP-Glucose} \xrightarrow{Toxin} \text{RhoA-Glucose + UDP}$$

With the exception of *C. novyi* α-toxin, which uses UDP-GlcNAc, all the large clostridial toxins use UDP-glucose as a substrate. The toxins split the activated nucleotide sugar and transfer only one sugar moiety onto the hydroxyl group of the acceptor amino acid threonine. Thus, they catalyze a mono-*O*-glucosylation or mono-*O*-*N*-acetylglucosaminylation of the eukaryotic target protein. In the absence of a protein substrate, the toxins exhibit glycohydrolase activity and split nucleotide sugars. The concentration of UDP-glucose in cells is ~100 μM and, therefore, not limiting for the glucosylation of target proteins. However, cell lines (e.g., Don cells) that are deficient in UDP-glucose are largely resistant to the toxins and regain sensitivity after microinjection of UDP-glucose, indicating the exclusive role of UDP-glucose for intoxication (e.g., by toxin B).

The Glucosyltransferase Domain of Large Clostridial Cytotoxins

Several findings indicate that the enzyme activity of large clostridial toxins is located at the N terminus (Fig. 1). In vitro the glucosyltransferase activities of the N-terminal 546-amino-acid residues of toxin B and lethal toxin are identical to those of the respective holotoxins, whereas additional deletion resulted in a dramatic loss of activity (about 1,000-fold reduction). Recently, some residues have been defined that are crucial for the enzyme activity of the N-terminal glucosyltransferase domain (3). A conserved DXD motif was identified that is conserved within all large clostridial cytotoxins and is also present in a great number of glycosyltransferases from prokaryotes and eukaryotes. Moreover, around the DXD motif, regions of high sequence similarity exist among these toxins. The DXD motif is essential for glucosyltransferase activity and also for glycohydrolase activity. This suggests a role of the DXD motif in catalysis and not in protein substrate recognition. Most likely, DXD is involved in the binding of the nucleotide sugar and/or of the divalent magnesium cation. Two findings support this view. First, mutation of the DXD motif causes loss of toxin labeling by azido-UDP-glucose and, second, a large excess of manganese ion partially compensates for the loss of activity due to mutations in the motif. It should be noted that the DXD motif is often found with α-retaining glycosyltransferases. The same type of reaction also holds true for large clostridial toxins. Thus, in contrast to ADP-ribosylating toxins, which generally cause inversion of the sugar acceptor amino acid bond (e.g., diphtheria toxin causes inversion from the α-anomeric ribose-nicotinamide bond to the β-anomeric ribose-diphthamide bond), glucosylation by toxin B results in retention of the α-anomeric sugar bond. In addition, Trp-102, which is located in a conserved region within the catalytic domain of all large clostridial cytotoxins, was important for nucleotide sugar binding.

The Family of Rho GTPases, Targets of Various Bacterial Protein Toxins

Rho GTPases are the preferred substrates of various bacterial toxins and effectors (Fig. 3). Therefore, this group of proteins will be briefly described. Mammalian Rho proteins comprise a family (>15 members) of small molecular size, monomeric GTPases, which act as molecular switches in various signaling processes (1). Rho, Rac, and Cdc42 subtypes and their respective isoforms are the best characterized. As Ras-related proteins, Rho GTPases cycle between an inactive GDP-bound state and an activated GTP-bound state. In the GDP-bound form, Rho GTPases are localized in the cytoplasm complexed to the guanine nucleotide dissociation inhibitor GDI. The Rho proteins are activated by extracellular signals (e.g., subsequent to activation of tyrosine kinase receptors or G-protein-coupled receptors) and dissociate from GDI, and GDP-GTP exchange is catalyzed by guanine nucleotide exchange factors (GEFs; more than 30 different exchange factors are known), resulting in the active GTP-bound form. In their active GTP-bound form, Rho GTPases interact with numerous effectors, which include serine/threonine kinases, lipid kinases, phospholipase D, and several adaptor proteins. Among the various described functions of Rho GTPases (e.g., control of secretion, endocytosis, transcriptional activation, cell cycle progression, transformation, and

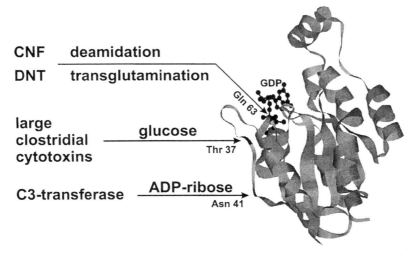

Figure 3 Rho GTPases are modified by various bacterial protein toxins. Large clostridial cytotoxins glucosylate Rho subfamily proteins at Thr-37 (RhoA) or at the equivalent Thr-35 (Rac, Cdc42). C3 transferases, including *Clostridium botulinum* C3 exoenzyme, ADP-ribosylate RhoA at Asn-41. CNFs from *Escherichia coli* and the DNT from *Bordetella* sp. deamidate and transglutaminate, respectively, Gln-63 of RhoA and Gln-61 of Rac and Cdc42 (2).

apoptosis), the involvement of Rho GTPases in the regulation of the actin cytoskeleton is best studied and appears to be of special importance. RhoA induces formation of stress fibers and focal adhesions in mammalian fibroblasts, whereas Cdc42 induces formation of microspikes and Rac leads to formation of lamellipodia and membrane ruffles. Conversion of the activated state of Rho proteins to the inactive GDP-bound form is catalyzed by GTPase-activating proteins (GAPs; again numerous [>30] proteins exist), which stimulate the GTP-hydrolyzing activity of Rho.

Glucosylation of Rho and Ras GTPases by the Large Clostridial Cytotoxins

Toxin B of *C. difficile* type 10463, the prototype large clostridial cytotoxin, monoglucosylates RhoA, B, and C at Thr-37. However, not only RhoA but also Rac and Cdc42 are substrates for glucosylation by toxin B (Table 1). Rac and Cdc42 are glucosylated by toxin B at the equivalent threonine, which is position 35. Also, toxin A from *C. difficile* and the hemorrhagic toxin from *C. sordellii* modify all three Rho family GTPases. However, the substrate specificity of large clostridial cytotoxins is not that simple. The lethal toxin of *C. sordellii* modifies Rac, and also Cdc42 to some extent, but not RhoA. In addition, the lethal toxin glucosylates Ras subfamily proteins including Ras, Rap, Ral, and R-Ras. These GTPases have different biological functions (e.g., Ras controls proliferation and differentiation) but also cross talk with signal pathways governed by Rho GTPases. The targeting specificity is even more complex, because various toxin B variants have been described that share the substrate specificity with the lethal

Table 1 Second substrates (cosubstrates) and protein substrates of the large clostridial cytotoxins[a]

Toxin	Cosubstrate	GTPase modified			Protein substrates		
Toxin A	UDP-glucose	Rho	Rac	Cdc42	–	(Rap)	–
Toxin B (10463)	UDP-glucose	Rho	Rac	Cdc42	–	–	–
Toxin B (1470)	UDP-glucose	–	Rac	–	(Ras)	(Rap)	Ral
HT	UDP-glucose	Rho	Rac	Cdc42	–	–	–
LT	UDP-glucose	–	Rac	(Cdc42)	Ras	Rap	Ral
α-Toxin	UDP-GlcNAc	Rho	Rac	Cdc42	–	–	–

[a] *C. difficile* toxin A (toxin A); *C. difficile* toxin B (toxin B) from strain 10463 (typical strain) and variant strain 1470; *C. sordellii* hemorrhagic toxin (HT); *C. sordellii* lethal toxin (LT); *C. novyi* α-toxin (α-toxin). Protein substrates given in parentheses are poor in vivo substrates.

toxin. Moreover, toxin A also has some specificity for Rap. In contrast to Rho/Ras GTPases, other low-molecular-mass GTP-binding proteins, e.g., from the Rab, Arf, or Ran subfamilies, are not modified by large clostridial cytotoxins.

Functional Consequences of the Glucosylation of Rho Proteins at Thr-37

Rho GTPases and other members of the Ras superfamily possess two loops, switch-I region (approximately amino acids 28 to 38 of RhoA) and switch-II region (amino acids 61 to 78), which undergo nucleotide-dependent conformational changes to define "on" and "off" states. Large clostridial toxins modify Rho/Ras proteins in the switch-I region by mono-O-glucosylating a specific threonine residue (Thr-37 in case of RhoA, which is equivalent to Thr-35 of Ras, Rac, or Cdc42). This threonine residue, which is highly conserved within the superfamily of Ras proteins, is essentially involved in coordination of the magnesium ion and nucleotide binding by the GTPase. The glucosylation site is within the "functional heart" of Rho GTPases, also called "effector region." In the active GTP-bound form of Rho, the hydroxyl group of Thr-37 participates in nucleotide binding and is located in the interior of the GTPase. In the inactive GDP-bound form, the hydroxyl group of threonine is directed to the solvent. In agreement with these conformational changes, GDP-bound Rho is a better substrate for glucosylation than the GTP-bound Rho. Note that glucosylation of Thr-37 by the large clostridial cytotoxins blocks the ADP-ribosylate Rho at Asn-41 by bacterial C3 transferases (Fig. 4).

What are the functional consequences of the glucosylation of RhoA (or Ras) by large clostridial cytotoxins? Probably the most important consequence of glucosylation is the inhibition of the interaction of Rho proteins with their effectors, which inhibits signaling mediated by the Rho proteins (Fig. 5). In addition, glucosylation inhibits the activation of Rho by GEFs. Further glucosylation targets Rho to the membrane. Thus, the cycling between the inactive cytosolic localization and the active membrane localization of Rho is blocked. Furthermore, glucosylation also inhibits the hydrolysis of GTP bound to Rho. Although this effect increases the half-life of the active state of Rho GTPases, it does not appear to have relevance for function, since the interaction of Rho with effectors is not

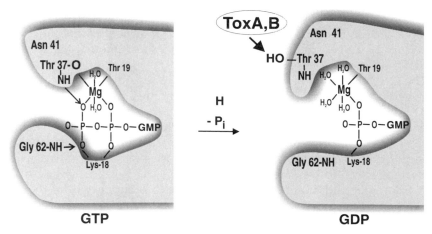

Figure 4 Toxins A and B from *C. difficile* modify GDP-bound RhoA. In GTP-bound RhoA, the hydroxyl group of Thr-37 coordinates a divalent cation directed into the protein and is not available for glucosylation by the toxins. In GDP-bound RhoA, the hydroxyl group of Thr-37 is directed to the solvent and is accessible for glucosylation by the toxins.

possible after glucosylation. Because the role of the threonine residue (e.g., Thr-37 of RhoA and Thr-35 of Ras), which is modified by glucosylation, is the same in all Rho/Ras family proteins, the glucosylation appears to cause the same consequences in all Rho substrates.

Figure 5 Functional consequences of the glucosylation of Rho GTPases. Inhibition of the interaction of glucosylated Rho with effectors is the most important effect of glucosylation. In addition, glucosylation inhibits Rho activation by guanine nucleotide exchange factors (GEF), the Rho cycling because glucosylated Rho locates at the membrane and does not bind to the guanine nucleotide dissociation inhibitor (GDI), and the stimulation of the GTP hydrolase activity by GTPase-activating proteins (GAP).

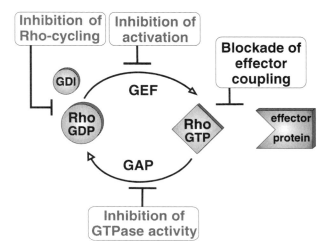

Binding and Uptake of the Toxins

The large clostridial cytotoxins possess three major structural and functional domains (Fig. 1): an N-terminal enzyme domain, a short internal hydrophobic region, and a C-terminal region.

The C-terminal region is composed of small repetitive oligopeptide motifs. Toxin A possesses 38 repetitive sequences containing 7 groups of 30 amino acids and 31 groups of 20 amino acid residues, whereas toxin B possesses 23 repetitive sequences containing 4 groups of 30 amino acids and 19 groups of 20 amino acid residues. These motifs show some similarity to structures of the carbohydrate-binding domain of glycosyltransferases from cariogenic *Streptococcus* species (*S. mutans* and *S. sobrinus*), suggesting a role for binding to carbohydrate structures at the eukaryotic cell membrane. This hypothesis is supported by the findings that the monoclonal antibody PCG-4, which blocks the binding and enterotoxicity of toxin A, interacts with these repeating units. In addition, toxin fragments that are missing the enzyme domain but include the C-terminal region block holotoxin intoxication and uptake. A minimal carbohydrate structure, consisting of the disaccharide Galβ1-4GlcNAc, binds toxin A. From studies with rabbit tissue, sucrase-isomaltase was suggested to be a receptor for toxin A; however, this enzyme is not expressed in most toxin A-sensitive cell lines. A human glycosphingolipid was described that binds to toxin A, supporting the possibility that the toxin has lectin-like binding capacity. Thus, the precise nature of toxin A remains to be elucidated, and the receptors for toxin B and the other glycosylating toxins are unknown.

Toxin translocation into cells occurs after binding and endocytosis of receptor-bound toxins from endosomes after their acidification, since inhibition of acidification by the vesicle proton pump inhibitor bafilomycin A blocks toxin translocation. Acidification may induce conformational changes of the toxin, causing exposure of hydrophobic regions that force the insertion into membranes. Recently, it was demonstrated that at low pH, toxin B forms pores. This was studied (after blocking the uptake from endosomes) by allowing toxin translocation through the cell membrane by acidification of the extracellular culture medium. Pore formation was suggested because eukaryotic target cells, which were preloaded with radioactive ^{86}Rb ions, released radioactivity after toxin B treatment of cells at low pH. Moreover, pore formation and increased ion permeability induced by toxin B were recognized in artificial membranes.

Glucosylating Toxins as Cell Biological Tools

The inactivating activity of large clostridial cytotoxin is widely used to study the participation of Rho GTPases in cell signaling. Cells or tissues are treated with these toxins and later tested to determine whether the signal or effector pathway studied remains active. For this purpose, a well-defined toxin should be used (e.g., toxin B from *C. difficile* 10463, which glucosylates Rho but not Ras subfamily proteins). Further characterization of the involvement of specific Rho GTPases is possible by application of additional toxins (e.g., C3 transferase to specifically ADP-ribosylate and inactivate RhoA, B, and C but not Rac and Cdc42). Further information about the signaling pathway can be obtained by using the lethal toxin from *C. sordellii* that modifies Rac but not Rho (note: lethal toxin also modifies Ras subfamily proteins). Through this regimen, Rho GTPases were implicated in the activation of mast cells via the high-affinity antigen receptor (FcεRI). These studies showed that Rac was essential for FcεRI signaling, because C3 transferase (used as a highly cell-accessible fusion toxin) had no effect on antigen-induced histamine release, whereas toxin B and lethal toxin blocked the signal-secretion coupling in mast cells.

METHODS TO STUDY GLUCOSYLATION OF RHO GTPases BY LARGE CLOSTRIDIAL CYTOTOXINS

Glucosylation of Rho GTPases in Cell Lysate or Recombinant Rho GTPases

To study the glucosylation of Rho GTPases by large clostridial cytotoxins in vitro, cell lysate is incubated in the presence of 30 μM of UDP[^{14}C]glucose with 1 μg/ml of toxin B for 45 min at 37°C. Thereafter, the proteins are separated by sodium dodecyl sulfate (SDS)–12% polyacrylamide gel electrophoresis (PAGE) and the ^{14}C-labeled proteins are analyzed by phosphorimaging for 6 to 24 h (for autoradiography, the exposure time is 5 to 14 days).

Toxin B Action in Intact Cells

Cultured cells are treated for 2 to 10 h with 10 to 100 ng/ml of toxin B (final concentration). Cell type-specific, morphological changes can be monitored. With phase-contrast microscopy, the cytotoxic features are characterized by cell shrinkage and formation of neurite-like extensions. Staining of the actin cytoskeleton, e.g., by rhodamine-phalloidin, shows the redistribution and disintegration of the actin cytoskeleton.

Differential Glucosylation

Differential glucosylation is used to measure the modification of Rho GTPases in intact cells. Toxin-modified Rho GTPase in intact cells blocks subsequent toxin-catalyzed labeling of Rho proteins in the cell lysate upon addition of UDP[^{14}C]glucose. Incorporation of [^{14}C]glucose into Rho GTPases is compared with controls and measured as described above.

DEAMIDATION AND TRANSGLUTAMINATION OF RHO GTPases BY BACTERIAL TOXIN

Covalent modification by bacterial protein toxins not only inhibits but also activates Rho GTPases. Members of the family of Rho-activating toxins include the cytotoxic necrotizing factors (CNFs) 1 and 2, which are produced by certain *Escherichia coli* strains, and the dermonecrotizing toxin (DNT) from *Bordetella* species.

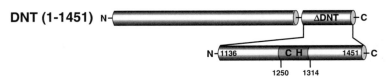

Figure 6 Structural comparison of CNF1 from *E. coli* with the DNT from *Bordetella* species. Alignment analysis identified a short region of high sequence identity at the C terminus. Indicated are the catalytic amino acids cysteine (C; Cys-1292 of DNT and Cys-866 of CNF1) and histidine (H, His-1307 of DNT and His-881 of CNF1). The minimal domain, possessing enzyme activity, is given as ΔCNF and ΔDNT, respectively.

CNF1 and -2 Deamidate Rho GTPases

In 1983, Caprioli and coworkers purified a toxin from *E. coli* isolated from a young patient with enteritis. The toxin caused necrotizing effects on rabbit skin and formation of large multinucleated cells and was termed CNF. Subsequently, CNF was identified in various *E. coli* strains isolated from domestic animals, and eventually, a second isoform called CNF2 was described. Both CNFs are single-chain toxins with molecular sizes of about 115 kDa, sharing more than 90% sequence similarity (Fig. 6). One difference between the toxins is the localization of their genes. Whereas CNF1 is chromosomally encoded, CNF2 is encoded by plasmids. Although their role as virulence factors is not clear, CNFs are toxic for a wide variety of mammalian cells, including CHO, Vero, HeLa, and Caco2 cells, and cause striking morphological changes (Fig. 6). CNF-intoxicated cells flatten and enlarge. Rapidly after intoxication (within a few hours), and often transiently, CNF-intoxicated cells form filopodia and membrane ruffles, while later (>6 h) a dense network of stress fibers is induced.

CNFs caused formation of filopodia, ruffles, and stress fibers, which suggested that Rho GTPases were involved in the action of the cytotoxins. This was supported by the findings that Rho proteins from CNF-treated cells exhibited an apparent larger molecular size in SDS-PAGE, indicating the covalent modification of Rho (Fig. 8). Important for the elucidation of the molecular mechanism of CNFs was the observation that the modification by CNF occurs with recombinant RhoA even in the absence of any additional factor, allowing a simple mass analysis of modified RhoA. Surprisingly, the difference in size turned out to be only 1 Da, resulting from deamidation of glutamine to glutamic acid (Fig. 8) (9). Thus, CNFs are deamidases, which specifically act on glutamine in position 63 of RhoA:

$$\text{RhoA-Gln-63} \xrightarrow{\text{CNFs}} \text{RhoA-Glu-63} + \text{NH}_3$$

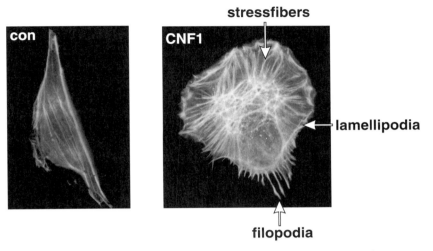

Figure 7 Cytotoxic effects of CNF1 on fibroblasts. CNF1 increases formation of lamellipodia, filopodia, and stress fibers.

Deamidation of Gln-63 of RhoA by CNFs

Gln-63 is essential for the GTPase activity of RhoA. Structure analysis suggests that Gln-63 stabilizes the transition state of GTP hydrolysis. Measurement of the GTP hydrolysis by CNF1-treated RhoA reveals an inhibition of the intrinsic and GAP-stimulated GTPase activity, indicating that deamidation of Gln-63 forms a constitutively active Rho protein. Thus, the "switch off" mechanism of the Rho GTPases is blocked by CNFs (Fig. 9). In fact, the morphological changes and

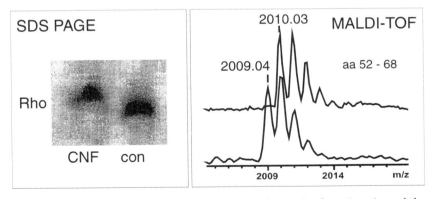

Figure 8 CNF treatment of RhoA induces a change in the migration of the GTPase on SDS-PAGE to an apparent larger molecular size. Deamidation of RhoA by CNF results in an increase in size of 1 Da. This can be analyzed by MALDI-TOF mass spectrometry. RhoA modified by CNF1 is digested by trypsin and subsequently the peptides formed are analyzed by mass spectrometry. One peptide, corresponding to amino acids 52 to 68, shows an increase in size of 1 Da, caused by deamidation of Gln-63.

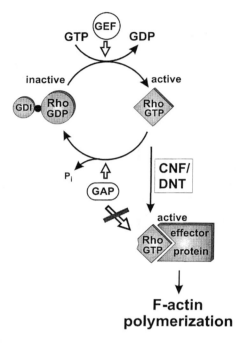

Figure 9 Deamidation or transglutamination of RhoA at Glu-63 by CNF and DNT, respectively, blocks the GTP hydrolysis stimulated by GAP. Rho-Q63E is constitutively active and induces activation of effectors, which eventually cause formation of stress fibers and a large array of other effects.

biological effects observed after CNF treatment can be explained by activation of Rho GTPases. Importantly, not only Rho but also Rac and Cdc42 are substrates of CNFs, with the deamidation of Rac and Cdc42 occurring at Gln-61. Modification of Rac and Cdc42 by CNFs and subsequent activation of the GTPases correspond with the observation that CNF-intoxicated cells show filopodia formation and membrane ruffling. Accordingly, CNF causes activation of Jun kinases, which is transient-like formation of membrane ruffles. Transient activation is most likely due to degradation of activated Rac protein. Blockade of Rac degradation by lactacystin causes permanent Jun kinase activation after CNF treatment, which is consistent with the constitutive activation of Rac by toxin-catalyzed deamidation of Gln-63. Thus, it appears that CNF-activated Rac is degraded by a proteasome-dependent pathway, thereby switching off the GTPase. Recently, it was shown that a small peptide, consisting of only 20 amino acid residues and resembling more or less the switch-II loop of RhoA, is a substrate for deamidation. However, Ras is not a substrate for CNFs.

Structure-Activity Analysis of CNF

Lemichez and coworkers suggested a structure of CNF consisting of the cell-binding component at the N terminus, a translocation domain with hydrophobic sequences in the middle of the protein, and the catalytic domain located at the C terminus. CNF appears to bind to a cell surface receptor and is taken up as an AB toxin, following the "diphtheria toxin" entry pathway. Accordingly, after endocytosis, receptor-bound CNF translocates from acidic endosomes into the cytosol. Whether the complete toxin or only the catalytic portion translocates is not clear. The putative binding region of CNF shows sequence similarity with the N terminus of *Pasteurella multocida* toxin, most likely sharing a similar fold of the domain involved in receptor binding.

A toxin fragment of amino acids 720 to 1,007 obtained from the C terminus of CNF possesses full deamidation activity but is not cytotoxic for cultured cells. After microinjection, this fragment induces the typical morphological changes induced by CNF holotoxin. In vitro, CNF possesses not only deamidating but also transglutamination activity. CNF is able to incorporate primary amines (e.g., putrescine, ethylenediamine, cadaverine) into RhoA at Gln-63, similar to eukaryotic transglutaminases, like tissue transglutaminase or factor XIII, which modify Rho GTPases. In general, transglutaminases are cross-linking enzymes that catalyze the exchange of the γ-carboxamide group of glutamine residues for primary amines (e.g., peptide-bound lysine residues or polyamines) to form ε-(γ-glutamyl)lysine or (γ-glutamyl)polyamine bonds. However, mammalian transglutaminases modify a large array of different eukaryotic protein substrates (e.g., fibronectin, actin, or casein), whereas the modification by CNF is specific for Rho GTPases. CNF and transglutaminases have little overt sequence homology but share a catalytic cysteine and histidine (Cys-866 and His-881 of CNF1), which are typical for transglutaminases or deamidases. The catalytic cysteine and histidine residues may form a thiolate-imidazolium ion pair, with the thiolate acting as the attacking nucleophile. Often transglutaminases and deamidases possess an additional catalytic aspartate as part of a catalytic triad, which is involved in orienting the histidine imidazole ring by hydrogen bonding. A catalytic aspartate residue of CNF has not been identified to date. The crystal structure of CNF1 suggests that the main chain carbonyl of Val-833 functions in place of a catalytic aspartate to align the catalytic His-881. The catalytic Cys-866 is located at the bottom of a deep pocket that most likely determines the target substrate specificity of CNF.

DNT

DNT, produced by *Bordetella bronchiseptica, Bordetella pertussis,* and *Bordetella parapertussis,* induces dermonecrotic lesions in mice and other laboratory animals. In pigs, DNT is a virulence factor for turbinate atrophic rhinitis. In cell culture, DNT induces enlargement of cells, polynucleation, and assembly of actin stress fibers similar, if not identical, to the CNFs. DNT is an ~160-kDa heat-labile protein that shares significant sequence similarity with CNF, which includes a small stretch of 260 amino acid residues (residues 1,160 to 1,420 of DNT). Sequences within this alignment are essential for enzyme activity in both CNF and DNT and include a catalytic cysteine and histidine. Accordingly, DNT also has deamidase activity and induces deamidation of Rho at position Gln-63, stimulating the formation of a constitutively active Rho protein in vitro and in intact cells. Similar to CNF, Rho, Rac, and Cdc42 are targets for DNT. However, differences exist between CNFs and DNT. One difference is the nucleotide dependency of the modification of Rho GTPases. The modification of Rho GTPases by DNT occurs predominantly with the GDP-bound form of Rho, while the modification of Rho by CNFs is not nucleotide dependent. Even more important is the difference in the enzyme activity. DNT possesses a high transglutaminase activity, whereas CNFs are primarily deamidases. DNT acts preferentially as a transglutaminase to attach primary amines onto Gln-63 of Rho GTPases:

$$\text{RhoA-Gln-63} + \text{NH}_2\text{-R} \xrightarrow{\text{DNT}} \text{RhoA-Gln-R1} + \text{NH}_3$$

In vitro, several primary amines are accepted as second substrates, e.g., ethylenediamine. In vivo, second substrates are polyamines like putrescine, spermine, and spermidine (8). Lysine may be an additional substrate for transglutamination of Rho by DNT. These primary amines, including lysine, are present in cells at high concentrations reaching the mM range. Like deamidation of Gln-63, transglutamination of Gln-63 inhibits GTP hydrolase activity of Rho proteins, although the precise function of the addition of primary amines onto Rho GTPases is not completely understood.

ASSAYS TO STUDY THE EFFECTS OF RHO-ACTIVATING TOXINS

Changes in Migration of Rho GTPases in SDS-PAGE

Deamidation and transglutamination by CNFs or DNT change the migration of GTPases by SDS-PAGE. This change in migration occurs with RhoA but not Rac or Cdc42 and depends on the deamidation of Glu-63 and, therefore, is also observed with the recombinant RhoA-Q63E. It is not clear why such a large change in the apparent molecular size (about 1-kDa upward shift) is observed, since deamidation increases the molecular size only by 1 Da. In contrast, transglutamination of RhoA with ethylenediamine as a cosubstrate causes a downward shift of the GTPase in SDS-PAGE. The extent of the downward shift depends on the second substrate and is not observed with dansylcadaverine.

Transglutamination with a Fluorescent Second Substrate

To screen recombinant proteins as toxin substrates, direct toxin-catalyzed labeling with the fluorescent amine derivative dansylcadaverine is possible. After incubation of Rho proteins with the toxins (CNFs or DNT) in the presence of high concentrations of dansylcadaverine (saturated solution) for up to 60 min at 37°C, the samples are applied to SDS-PAGE and the labeling analyzed under UV light before the gel is stained and dried.

Mass Spectrometry

Mass spectrometric analysis of proteolytic peptides of Rho GTPases allows the detection of the 1-Da shift caused by CNF1-induced deamidation of Gln-63 of RhoA (Gln-61 of Rac and Cdc42). Similarly, the mass shifts of 43 Da and 71 Da after treatment of the GTPase with the toxins in the presence of putrescine and ethylenediamine, respectively, are detected by mass analysis. For this purpose, Rho proteins are modified by the toxins and analyzed by SDS-PAGE. The proteins are stained, and the Rho protein band is excised. The excised gel plugs of RhoA are distained, dried in a vacuum centrifuge, and digested with trypsin for 12 h at 37°C. The sample is then subjected to matrix-assisted laser desorption ionization–time of flight (MALDI-TOF) mass spectrometry.

Measurement of Ammonia Released

Deamidation, as well as transglutamination, of Rho releases ammonia, which can be measured in a coupled enzymatic reaction based on the Ammonia Test Combination for Food Analysis. In the presence of glutamate dehydrogenase and NADH, ammonia reacts with 2-oxoglutarate to form L-glutamine and NAD^+. The decrease in NADH-fluorescence is a measure of ammonia release and can be monitored in a luminescence spectrometer.

Study of the Influence of Deamidation or Transglutamination on GTPase Activity of Rho Proteins

Deamidation and transglutamination of Gln-63 of Rho (Gln-61 of Rac and Cdc42) block the intrinsic and GAP-stimulated GTPase activity. Thus, toxin-induced modification of the small GTPases can be analyzed by measuring their GTPase activity. Rho proteins are loaded with $[\gamma^{-32}P]GTP$ in the presence of EDTA to first release GDP by chelating Mg^{2+}, which is essential for GTP binding by the Rho protein. Subsequently, excess of $MgCl_2$ enables Rho proteins to then bind $[\gamma^{-32}P]GTP$. GTPase activity as remaining $[\gamma^{-32}P]GTP$ is analyzed by filter binding assay.

REFERENCES

1. **Bishop, A. L., and A. Hall.** 2000. Rho GTPases and their effector proteins. *Biochem. J.* **348:**241–255.
2. **Boquet, P., P. Munro, C. Fiorentini, and I. Just.** 1998. Toxins from anaerobic bacteria: specificity and molecular mechanisms of action. *Curr. Opin. Microbiol.* **1:**66–74.
3. **Busch, C., and K. Aktories.** 2000. Microbial toxins and the glucosylation of Rho family GTPases. *Curr. Opin. Struct. Biol.* **10:**528–535.
4. **Farrell, R. J., and J. T. LaMont.** 2000. Pathogenesis and clinical manifestations of *Clostridium difficile* diarrhea and colitis. *Curr. Top. Microbiol. Immunol.* **250:**109–125.
5. **Fiorentini, C., and M. Thelestam.** 1991. *Clostridium difficile* toxin A and its effects on cells. *Toxicon* **29:**543–567.
6. **Just, I., J. Selzer, M. Wilm, C. Von Eichel-Streiber, M. Mann, and K. Aktories.** 1995. Glucosylation of Rho proteins by *Clostridium difficile* toxin B. *Nature* **375:**500–503.
7. **Lemichez, E., G. Flatau, M. Bruzzone, P. Boquet, and M. Gauthier.** 1997. Molecular localization of the *Escherichia coli* cytotoxic necrotizing factor CNF1 cell-binding and catalytic domains. *Mol. Microbiol.* **24:**1061–1070.
8. **Masuda, M., L. Betancourt, T. Matsuzawa, T. Kashimoto, T. Takao, Y. Shimonishi, and Y. Horiguchi.** 2000. Activation of Rho through a cross-link with polyamines catalyzed by *Bordetella* dermonecrotizing toxin. *EMBO J.* **19:**521–530.
9. **Schmidt, G., P. Sehr, M. Wilm, J. Selzer, M. Mann, and K. Aktories.** 1997. Gln63 of Rho is deamidated by *Escherichia coli* cytotoxic necrotizing factor 1. *Nature* **387:**725–729.
10. **Von Eichel-Streiber, C., P. Boquet, M. Sauerborn, and M. Thelestam.** 1996. Large clostridial cytotoxins—a family of glycosyltransferases modifying small GTP-binding proteins. *Trends Microbiol.* **4:**375–382.

Bacterial Protein Toxins
D. Burns et al., Editors
©2003 ASM Press, Washington, DC 20036

Chapter 17

Plant and Bacterial Toxins as RNA *N*-Glycosidases

Angela Melton-Celsa and Alison D. O'Brien

The knowledge that some plants or plant extracts are toxic to viruses, fungi, and animals or could be used as abortifacients began before modern recognition that those plants contained enzymes that inhibit protein synthesis. Indeed, the toxic factor in castor bean seeds, ricin, was identified in the late 1800s, but the definition of its enzymatic activity did not occur until nearly a century later. Ricin was known to be one of a large number of plant proteins that could injure host cells by somehow preventing the action of the host ribosomes. Thus, this group of plant proteins was called ribosome-inactivating proteins (RIPs). The RIPs were later defined as RNA *N*-glycosidases. Along with identification of the enzymatic function of RIPs came the recognition that certain bacterial toxins, called Shiga toxins (Stxs), also fit into this category of molecules. We now know that these plant and bacterial poisons have the broader specificity of polynucleotide:adenosine glycosidases, at least in vitro. This review will refer to the enzymes interchangeably as RIPs or RNA *N*-glycosidases.

ENZYMATIC ACTIVITIES OF THE RIPs

In 1987 Endo and colleagues demonstrated that ricin is an RNA *N*-glycosidase that cleaves an adenine residue (A_{4324}) in rat 28S rRNA. The deadenylation of the rRNA reduces or prevents interaction with the elongation factors and, thus, effectively halts protein synthesis in the target cell. The Stxs deadenylate the rRNA at the same site as the RIPs. The cleaved adenosine is found in a hairpin loop, often referred to as the sarcin/ricin or S/R loop, in which the sequence GAGA appears (Fig. 1). The first adenosine in the GAGA sequence is the substrate for the RIPs, whereas the *Aspergillus* α-sarcin ribotoxin cleaves after the second guanosine. The activity of the RIPs is often measured in vitro with rabbit

Angela Melton-Celsa and Alison D. O'Brien • Department of Microbiology and Immunology, Uniformed Services University of the Health Sciences, F. Edward Hebert School of Medicine, Bethesda, MD 20814-4799.

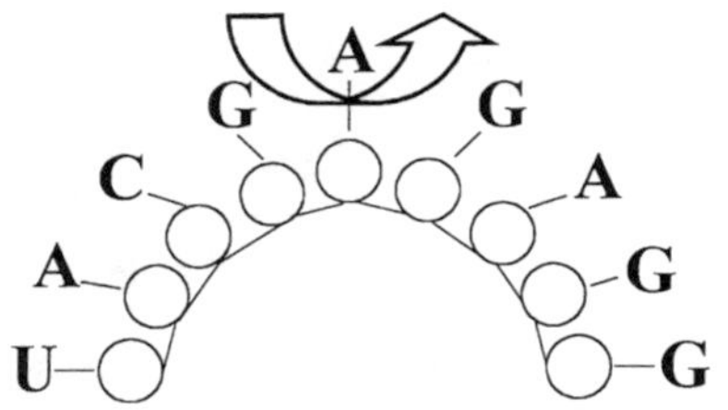

Figure 1 Schematic of a portion of the S/R loop in 28S rRNA that serves as the substrate for the RIPs. The RIP catalyzes release of the adenine residue from the loop without cleavage of the sugar backbone (shown as open circles) of the rRNA.

reticulocyte assays in which the incorporation of tritiated leucine allows quantitation of the protein synthesis inhibition capacity of the enzymes.

Recently, additional enzymatic activities have been attributed to the RIPs. However, when four different RIPs were highly purified, the only enzymatic activity they retained was the capacity to remove adenine from rRNA or DNA (1). This observation indicates that the RIPs, originally defined as RNA *N*-glycosidases, are actually polynucleotide:adenosine glycosidases, at least in vitro.

TYPES OF RIPs AND THEIR GENERAL STRUCTURES

There is a consensus in the field that the RIPs fall into two major classes (called type 1 and type 2 RIPs). However, some researchers argue that a third category should be established to reflect the unique structural features of certain atypical RIPs. The largest class of RNA *N*-glycosidases, the type 1 RIP group, consists of a single polypeptide chain that exhibits RNA *N*-glycosidase activity. Most type 1 RIPs have a signal sequence and a C-terminal extension that must be cleaved for maximal enzymatic activity. The type 1 RIPs are generally basic proteins with a molecular size of approximately 30 kDa. The first type 1 RIP to be identified was pokeweed antiviral protein (PAP). As early as 1925, PAP was found to have antiviral properties, but it was not shown to be a protein-synthesis inhibitor until the 1980s. Table 1 lists some type 1 RIPs.

In contrast to the single-chain type 1 RIPs, type 2 RIPs, such as ricin, have two polypeptide chains or subunits. These subunits include one polypeptide that has the RNA *N*-glycosidase activity and one polypeptide that has lectin-like activity. The enzymatic subunit of the type 2 RIPs is called the A chain. The second

Table 1 Examples of type 1 RIPs

Name	Source	Location
Asparin 1	*Asparagus officinalis* (asparagus)	Seeds
Bryodin 1	*Bryonia dioica*	Roots
Dianthin 30	*Dianthus caryophyllus* (carnation)	Leaves
Maize RIP	*Zea mays*	Seeds
PAP-R	*Phytolacca americana* (pokeweed)	Roots
Saporin-L1	*Saponaria officinalis*	Leaves
Saporin-R1	*S. officinalis*	Roots
Trichosanthin	*Tricchosanthes kirilowii* (Chinese cucumber)	Roots
Tritin	*Triticum aestivum*	Germ

subunit found in type 2 RIPs allows efficient entry into a target cell and is called the B chain. The presence of the B chain in the type 2 RIPs turns these RNA N-glycosidases into potent cytotoxins. Each subunit of the type 2 RIPs has an approximate molecular size of 30 kDa. The B chain of ricin, like that of most type 2 RIPs, has a nonspecific affinity for terminal galactose residues on target cells. The type 2 RIPs are synthesized as a single polypeptide that has a signal sequence and a small peptide linker separating the A and B domains. During processing of the protein the signal sequence and the peptide linker are removed. A single disulfide bridge connects the A and B chains. Table 2 lists some examples of type 2 RIPs.

A few RIPs, such as maize b-32, are synthesized as proRIPs from which an internal peptide linker must be removed before enzymatic activity can be detected. Some authors have classified such proteins as type 3 RIPs, while others prefer to treat them as specialized type 1 RIPs since the ultimate enzyme has a single chain. Barley plants produce yet another RIP, JIP60, that, similar to this latter group, must have an internal linker removed from the proenzyme. However, in JIP60 the linker separates the enzymatic domain from a domain of unknown function. Because JIP60 has this second domain of unknown function, it has been considered a different type of RIP by some authors. Future research will reveal if these rare examples of RIPs will be found in other plants.

The Stxs Are a Subclass of Type 2 RIPs

The Stxs can be classified as type 2 RIPs, but they are unique in several ways. The Stxs have an A polypeptide with N-glycosidase activity but, in contrast to type 2 RIPs such as ricin, have a pentameric B subunit that facilitates specific binding to the terminal galactose residues on globotriaosylceramide (Gb_3). The Stxs are generally encoded on phages or defective phages in *Shigella dysenteriae* or *Escherichia coli*. There are two groups of Stxs, Stx/Stx1 and the Stx2s. Stx was

Table 2 Examples of type 2 RIPs

Name	Source	Location
Plant		
Abrin	*Abrus precatorius*	Seeds
Ebulin[a]	*Sambucus ebulus*	Leaves
Modeccin	*Adenia digitata*	Leaves
Nigrin b[a]	*Sambucus nigra* (elderberry)	Bark
Ricin	*Ricinus commnis* (castor bean)	Seeds
Viscumin	*Viscum album* (mistletoe)	Leaves
Volkensin	*Adenia volensii*	Roots
Bacterial		
Stx[b]	*S. dysenteriae*	Chromosome[c]
Stx1[b]	*E. coli*	Phage or chromosome
Stx2	*E. coli*	Phage

[a] Ebulin and nigrin b are unusual, since they are not toxic to mice and cultured cells.
[b] Stx and Stx1 are essentially identical, possessing one amino acid difference.
[c] DNA surrounding the *stx* gene in *S. dysenteriae* appears to encode an inactivated phage.

originally found in *S. dysenteriae*, but an almost identical toxin, Stx1, was identified from *E. coli* in the 1980s. The Stx2 toxins are also found in *E. coli* and have the same mode of action as the Stx/Stx1 group, but have biological or immunological differences that are generally attributable to the B subunit of the toxin. The Stxs from *E. coli* have also been called verotoxins and were previously known as Shiga-like toxins. Even though Stx1 and Stx2 have approximately 56% sequence similarity, polyclonal antiserum raised against Stx1 does not cross-neutralize Stx2, and vice versa. Sequence variations in the B subunit of Stx2-type toxins have yielded variants in the Stx2 group, whereas the Stx1 group remains homogeneous. The Stxs are different from other type 2 RIPs in that they do not have a disulfide bridge that links the A and B subunits. Instead, the A subunit has an internal disulfide linkage with a protease-sensitive site between the linked cysteine residues that, when cleaved, separates the A chain into two domains. The N-terminal segment retains the RNA N-glycosidase activity and is called the A_1 subunit. The second, smaller segment of the A chain is called the A_2 peptide and serves to join the A_1 and B subunits. The disulfide bridge, in addition to other noncovalent forces, keeps the A_1 and A_2/B portions of the toxin together. The A_2 peptide further serves to block the active site in the A_1 subunit until the A_1 and A_2/B components are separated inside the target cell.

Localization of the RIPs

RIPs have been isolated from most plant organs and tissues. A particular RIP may be found at many sites in a single plant, and a plant tissue may contain more than one RIP, generally isoforms of the same RIP. Ricin, for example, is encoded within a multigene family that produces several isoforms, such as ricin D and ricin E, in the endosperm of seeds. In another example, dianthin 30 is found throughout the carnation plant but dianthin 32 exists only in leaves and shoots. The type 1 RIP PAP is localized primarily to the cell wall matrix within leaves, whereas saporins are found in intercellular spaces of both leaves and seeds. The Stxs are found associated with the bacteria or in the culture supernatant.

Expression and Glycosylation of the RIPs

The expression of some of the RIPs has been shown to be inducible. For example, jasmonic acid induces expression of JIP60, whereas the production of some other RIPs is enhanced when the plant is exposed to hydrogen peroxide or salicylic acid. The bacterial RIPs Stx and Stx1 are induced when iron is limited in the growth culture. Additionally, the expression of Stxs that are bacteriophage encoded increases when the prophage itself is induced, often in response to stress. Similarly, expression of some of the RIPs increases in response to factors such as wounding, heat shock, salt stress, osmotic stress, or the presence of microorganisms in the soil. A RIP may also be expressed temporally within the plant tissues or fluctuate with changing light levels.

Most plant RIPs, including ricin, are N-glycosylated. However, the sugar component of the enzyme does not appear to be required for enzymatic activity because when ricin is produced in *E. coli*, and thus not glycosylated, it is still active. Additionally, some RIPs that are not synthesized as glycoproteins, such as trichosanthin, are fully functional enzymes.

RIP Binding, Entry, and Movement in Host Cells

Type 1 RNA *N*-glycosidases may use several different mechanisms to enter target cells. Fluid-phase endocytosis likely is responsible for much of the uptake of the type 1 RIPs. However, many type 1 RIPs are glycosylated and thus may have the capacity to enter cells by a carbohydrate-dependent mechanism. Alternatively, some nonglycosylated type 1 RIPs interact with low-density lipoprotein receptors. In fact, α-macroglobulin, megalin, and some chemokine receptors bind trichosanthin specifically.

As described previously, Stx uses a specific receptor, Gb_3, to enter cells. However, some variants of Stx2 show marked preferences for Gb_3 molecules with shorter fatty acid chains compared to those preferred by Stx1. Another Stx2 variant, Stx2e, which causes edema disease of swine, binds globotetraosylceramide in preference to Gb_3. Most other type 2 RNA *N*-glycosidases bind nonspecifically to terminal galactose residues on host cells. Ricin can gain access to host cells through a mannose-dependent pathway as well as a galactose-dependent pathway. Elegant studies by Sandvig and coworkers showed that after binding to the target cell, ricin and Stx enter target cells via endocytosis, generally through clathrin-coated pits. However, ricin also uses clathrin-independent mechanisms to enter host cells. Once inside the cell, ricin and Stx traffic through the cell in a retrograde manner: from the entry vacuole, to the Golgi, to the endoplasmic reticulum, and into the cytosol to deadenylate the ribosomes. Only a few molecules of ricin are needed to kill the target cell. It is presumed that since the target cell DNA is contained within the nucleus, the ricin and Stx do not act directly on host cell DNA. In any case, host cells intoxicated with ricin or Stx undergo apoptosis, a process that involves degradation of nuclear DNA by host cell factors.

The Active Site and Three-Dimensional Structure of the RIPs

The active site of ricin has been studied both biochemically and genetically. The Stx active site has also undergone extensive mutational analysis. These studies reveal that a glutamic acid at position 177 in ricin and 167 in Stx is required for enzymatic activity. Several other residues, such as the arginine at position 180 in ricin and 170 in Stx, are important for full enzymatic activity as well. Alignment of the sequences of several RIPs reveals that, as a group, they have very little sequence homology, except for residues determined to be involved in the active site. These active-site residues include, for ricin, Tyr-80 and Tyr-123, Glu-177, Arg-180, and Trp-211. Robertus and coworkers proposed the following mechanism of action for ricin. The targeted adenosine becomes stacked in the active-site cleft of ricin between two conserved tyrosines, where the adenosine undergoes protonation by the arginine at position 180. This protonation results in formation of an oxycarbonium ion on the ribose of the adenine that is stabilized through ion pairing with glutamic acid at position 177. The oxycarbonium ion then undergoes nucleophilic attack by a nearby water molecule, which, in turn, causes the release of the adenine from the RNA.

The three-dimensional structure of several RIPs has been solved alone or with an analog bound in the active site. The active site of the RNA N-glycosidases appears to be conserved, even though only a few conserved residues exist among the RIPs. Figure 2 shows the crystal structures of three of the RIPs.

Substrate Specificity

One unusual feature of the RIPs is that, although they share the same enzymatic activity and their preferred target is intact mammalian ribosomes, they are not equally active on target ribosomes in vitro. There appear to be at least two explanations for this phenomenon. One reason that some RNA N-glycosidases do not appear to be as active as others in cell-free assays is that they require cofactors such as ATP. The activity of other RIPs seems to be enhanced in the presence of some tRNAs. The second reason that some RIPs have varying specificities is that they appear to interact with ribosomal proteins. Therefore, if those proteins are removed from the rRNA or a ribosome with different ribosomal proteins is used as a substrate, the enzyme may not work as efficiently or at all. In other cases, removal of the ribosomal proteins may allow a particular RIP to act on that rRNA. So, for example, even though E. coli has the GAGA target sequence in its rRNA, only a few RIPs attack the E. coli ribosome when they are expressed in the E. coli cytosol. For that reason, E. coli can usually be used to maintain clones of the RIPs. However, E. coli 23S rRNA can be deadenylated by high concentrations of ricin in vitro, if the rRNA has first been deproteinated, a fact that suggests that the ribosomal proteins protect the E. coli rRNA from the N-glycosidase activity in vivo.

Type 1 RIPs efficiently inhibit protein synthesis in a cell-free rabbit reticulocyte system with 50% inhibition concentrations (IC_{50}s) ranging from 0.03 to 4

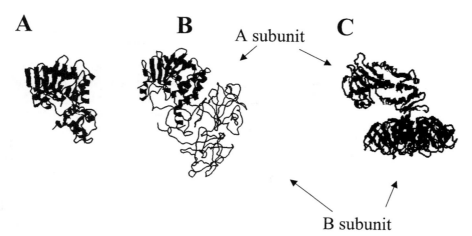

Figure 2 Crystal structures of the (A) type 1 RIP bryodin 1 (Protein Data Bank [PDB] ID: 1BRY, Gawlak et al. [3]) and the type 2 RIPs (B) ricin (PDB ID: 2AAI, Rutenber et al. [4]) and (C) Stx (PDB ID: 1DMO, Fraser et al. [2]).

nM, with most having IC_{50}s of less than 1 nM. Intact type 2 RIPs are less active in the cell-free assays, with IC_{50}s near 100 nM. However, if the type 2 RIP A chain is tested in the absence of the B chain, the cell-free IC_{50} drops to 0.1 to 4 nM.

Protection from Autologous RNA *N*-Glycosidases

How does a cell protect itself from its own RNA *N*-glycosidase(s)? Many of the RIPs have N-terminal signal sequences that target the protein to the endoplasmic reticulum, a location that may segregate the RNA *N*-glycosidase from the target ribosomes within the producing cell. The C-terminal extensions on the RIPs may also target them to vacuolar locations. However, since the RIPs can move in a retrograde fashion, additional protective features may be required to protect autologous ribosomes from inactivation. There are at least six possible mechanisms by which a cell might protect its ribosomes from the autologous RNA *N*-glycosidase: (i) production of the enzyme in an inactive form (saporin, ricin); (ii) segregation of the enzyme in storage vacuoles within the cell (PAP, ricin) and/or within structures such as seeds (ricin); (iii) insensitivity to the autologous RNA *N*-glycosidase, perhaps through an inability to interact with autologous ribosomal proteins (tritin, cereal RIPs); (iv) absence of a cofactor required by the enzyme (gelonin); (v) blockage of the active site of the enzyme (Stxs); and (vi) secretion of the enzyme (PAP, Stxs). Figure 3 illustrates these mechanisms.

Pathogenesis of RNA *N*-Glycosidases in Animals

The plant type 2 RIPs and the bacterial Stxs are similar in that, for animals, natural exposure to the toxin begins with oral ingestion. In the case of the plant RIPs, the item eaten might be seeds, leaves, or roots. RIPs have also been administered to animals either purposefully or accidentally, at both lethal and sublethal doses. Following either oral consumption or injection, a RIP would be adsorbed into the bloodstream, circulate, and bind to cells with terminal galactose residues. Once internalized into those cells, the RIP would be trafficked as described earlier, halt protein synthesis, and kill the intoxicated cell. In an animal, low doses of type 1 RNA *N*-glycosidases cause rashes and fever. At higher doses of type 1 RIPs, necrotic lesions are observed in the liver, kidney, and spleen. The more potent type 2 RNA *N*-glycosidases cause damage in different locations in the animal, presumably reflecting the binding specificity of the B chain. For example, ricin, modeccin, and volkensin affect the liver, whereas abrin does not. Ricin and abrin can travel along peripheral nerves, while modeccin and volkensin attack the central nervous system. The mouse 50% lethal dose (LD_{50}) ranges from about 1 to 40 mg/kg for type 1 RIPs and from about 1 to 80 μg/kg for the type 2 RIPs.

For the Stxs, the delivery vehicle is food or water contaminated with *S. dysenteriae* type 1 or *E. coli* of certain serotypes, such as O157:H7, that encode one or more types of Stx. The Stx-producing bacteria have the capacity to set up an intestinal infection and deliver the Stx(s) into the intestines and bloodstream of the infected person. The consequences of infection with these Stx-producing bacteria can range from a mild, watery diarrhea to severe bloody diarrhea (popularly called hamburger disease), to the hemolytic-uremic syndrome (HUS), or

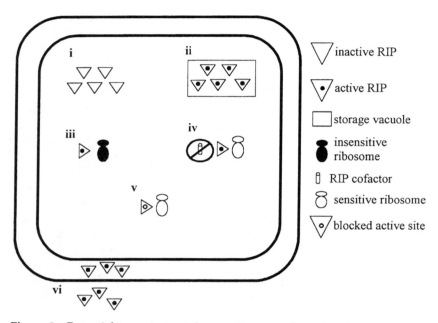

Figure 3 Potential to protect autologous ribosomes from the plant's own RIP. (i) The RIP is produced in an inactive form. (ii) The RIP is active but stored so that it does not have access to the ribosome. (iii) The ribosome is refractory to the action of the RIP. (iv) The RIP requires a cofactor that is absent in the cytosol. (v) The active site of the enzyme is blocked. (vi) The active RIP is secreted into the cell wall or outside the cell.

even death. It is important to point out, however, that the Stx-producing bacteria have other pathogenic determinants besides Stx(s) that are important for such early events as colonization and, for some Stx-producing *E. coli*, attach and efface lesion formation at the gut mucosa. Even so, the Stxs appear to play a critical role in the development of the more severe sequelae of infection from Stx-producing bacteria. The Stxs are quite toxic to mice, with a mouse LD_{50} for Stx1 of about 6 μg/kg and approximately 0.05 μg/kg for Stx2. In fact, Stx2 is likely the most potent RNA N-glycosidase, with a mouse LD_{50} about 100-fold lower than that of the other RIPs. Furthermore, epidemiological evidence suggests that Stx2-producing strains of *E. coli* are more likely to cause HUS than are strains that produce only Stx1. The pathogenesis of HUS is not that straightforward but may reflect that the Stxs (as well as some of the plant RIPs) induce the production of cytokines such as tumor necrosis factor α and interleukin 1β. Lipopolysaccharide (LPS) also plays a role in the pathogenesis of HUS due to Stx-producing bacteria. Both LPS and the cytokines induced by the presence of Stx and LPS may cause target cells to up-regulate expression of Gb_3 and thus enhance sensitivity of those cells to Stx. This up-regulation of Gb_3 may result in the unfortunate situation for the host wherein the presence of Stx causes target cells to become even more sensitive to this bacterial poison.

In Vivo Role(s) of the RIPs

The major role of the plant RIPs is presumed to be a defensive one. Indeed, one of the first activities attributed to the RIPs was viral inhibition. For example, the type 1 RIP PAP prevents tobacco mosaic virus transmission to other plants and inhibits influenza virus replication in monkey kidney cells. Indeed, PAP inhibits several different viruses, including both DNA and RNA viruses. Furthermore, PAP expression within tobacco or potato plants (which lack native RIPs) confers a broad-spectrum viral resistance upon those recombinant plants. RIPs that inhibit viral replication in cells do so at lower concentrations than are required to halt protein synthesis in uninfected cells. One of the hypotheses from these studies was that the infected cells are more permeable to the RIPs, so lower concentrations of the RNA N-glycosidase are required for viral inhibition. Presumably, once a host plant cell is invaded by the virus, the plant's own RIP could enter the damaged cell or be released from an internal storage compartment and halt protein synthesis inside the cell. The virus would thus be prevented from usurping host cell functions for replication and infection of neighboring cells.

Understanding of the antiviral role of the RIPs is complicated by the fact that a deletion mutant of PAP that was unable to deadenylate rRNA retained the capacity to inhibit *Potato virus X* in tobacco plants. However, mutant PAP, as well as wild-type PAP, was later shown to be able to depurinate capped mRNAs, a finding that suggests that an RNA N-glycosidase might be able to attack viral RNA directly. Interest in the RIPs as antiviral agents has increased since the recognition that PAP, trichosanthin, and saporin, among others, could inhibit replication of human immunodeficiency virus (HIV). Several RIPs, including trichosanthin, are under study as potential therapeutics for HIV treatment.

Besides antiviral activity, a few RNA N-glycosidases demonstrate antifungal, antibacterial, and/or insecticidal properties. Some type 1 RIPs also kill tumor cell lines, mouse macrophages, and human fibroblasts in vitro. However, much higher levels of type 1 RIPs are required for cytotoxicity (μM) compared to the type 2 RIPs, which are generally toxic in the pM range. The exception to this general rule is that the type 1 RIPs demonstrate a high toxicity for trophoblasts and macrophages. Indeed, the IC_{50} of the type 1 RIPs for macrophages is about 10 nM.

Consistent with a protective function for the plant RNA N-glycosidases is that the RIPs are both lethal and often stored in large quantities. However, the definitive function of the plant RNA N-glycosidases remains a controversial question within the RIP field, in part because plants that express RIPs may still be susceptible to viruses. Studies with defined RIP mutants would further define the in vivo role of RNA N-glycosidases for plants.

The function of the Stxs for the bacteria that carry them is more speculative than that of the RIPs for plants. Certainly, people who have diarrhea shed and spread the infecting organism. But people may be an accidental host of the *E. coli* that makes Stxs. Cattle, deer, and sheep appear to carry Stx-producing *E. coli* at least transiently in their intestines without harm. That leads to the question of whether there is another role for the Stxs for the bacteria. Perhaps the Stxs serve as a colonization factor in certain situations or enhance environmental survival of the host bacteria.

POTENTIAL USES OF RIPs IN MEDICINE

Because the type 2 RIPs are potent cytotoxins and the type 1 RIPs are efficient inhibitors of protein synthesis in a whole cell if they are delivered into the target cell cytoplasm, much effort has been put into developing various RIPs as immunotoxins. Proposals have been made to use RIPs in ex vivo and in vivo situations for the treatment of cancer. Ricin, PAP, bryodin 1, and saporin, to name a few, have been used to make immunotoxins. To increase the target specificity of type 2 RIPs, the A domain only is generally conjugated to a carrier, such as a monoclonal antibody. RIP immunotoxins have been used to deplete bone marrow of T cells before transplantation into a recipient to reduce the possibility of graft-versus-host disease. Attempts have been made to use RIP immunotoxins to eliminate malignant cells from bone marrow removed from a patient for autologous transplantation following radiation treatment. The RIP immunotoxins have not been as successful as originally hoped, because the target cell type was not entirely eliminated from the bone marrow. Another problem with the RIPs is that they are potent immunogens, a difficulty that might reduce the number of times a therapy could be utilized.

CONCLUDING REMARKS

The plant and bacterial RNA N-glycosidases represent a class of highly potent protein toxins whose natural role for the plants or bacteria that express them is still under investigation. For plants, certainly, the most likely function of these polynucleotide:adenosine glycosidases is for self-defense. However, the question of the precise roles of the RIPs for the plant or bacteria that produce them may not be answered soon, since much of the current RIP research is focused on how to exploit these toxic molecules for treatment of such human illnesses as HIV infection and cancer.

REFERENCE
1. **Barbieri, L., P. Valbonesi, F. Righi, G. Zuccheri, F. Monti, P. Gorini, B. Samori, and F. Stirpe.** 2000. Polynucleotide:adenosine glycosidase is the sole activity of ribosome-inactivating proteins on DNA. *J. Biochem.* **128:**883–889.
2. **Fraser, M. E., M. M. Chernaia, Y. V. Kozlov, and M. N. James.** 1994. Crystal structure of the holotoxin from *Shigella dysenteriae* at 2.5 Å resolution. *Nat. Struct. Biol.* **1:**59–64.
3. **Gawlak, S. L., M. Neubauer, H. E. Klei, C. Y. Chang, H. M. Einspahr, and C. B. Siegall.** 1997. Molecular, biological, and preliminary structural analysis of recombinant bryodin 1, a ribosome-inactivating protein from the plant *Bryonia dioica. Biochemistry* **36:**3095–3103.
4. **Rutenber, E., B. J. Katzin, S. Ernst, E. J. Collins, D. Mlsna, M. P. Ready, and J. D. Robertus.** 1991. Crystallographic refinement of ricin to 2.5 Å. *Proteins* **10:**240–250.

SUGGESTED READING
1. **Barbieri, L., M. G. Battelli, and F. Stirpe.** 1993. Ribosome-inactivating proteins from plants. *Biochim. Biophys. Acta* **1154:**237–282.
2. **Endo, Y., K. Mitsui, M. Motizuki, and K. Tsurugi.** 1987. The mechanism of action of ricin and related lectins on eukaryotic ribosomes. *J. Biol. Chem.* **262:**5908–5912.

3. **Hartley, M., J. Chaddock, and M. Bonness.** 1996. The structure and function of ribosome-inactivating proteins. *Tr. Plant Sci.* **1:**254–260.

4. **Lord, J. M., M. R. Hartley, and L. M. Roberts.** 1991. Ribosome inactivating proteins of plants. *Semin. Cell Biol.* **2:**15–22.

5. **Melton-Celsa, A. R., and A. D. O'Brien.** 2000. Shiga toxins of *Shigella dysenteriae* and *Escherichia coli*, p. 385–406. *In* K. Aktories and I. Just (ed.), *Handbook of Experimental Pharmacology*, vol. 145: *Bacterial Protein Toxins*. Springer Verlag, Berlin, Germany.

6. **Melton-Celsa, A. R., J. E. Rogers, C. K. Schmitt, S. C. Darnell, and A. D. O'Brien.** 2001. Virulence of Shiga toxin-producing *Escherichia coli* (STEC) in orally-infected mice correlates with the type of toxin produced by the infecting strain. *Jpn. J. Med. Sci. Biol.* **50:**S108–S114.

7. **Nielsen, K., and R. S. Boston.** 2001. Ribosome-inactivating proteins: a plant perspective. *Annu. Rev. Physiol. Plant Mol. Biol.* **52:**785–816.

8. **Paton, J. C., and A. W. Paton.** 1998. Pathogenesis and diagnosis of Shiga toxin-producing *Escherichia coli* infections. *Clin. Microbiol. Rev.* **11:**450–479.

9. **Peumans, W. J., Q. Hao, and E. J. M. Van Damme.** 2001. Ribosome-inactivating proteins from plants: more than RNA N-glycosidases? *FASEB J.* **15:**1493–1506.

10. **Robertus, J.** 1991. The structure and action of ricin, a cytotoxic N-glycosidase. *Semin. Cell Biol.* **2:**23–30.

11. **Sandvig, K., and B. van Deurs.** 2000. Entry of ricin and Shiga toxin into cells: molecular mechanisms and medical perspectives. *EMBO J.* **19:**5943–5950.

12. **Stirpe, F., L. Barbieri, M. G. Battelli, M. Soria, and D. A. Lappi.** 1992. Ribosome-inactivating proteins from plants: present status and future prospects. *Biotechnology* **10:**405–412.

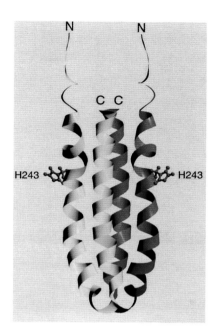

Figure 3 [chapter 1] NMR solution structure of the dimeric core domain of EnvZ. The protein structure is depicted in ribbon format showing the four-helix bundle formed from two identical monomers. The catalytic histidine at position 243 is shown in ball-and-stick format. Based on structural coordinates deposited in the Research Collaboratory for Structural Bioinformatics (RCSB) Protein Data Bank (PDB), as described in reference 10.

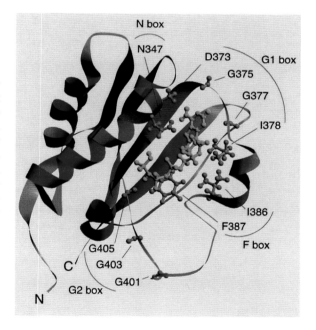

Figure 4 [chapter 1] NMR solution structure of the catalytic kinase domain of EnvZ. The protein structure is depicted in ribbon format. The segments corresponding to the conserved N, G1, F, and G2 boxes are colored blue, and key residues from these are shown in ball-and-stick format. This structure was determined in the presence of the nonhydrolyzable ATP analog AMP-PNP, which is shown in green ball-and-stick format. Based on structural coordinates deposited in the RCSB PDB, as described in reference 8.

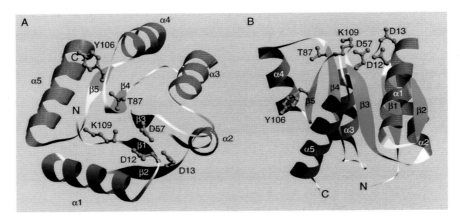

Figure 5 [chapter 1] CheY structure determined by X-ray crystallography. The structure of CheY is presented as representative of the common structure of receiver domains from many sources. The protein structure is shown in ribbon format, and the α-helical and β-strand segments are numbered. Key residues introduced in Fig. 2 are shown in blue ball-and-stick format. Two views are shown for clarity. Based on structural coordinates deposited in the RCSB PDB, as described in reference 7.

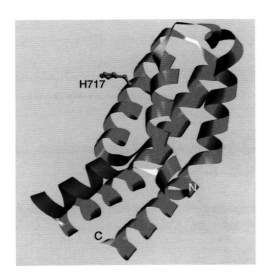

Figure 6 [chapter 1] Structure of the monomeric HPt domain from the ArcB sensor kinase of *E. coli* determined by X-ray crystallography. Protein structure is shown in ribbon format, with the catalytic histidine at position 717 shown in blue ball-and-stick format. Based on structural coordinates deposited in the RCSB PDB, as described in reference 3.

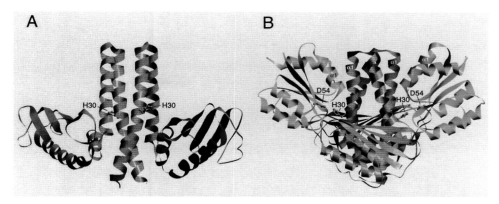

Figure 7 [chapter 1] Structure of Spo0B (A) and a complex between Spo0F and Spo0B (B) determined by X-ray crystallography. (A) The structure of a Spo0B dimer is presented as representative of the overall structure of histidine kinase domains. Protein structure is shown in ribbon format while catalytic histidines are shown in blue ball-and-stick format. Based on structural coordinates deposited in the RCSB PDB, as described in reference 11. (B) The structure of a complex between a Spo0B dimer and two cognate receiver domains of Spo0F. Based on structural coordinates deposited in the RCSB PDB, as described in reference 13.

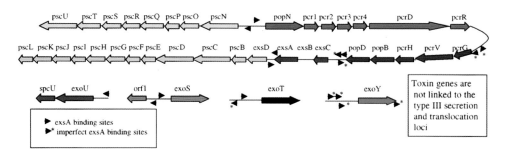

Figure 3 [chapter 3] The operons of *Pseudomonas aeruginosa* type III secretion system and effector toxins ExoU, ExoS, ExoT, and ExoY are shown. Individual gene products are indicated as colored arrows (the sizes of protein products are approximate and not to scale). Different colors are used to identify operons based on promoter mapping, transcriptional fusion, or DNA-binding analyses. The toxins and known chaperone proteins are located in different chromosomal locations than the set of operons that encode the secretory, translocation, and regulatory proteins. ExsA consensus and imperfect binding sites are illustrated.

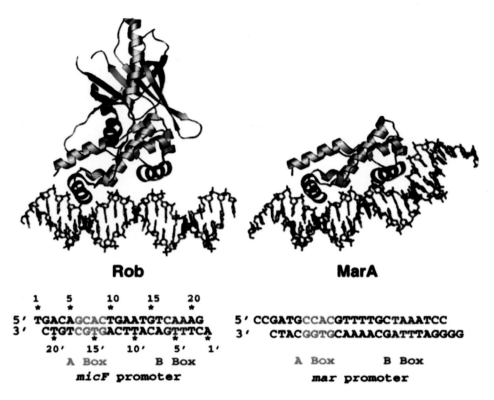

Figure 1 [chapter 3] Structures of Rob and MarA in complex with target DNA sequences. The structurally similar N-terminal DNA binding domains are colored orange in the ribbon diagrams. The C-terminal domain unique to Rob is blue. MicF and MarA sequences used for cocrystallization are shown with A-box sequences in orange and B-box sequences in light blue type. Only the N-terminal HTH motif of Rob directly contacts bases of the binding sequence. MarA bends the DNA target so that both HTH motifs are located in adjacent major groove surfaces on one side of the DNA. From reference 8 with permission.

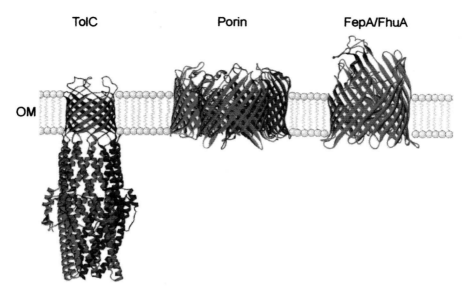

Figure 4 [chapter 5] The structures of TolC porins (e.g., OmpF and SecY) and the siderophore transporters FhuA and FepA.

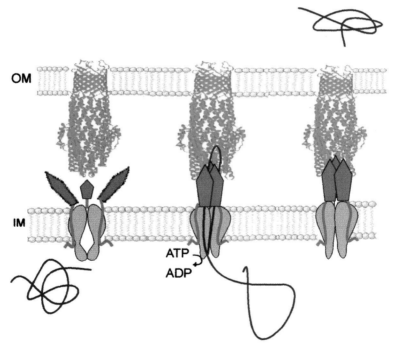

Figure 5 [chapter 5] Schematic presentation of the proposed interaction of trimeric TolC (blue) with substrate-specific inner membrane complexes containing an adaptor protein (red) and a traffic ATPase (green). When the substrate binds to the inner membrane complex, the trimeric adaptor protein contacts the periplasmic tunnel, possibly via the predicted coiled-coil structures, triggering the conformational change that opens the entrance and presents the exit duct. Following export, the components revert to resting state. An animated model of protein export is available at http://archive.bmn.com/supp/ceb/ain1.html.

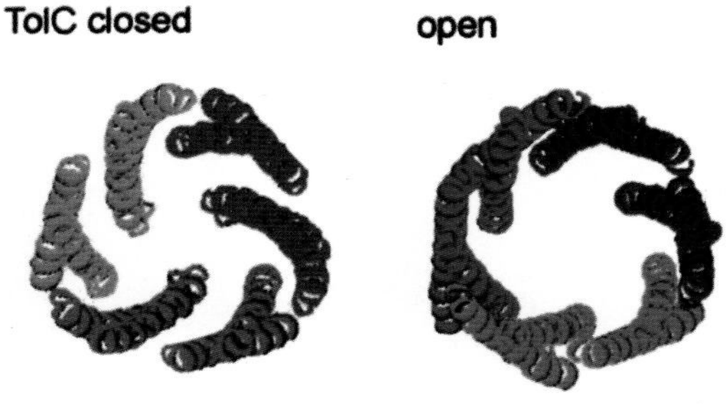

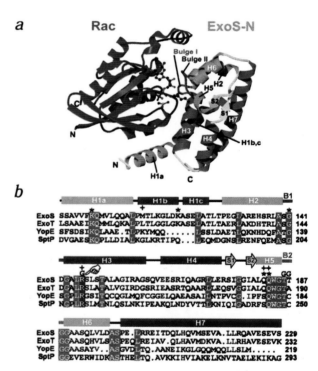

Figure 6 [chapter 5] The TolC tunnel entrance closed and modeled open state showing the monomers colored in green, blue, and red.

Figure 4 [chapter 20] Secondary and tertiary structure of the GAP domain of ExoS. (a) Ribbon diagram of the GAP domain of ExoS in a complex with Rac1, GDP, and aluminum fluoride (AlF3). AlF3 is required to trap the ExoS-Rac1 complex in a transition state. Rac1 is colored blue. Bulge I is colored red, and bulge II is colored green in ExoS. The catalytic arginine in ExoS (Arg-146), the GDP, and the AlF3 are shown as ball-and-stick models. (b) Secondary structure assignments for ExoS GAP domain and sequence alignments of ExoS, ExoT, YopE, and SptP. Invariant residues are shown as white on green or red (catalytic Arg). Asterisks and plus signs indicate residues involved in hydrophilic or hydrophobic interactions, respectively. G indicates residues that contact GDP. Reprinted from reference 15 with permission.

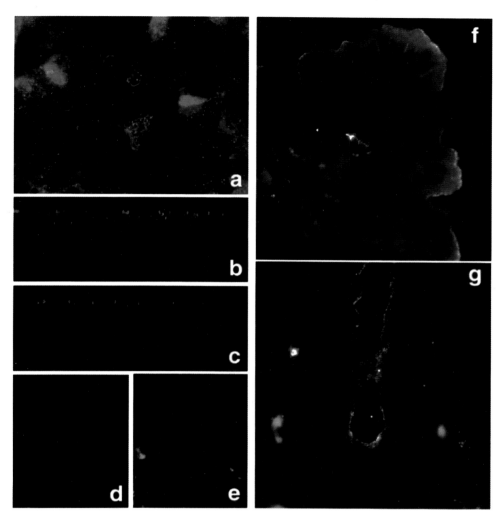

Figure 3 [chapter 12] Immunolocalization of FcRn in polarized T84 monolayers and normal adult human small intestinal mucosa. (a) Whole-mount T84
monolayers visualized en face show a diffuse, punctate staining pattern. The
Z0-1 image was captured slightly above the focal plane of FcRn. (b) FcRn
staining of whole-mount T84 monolayers visualized as confocal vertical sections. (c) FcRn staining was absent in the presence of an isotype-matched,
irrelevant antiserum. (d and e) FcRn staining was absent in the presence of
an irrelevant anti-serum or with secondary antibody alone. (f) Villous enterocytes of normal adult human small intestine show delicate linear staining in
the region of the apical cytoplasmic membrane. (g) Crypt enterocytes show
an apical and punctate staining pattern visible not only at the level of the
apical cytoplasmic membrane, but also in the apical cytoplasm below the level
of the apical membrane.

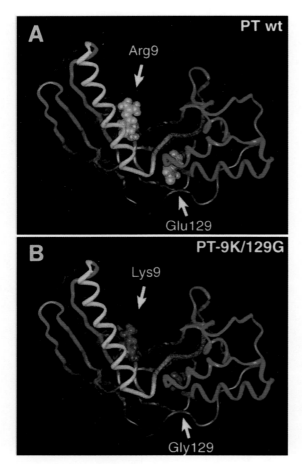

Figure 3 [chapter 22] Computer modeling of the S1 subunit of (A) wild-type PT and of (B) PT-9K/129G mutant represented as an α-carbon trace. The NAD-binding site is highlighted in yellow; the catalytic residues in position 9 and 129 are shown in yellow for the wild-type PT and in red for the PT-9K/129G mutant, respectively.

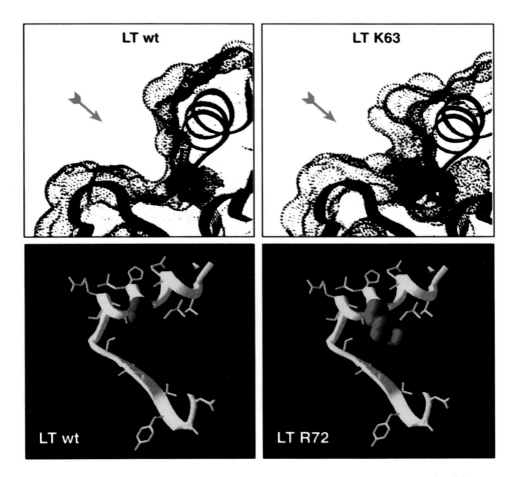

Figure 6 [chapter 22] Three-dimensional structure of the A subunit of wild-type LT (top left panel) and LTK63 mutant (top right panel) represented as α-carbon trace and as solvent-exposed surfaces. The residues in position 63 (serine in the wild-type LT and lysine in the LTK63 mutant) are indicated by the arrow. Computer modeling of the NAD-binding site of wild-type LT (top left panel) and LTR72 (bottom right panel). The residues in position 72 (alanine in the wild-type LT and arginine in LTR72) are shown in green and red, respectively.

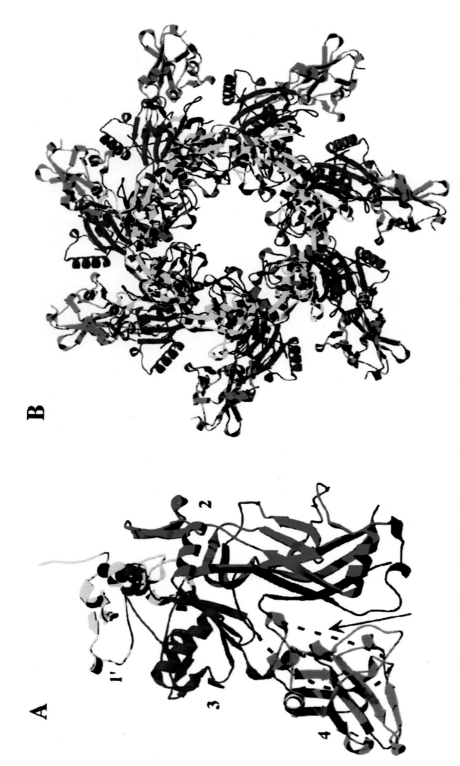

Figure 1 [chapter 23] Three-dimensional structure of the PA of anthrax toxin. (A) Three-dimensional structure of monomeric PA. The four domains of the protein are indicated. The dashed line indicated by the arrow represents a loop that is believed to form a segment of a transmembrane β-barrel upon pore formation. (B) Modeled heptamer of PA depicting its predicted ring-like structure. Adapted from reference 8 with permission.

x

Bacterial Protein Toxins
D. Burns et al., Editors
©2003 ASM Press, Washington, DC 20036

Chapter 18

Cytolethal Distending Toxin

Lawrence A. Dreyfus

Cytolethal distending toxin (CDT) was first recognized in 1987 when it was found associated with an *Escherichia coli* O55:K59:H4 isolate recovered from a pediatric case of gastroenteritis and encephalopathy. Soon after its initial identification in *E. coli*, CDT production was observed in several *Campylobacter jejuni* isolates. CDT was named by Johnson and Lior to reflect the dramatic and progressive cellular distension and eventual death of tissue culture cells treated with the newly described toxic factor (10). At the time, CDT appeared to be associated primarily with *Campylobacter* species and scattered isolates of *E. coli* and *Shigella dysenteriae*. Today, a growing list of unrelated bacterial pathogens are known to produce CDT activity or at least encode CDT-related genes (Table 1). Because of the diverse collection of bacterial species known to produce CDT, it is likely that the list will continue to expand as additional strains are examined for the *cdt* gene cluster.

Cells treated with CDT elongate within the first 24 h following the addition of toxin, then progressively distend over the next 48 to 72 h to several times their normal size. Although distended cells remain viable, cell division ceases early after toxin treatment due to an irreversible block at the G_2/M boundary (Fig. 1). Cessation of cell division is accompanied by an accumulation of the phosphorylated (inactive) form of Cdc2, a key regulatory kinase that, along with the mitotic cyclin, cyclin B, mediates progression of the cell cycle from G_2 to M. The irreversible inactivation of Cdc2, a hallmark of CDT action, blocks the cell's entry into mitosis. Recent studies indicate that Cdc2 phosphorylation in response to CDT results from activation of the mitotic checkpoint cascade, the cellular response to DNA injury (3). These observations suggest that the action of CDT on mammalian cells is similar to that of ionizing radiation (IR) and probably results in chromosomal damage. In this chapter the molecular biology and biochemistry of CDT and its molecular mode of action will be examined.

Lawrence A. Dreyfus • Division of Cell Biology and Biophysics, University of Missouri-Kansas City, Kansas City, MO 64110.

Table 1 CDT-producing bacterial species and CDT genes

Species	Designation	Accession number[a]
A. actinomycetemcomitans	Aa-CDT	AF102554
C. jejuni	Cj-CDT	U51121
E. coli (strain 6468/62)	Ec-CDT-I	U03293
E. coli (strain 9142-88)	Ec-CDT-II	U04208
E. coli (strain S5)	Ec-CDT-III	U89305
Haemophilus ducreyi	Hd-CDT	U53215
Helicobacter hepaticus	Hh-CDT	AF243076
Helicobacter pullorum	Hp-CDT	AF220065
Salmonella enterica serovar Typhi	St-CdtB	NC-003198 (STY1886)

[a] GenBank (nucleotide sequence database) accession numbers.

GENETICS OF CDT PRODUCTION

The genes encoding CDT have been cloned and sequenced from several bacterial strains including *C. jejuni*, *Haemophilus ducreyi*, *Actinobacillus actinomycetemcomitans*, certain isolates of *E. coli* and *Shigella dysenteriae*, and a number of isolates of *Helicobacter* species pathogenic for animals (Table 1). In addition, the recent completion of the *Salmonella enterica* serovar Typhi genomic sequence reveals the presence of a *cdtB* gene, though orthologs of the *cdtA* and *cdtC* genes were not found (4, 10). Although it is likely that all of the CDTs will fundamentally possess the same biological activity, there are subtle but significant differences that will be discussed in a later section. Recently, Cortes-Bratti et al. (3) proposed a strain-specific nomenclature to differentiate CDTs when referring to a specific toxin; this nomenclature will be used (Table 1). In addition to the variety of closely related CDTs from the unrelated group of organisms listed in Table 1, there are notable strain-specific differences in the CDTs produced by different *E. coli* isolates. The three *E. coli* CDTs, designated Ec-CDT-I, Ec-CDT-II, and Ec-CDT-III, have been identified and differentiated based on sequence relatedness and gene location. Like all other CDTs, the *cdt* genes specifying Ec-CDT-I and Ec-CDT-II are chromosomally located and encode similar but not identical CdtA, B, and C proteins. For instance, the CdtBs encoded by Ec-CDT-I and Ec-CDT-II display 54.5% identity (60.8% similarity) that is typical of the similarity among all CdtBs characterized, whereas CdtB-III is 93.7% identical (94.4% similar) to CdtB-II. In addition, CDT-III of *E. coli* strain S5 (O15:H21) is encoded on a large virulence plasmid (pVir), which also contains genes for the cytotoxic necrotizing factor type 2 (CNF2) and the F17b fimbrial adhesin. The plasmid-encoded CDT-III may be part of a pathogenicity island, which would represent the first described to encode CDT. This observation provides indirect evidence for the obvious conclusion that the *cdt* genes have disseminated among gram-negative bacterial species by horizontal gene transfer. Additional evidence supporting horizontal movement of *cdt* genes includes the findings that *cdt* genes are, in some cases, flanked by bacteriophage and insertion sequence-related elements, transposase-like genes, and a bacteriophage att site (4, 10). The increasing spectrum of CDT gene dissemination among the largely unrelated group of gram-negative bacterial

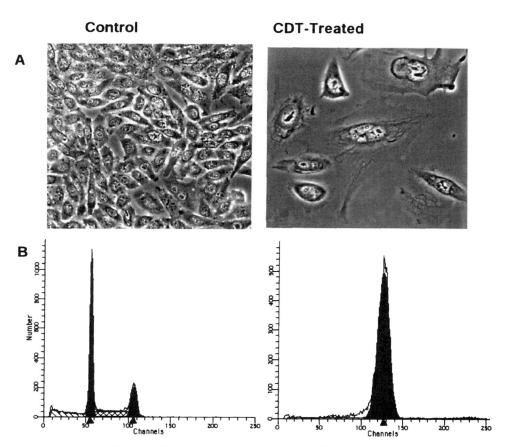

Figure 1 Effects of CDT treatment on growing cells. (A) Phase-contrast microscopy and (B) cell cycle distribution of Chinese hamster ovary cells treated with Ec-CDT-I. Microscopic examination reveals the dramatic increase in cell size following 48 h of CDT treatment. Control cells were treated with a cell extract from a non-CDT-producing *E. coli* strain. Cell cycle distribution analysis by flow cytometry reveals that CDT cells are completely blocked at the G_2/M boundary whereas control cells exhibit a normal distribution for cycling cells.

pathogens is a unique phenomenon for a bacterial toxin and may reflect a yet-to-be-determined selective advantage for CDT production.

All CDTs thus far characterized are encoded by three nearly or slightly overlapping open reading frames designated *cdtA*, *cdtB*, and *cdtC*. Transcript analysis of the *Haemophilus ducreyi cdt* gene cluster (*Hd-cdtABC*) indicated that the *cdt* genes are organized in an operon. On the basis of sequence analysis of other gene clusters, this fact probably holds true for all *cdt* clusters characterized to date. The prevalence of *cdt* genes among isolates of the various bacterial species is somewhat variable. Virtually all *C. jejuni* strains possess *cdt* genes and express CDT toxicity. A slightly lower percentage of *Campylobacter fetus* and *Campylobacter coli* harbor *cdt* sequences; however, when a larger sample of isolates is surveyed, it is likely that *C. fetus* will be similar to *C. jejuni* in *cdt* carriage. Studies investigating the prevalence of *cdt* in clinical isolates of *Haemophilus ducreyi, A. actinomycetemcomitans*, and the enteric and hepatic isolates of *Helicobacter* species examined thus far indicate that 80 to 90% carry the *cdt* genes. In contrast to the relatively high frequency of *cdt* in the aforementioned species, the presence of *cdt* in *E. coli* and *Shigella* species, although notable, is sporadic (3, 4, 10).

Biochemistry of CDT Polypeptides

The *cdtABC* operon encodes three CDT polypeptides, CdtA, CdtB, and CdtC (Fig. 2). All three proteins bear apparent signal peptide coding sequences consistent with secretion across the inner membrane by the general export pathway. The calculated molecular weights of the three proteins are approximately 28,500 (CdtA), 30,000 (CdtB), and 22,000 (CdtC). Processing by signal peptidase I cleavage would result in proteins with molecular weights of approximately 26,500 (CdtA), 28,000 (CdtB), and 20,000 (CdtC), which is consistent with published molecular weight estimates based on sodium dodecyl sulfate-polyacrylamide gel electrophoresis (SDS-PAGE) analysis (Fig. 2). Early reports describing the cloning of *cdt* genes provided genetic evidence suggesting that all three proteins were required for biological activity. However, ascertaining the role of each of the three CDT polypeptides in biological activity has been hampered by difficulties in purifying the CDT holotoxin. Difficulties in separating CDT from outer membrane components and the low level of CDT production have hindered purification. Biochemical evidence for the requirement of CdtA, CdtB, and CdtC in biological activity has recently been unequivocally demonstrated. Lara-Tejero and Galán recently purified and characterized the individual *C. jejuni* CDT polypeptides and reconstituted an active CDT holotoxin (9). When mixed in

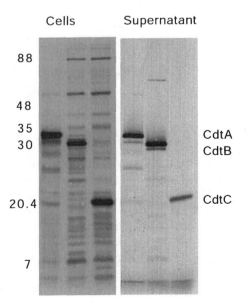

Figure 2 T7 expression and exclusive radiolabeling of CDT polypeptides. The Ec-CDT-II genes were separated by cloning and expressed under a T7 promoter in the presence of rifampicin and ^{35}S-methionine/cysteine. Total cell extract and supernatant fractions were then prepared and analyzed by SDS-PAGE and autoradiography. The expressed forms of Ec-CDT-II migrated with molecular sizes of approximately 32, 30, and 20 kDa. These results coincide with the calculated molecular sizes for CDT. As expected, CdtA, CdtB, and CdtC also appeared in the supernatant fraction of the cell culture.

equal molar amounts, CdtA, CdtB, and CdtC formed a biologically active complex that was separated from the individual subunits by gel filtration chromatography. The complex contains CdtA, CdtB, and CdtC in an apparent subunit stoichiometry of 1:1:1. Any combination of CDT subunits other than all three fails to exhibit biological activity. Recent studies linking CdtB to mammalian type I DNase (described in the next section) suggest that CdtB may be the active CDT subunit. In addition, CdtA appears to possess a protein fold similar to that of the carbohydrate-binding domain of the ricin A chain (9). However, a functional carbohydrate-binding activity of CdtA has not been reported. It has also been suggested that Hd-CdtA may covalently modify Hd-CdtB and/or Hd-CdtC to mediate formation of an active CDT tripartite complex, although this has not been conclusively demonstrated or reported for purified Cj- or Aa-CDT (3). On the basis of the activity of CdtB (below) and the apparent requirement for CdtA and CdtC to mediate toxicity, Lara-Tejero and Galán have proposed that CDT is a trimeric AB-type toxin composed of equimolar CdtA, CdtB, and CdtC in which CdtB is the active A component and a heterodimer of CdtA and CdtC makes up the B component (9).

Role of CdtB in the Mode of CDT Action

Protein sequence alignments often provide insights into the potential function of otherwise uncharacterized proteins. Initial alignments of the CDT polypeptides against the protein sequence data bank indicated that CdtA, CdtB, and CdtC were unlike any known proteins. This is not an uncommon occurrence since many functionally and evolutionarily important protein similarities are only recognizable through structural comparison. Examination of protein primary sequence data by iterated profile search methods has improved the chances of identifying related proteins when the primary sequence identity is beneath the threshold of what might otherwise be categorized as a related protein by a simple BLAST alignment. Recent improvements in the speed and efficiency of iterated profile search methods have greatly enhanced the ability to identify proteins distantly related to a query sequence. When a query sequence is compared against the protein database using the Position-Specific Iterated BLAST (PSI-BLAST) search method, a position-specific scoring matrix is generated from the multiple alignments of sequences identified in that search. The position-specific scoring matrix is used to search the database for additional similar sequences, and these are now used to update the matrix. The process is repeated until the searches converge, that is, no new sequences are determined. The procedure has allowed the identification of weak but biologically relevant similarities that would have otherwise gone undetected. When Ec-CdtB-II was subjected to PSI-BLAST analysis, a striking sequence similarity to mammalian type I DNase was observed (6). A similar finding using an analogous search strategy was observed with Cj-CdtB (8). PSI-BLAST and motif-based methods can be optimized to identify the most distantly related protein sequences. This, however, may permit the identification of nonrelated artifacts due to highly biased amino acid composition. The proof of a pattern homology match lies in conservation of active-site matches and ultimately similar protein function. In the case of CdtB, both of these criteria have been met. The PSI-BLAST alignment of Ec-CdtB-II and human

DNase I (Fig. 3A) reveals that the best-conserved portions of the alignment correspond to DNase I catalytic residues, including a conserved carboxy-terminal motif found in all phosphohydrolases (Fig. 3B). On the basis of iterated sequence alignments, CdtB is classed in a heterogeneous superfamily of Mg^{2+}-dependent phosphohydrolases including sphingomyelinases, exonucleases, endonucleases, and inositol 5-phosphatases (4). Subdivision of the superfamily into subgroups places CdtBs in a separate grouping, with type I endonucleases being the nearest neighboring group. Moreover, analysis of CdtB from *E. coli*, *C. jejuni*, and *Haemophilus ducreyi* indicates that CdtB possesses DNase activity (4). DNase activity associated with purified Ec-CdtB-II is Mg^{2+} or Mn^{2+} dependent, stimulated by Ca^{2+}, and inhibited by Zn^{2+} (Chao and Dreyfus, unpublished data). In addition, Ec-CdtB-II forms a tight complex with DNA in the presence of Zn^{2+}; but, unlike mammalian type I DNase, Ec-CdtB-II is not inhibited or bound by G-actin (Chao and Dreyfus, unpublished). Mutation analysis of (five) DNase-related residues

```
                                                          M
A   CdtB      ADLTDFRVATWNLQGASATTESKWNINVRQLISGENAVDILAVQEAG-S----------- 65
              +  F + T+     S    T + + + +        DI+ +QE
    DNASE I   LKIAAFNIRTFGETKMSNATLASYIVRIVRRY------DIVLIQEVRDSHLVAVGKLLDY 54
                                      C
    CdtB      PPSTAVDTGTLIPSPGIPVRELIWNLSTNSRPQQVYIYFSA-VDALGGRVNL-ALVSNER 124
              +T   + S  +       `.    RP +V + +  D        S R
    DNASE I   LNQDDPNTYHYVVSEPLG-RNSYKERYLFLFRPNKVSVLDTYQYDDGCESCGNDSFS-RE 112
                                               C
    CdtB      DEVFVLSPVRQGGRPLLGIRIGNDAFFTAHAIAMRNNDAPALVEEVYNFFRDSRDPVHQA 184
                  V +    ++    F   A+    +DA A +  +Y+ + D +   H
    DNASE I   ------PAVVKFSSHSTKVKE-----FAIVALHSAPSDAVAEINSLYDVYLDVQQKWHLN 161
                        M   *                                          C
    CdtB      LNWMILGEFVREPADLEMNLTVPVRRASE------IISPAAATQTSQRT-LD-------- 229
              + M++GDFN + + +  +    +R  +       I   A  TS    D
    DNASE I   -DVMLMGDFNADCSYVTSSQWSSIRLRTSSTFQWLIPDSADTTATSTNCAYDRIVVAGSL 220
                                                              MC
    CdtB      --YAVAGNSVAFRPSPLQAGIVYGARRTQISSDHFPVGVSR 268
                +V   S A P    QA          SDH+PV V+
    DNASE I   LQSSVVPGSAA--PFDFQAAYGLSNEMALAISDHYPVEVTL 259
```

```
                                      **
B   A.c. CdtB        259   ASLMLNQLRSQITSDHFPVSFVHDR        283
    C.j. CdtB        242   AILMLASLRSHIVSDHFPVNFRKF         265
    H.d. CdtB        259   ASLMLNQLRSQITSDHFPVSFVRDR        283
    E.c. CdtB-I      245   ASLLFGLLRGQIASKHFPVGFIPGRGARR    273
    E.c. CdtB-II     246   AGIVYGARRTQISSDHFPVGVSRR         269
    E.c. CdtB-III    246   AGIVYGARRTQMSSDHYPVGVSRR         269
    Hum. DNaseI      237   AAYGLSKQLAQAISDHYPVEVMLK         282
    B.c. SMase       208   VTSWFQKYTYNDYSDHYPVEATISMK       333
    S.a. Smase       205   VYAFPYYYVYNDFSDHYPIKAYSK         330
```

Figure 3. Homology of CdtB to mammalian type I DNase. (A) PSI-BLAST alignment of Ec-CDT-II was performed through the National Center for Biotechnology Information BLAST site (http://www.ncbi.nlm.nih.gov/ BLAST/). Iterations were repeated until the sequences converged. Shaded residues correspond to catalytic or metal ion-binding residues. The underlined sequence represents the phosphohydrolases-related motif found in all CdtBs, mammalian type I endonucleases, and sphingomyelinases (B). The CdtBs aligned are designated as shown in Table 1. Hum. DNaseI, human DNase I; B.c. SMase, *Bacillus cereus* sphingomyelinase; S.a. SMase, *Staphylococcus aureus* sphingomyelinase.

found in Ec-CdtB-II resulted in either a dramatic reduction or a total loss in DNase activity and cell cycle arrest activity (5, 6). These data suggest that the DNase activity associated with CdtB is, at the very least, involved in CDT-mediated cell cycle arrest, if not directly responsible following damage to target cell chromosomal DNA.

BIOLOGICAL ACTIVITY OF CDT

Cell Cycle Arrest

CDT was first described as a product that induced cellular elongation and massive swelling followed by cell death in 48 to 96 h following toxin treatment. A curious observation that provided clues to the action of CDT was that confluent cell monolayers were apparently resistant to CDT whereas cells in sparsely seeded culture wells were readily susceptible. It was apparent from this observation that the effect of CDT was to block cell division in actively dividing cells. Several groups soon identified that CDT-treated cells were irreversibly arrested at the G_2/M boundary (10). Control of the cell cycle at this position is regulated by the activity of the cyclin-dependent kinase Cdc2 (CDK1) (13). Before mitosis, Cdc2 is phosphorylated on two regulatory sites, Thr-14 and Tyr-15, by the action of the Myt1 and Wee1/Mik1 kinases, respectively. Activation of Cdc2 at the mitotic boundary is a consequence of dephosphorylation of Tyr-15 and Thr-14 by the action of the dual-specificity phosphatase Cdc25 (Fig. 4) (13). Dephosphorylated Cdc2 then forms an active complex with cyclin B to promote entry into mitosis. CDT treatment of cells results in persistent phosphorylation of Cdc2 consistent with either Myt1/Wee1 activation or inactivation of Cdc25. Examination of Cdc2 reactivation in CDT-treated cells indicated that, indeed, removal of inhibitory phosphates from Cdc2 was sufficient to promote Cdc2 activity. In addition, CDT intoxication could be overridden in cells following overexpression of Cdc25B or Cdc25C (3, 4). These findings provided the evidence to support a hypothesis that CDT action targets components upstream of Cdc25 regulation, namely, the mitotic checkpoint pathway (14).

   Progression through the eukaryotic cell cycle is controlled by a series of cell cycle regulators, predominantly the cyclin-dependent kinases (CDKs). Sequential activation of different CDKs is responsible for the timing of the onset of S phase and the transition into mitosis. The transition to mitosis is controlled by the CDK Cdc2 as described above. The integrity of cycle progression is monitored by checkpoints that assess the status of DNA synthesis and/or the presence of unrepaired DNA damage. Recognition of DNA damage or unreplicated DNA triggers a cascade of events that lead to cycle arrest during G_1, before the end of S or at the G_2/M boundary. The mitotic checkpoint (G_2/M boundary) is characterized by the persistence of inactive (Tyr-15-phosphorylated) Cdc2, arrest of the cycle in G_2, and rescue from the block by overexpression of Cdc25 (Fig. 4) (13). These events are identical to those following CDT intoxication and thus support a hypothesis that CDT triggers the mitotic checkpoint similar to that observed in cells following an assault of IR (Fig. 4) (3). It is tempting to speculate that CDT directly damages chromosomal DNA following entry into the cell and transit to the nucleus. The DNase activity associated with CdtB may gain access to

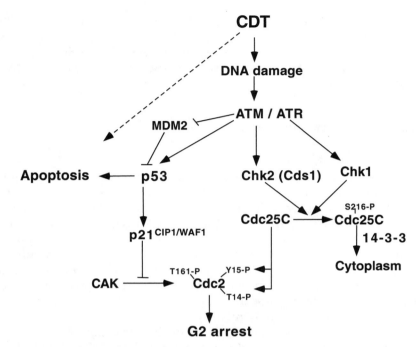

Figure 4 Mitotic checkpoint pathway following DNA damage. DNA damage following exposure to IR, UV light, or genotoxic chemical is sensed by either ATM or the ATM-related kinase ATR. Once activated, ATM and ATR phosphorylate several substrates that affect growth arrest. Chk2 (Cds1) is phosphorylated by ATM, which in turn phosphorylates the Cdc25 phosphatase on Ser-216. Phosphorylation of Cdc25 on Ser-216 creates a binding site for a 14-3-3 protein that binds to the phosphatase and escorts it to the cytoplasm. Cdc25 is required by the cell to mediate the transition between G_2 and M phase by activating Cdc2. Cdc2 is dephosphorylated (activated) by Cdc25 in a cycle-regulated manner to allow formation of the active Cdc2/cyclin B complex, thus directing the onset of mitosis. The persistent phosphorylation of Cdc2 following CDT treatment is a result of the Chk2-dependent inactivation of Cdc25 (13). ATM activation also is responsible for stabilization of the tumor suppressor p53. Both direct phosphorylation of p53 and phosphorylation of MDM2 (a protein that normally escorts p53 to the cytoplasm for degradation) result in a dramatic elevation and activation of p53. Activation of p53 in turn induces expression of p21, a cyclin-dependent kinase inhibitor that blocks activation of Cdc2, thus enforcing the G_2 arrest. In addition, p53 activation results in apoptosis in some cell types (see text and suggested reading for additional background).

the chromosome during S phase and impose a G_2 block following detection of single- and possibly double-stranded breaks. Thus far, CDT-mediated fragmentation of chromosomal DNA is controversial and has not been directly observed in CDT-treated mammalian cells. Microinjection or electroporation of wild-type but not DNase-deficient CdtB into mammalian cells results in chromatin degradation indicative of DNA damage (5, 8). Recently, Hassane et al., using *Saccharomyces cerevisiae* as a model for CDT action, demonstrated DNA degradation in yeast expressing wild-type CdtB but not a CdtB mutant lacking a key phosphodiesterase-related residue (7). CdtB expression in yeast was accompanied by an up-regulation of the DNA damage-inducible ribonucleotide reductase (*RNR2*). The resulting DNA degradation was independent of an elevation of reactive oxygen species and, thus, not a function of the yeast apoptotic-like response. However, DNA fragmentation has not been directly observed by other groups (4). In addition, the S phase of CDT-treated cells was not delayed as is often detected in cells undergoing a DNA-damaging assault (4). These latter observations suggest that CDT induces cell cycle arrest without introducing DNA damage in treated cells. This conclusion argues that either the relationship of the CdtB-associated DNase activity to CDT action is coincidental or that the in vivo CdtB substrate is a phosphodiester-containing molecule other than DNA. Alternatively, the threshold of DNA strand breakage tolerated by the cell before induction of the damage cascade may be extremely low. Quantitative studies indicate that a single double-stranded break is sufficient to initiate DNA damage-mediated cell cycle arrest. Unlike treatment with IR, which is a single assault, CDT treatment lasts for many hours, during which it is likely that CdtB remains within the nucleus, continually inflicting a low but persistent level of strand breakage.

Apoptosis

Mammalian cells undergo growth restriction following genotoxic stress either by cell cycle arrest or apoptosis. At the top of the DNA damage recognition pathway are sensor kinases that are characterized by ataxia telangiectasia-mutated (ATM) and ataxia telangiectasia-related (ATR) (11, 14). Damage to chromosomal DNA, such as that inflicted by IR or UV light, initiates a cascade consisting of a twofold approach to arresting cell growth (1, 14). One arm of the pathway stops the cell from entering mitosis by indirectly inhibiting the Cdc25C regulatory phosphatase (see earlier and Fig. 4). In addition to the blockage of the cell cycle by inactivation of Cdc25, an equally important response to DNA damage is the ATM/ATR-dependent growth restriction mediated by the tumor suppressor gene product p53 (1). Following DNA damage, p53, a sequence-specific transcription factor, undergoes several posttranslational modifications that result in an increase in its steady-state level and the induction of a number of downstream effectors that mediate cell cycle arrest and apoptosis. The level of p53 is maintained largely by regulation of its degradation. Under normal circumstances, the murine double minute (Mdm2) oncoprotein binds p53 and escorts it to the cytoplasm, where it is targeted for ubiquitin-dependent degradation. Following DNA damage the association of Mdm2 for p53 is dramatically decreased and the nuclear level of p53 rises. Coincident with p53 stabilization is the modification-specific increase in its activity as a transactivator of cell cycle arrest and apoptosis (1).

Activation of p53 induces myriad additional factors that may induce cell cycle arrest at the G_1/S and the G_2/M boundaries and/or apoptosis. Activation of p53 following DNA damage- and CDT-dependent stabilization may thus be central to the decision of a particular cell type to undergo apoptosis versus cycle arrest. Given the similarity of the cellular response to CDT and that following an assault with IR, it is no surprise that certain CDT-treated cells undergo apoptosis. Aa-CDT and Hd-CDT induce apoptosis in human peripheral blood mononuclear cells (HPBMCs), Jurkat human acute T-lymphoblastic cells, and Epstein-Barr virus-induced B-cell lymphoblast lines (3). Stabilization of p53 and induction of the downstream cyclin-dependent kinase inhibitors p21 and p27 were observed in cells treated with Hd-CDT and IR (3). Elevation of p21 in response to CDT might be expected if CDT does, in fact, induce DNA damage. Although associated with arrest in G_1, p21 has also been shown to enforce arrest in G_2 by interfering with the Thr-161-mediated phosphorylation (activation) of Cdc2. Thus CDT-mediated elevation of p21 will result in some cells being blocked in G_1 while at the same time enforcing the Chk2-mediated block in G_2. Together, these data provide further evidence that CDT acts through the ATM-dependent DNA damage regulatory cascade or a redundant cascade such as that mediated by ATR.

ATM and ATR directly phosphorylate p53 on Ser-15 in response to DNA damage (14). Examination of p53 following treatment with Hd-CDT or IR indicated that p53^{Ser15} is phosphorylated with similar kinetics by both treatments (3). Apoptosis in human T cells by Aa-CDT is accompanied by mitochondrial permeability; activation of caspases 8, 9, and 3; and expression of the mitochondrial membrane protein Apo 2.7, a marker of apoptosis. In addition, T cells were reduced in size and displayed an increase in nuclear condensation and DNA fragmentation (3). Overexpression of the antiapoptosis proto-oncogene bcl-2 in Aa-CDT-treated T cells significantly diminished the development of apoptosis in these cells. Development of apoptosis was somewhat delayed, in that the majority of apoptotic cells were observed 72 to 96 h after the addition of toxin. This same observation was made with B cells treated with Hd-CDT. The potential involvement of the CD95 death receptor in CDT-mediated apoptosis is presently unclear. Recent studies, however, indicate that DNA damage-mediated apoptosis occurs independent of CD95 or fatty acid synthase involvement. In addition, the antiapoptotic activity of Bcl-2 appears to provide protection against radiation-induced apoptosis but does not influence CD95- or TRAIL-mediated apoptosis.

The decision of a cell to undergo cycle arrest versus apoptosis is complicated by the interconnectivity of these pathways and the responses of different cell types to identical stimuli. It is clear that CDT treatment of cells mimics genotoxic stress in a number of ways discussed above, including the dependence on ATM (and perhaps ATR), activation of Chk2, stabilization of p53 and elevation of downstream p53-regulated target genes, inactivation of Cdc25 by phosphorylation on Ser-216, the capacity of elevated Cdc25^{wt} and Cdc25^{S216S} to override a CDT block, and the accumulation of an inactive phosphorylated form of Cdc2 (Fig. 4). Like other forms of DNA damage, CDT treatment of cells leads to apoptosis in certain cell types and cell cycle arrest in others. Thus far, apoptosis in response to CDT has only been observed in cells of lymphoid and myeloid lineages. This may reflect the sensitivity of these cells to apoptosis and/or the resistance of transformed cells of epithelial and fibroblast lineages to apoptosis. To further complicate matters, apoptosis has thus far only been reported in cells treated with Aa-CDT and Hd-CDT. These two CDTs are highly related to one another, yet reasonably different from CDTs that are more closely related to the Cj-CDT cluster (4, 10). Induction of apoptosis may, therefore, be specific to Aa-

CDT and Hd-CDT. Although it is more likely that all CDTs will be found to induce apoptosis in cells of lymphoid and myeloid lineage, this is yet to be proved. It has been reported that Aa-CdtB is sufficient to induce apoptosis in HPBMCs (12). However, induction of apoptosis with Hd-CDT required the presence of Hd-CdtB and Hd-CdtC purified from a strain producing all three proteins. Whether this discrepancy represents a difference between Aa-CDT and Hd-CDT and the other CDTs or differential susceptibilities of different target cells remains to be determined.

Cell Death

Unless CDT-treated cells undergo apoptosis, death occurs over several days following toxin treatment. Specifically, HeLa cells treated with a cytostatic dose of CDT (the amount of CDT that will arrest 100% of cells in G_2 after 24 h) are unchanged with respect to viability for 48 h, after which viability steadily decreases over the next 3 to 4 days. During this time, cells continue to grow in size and synthesize protein. During the period of decline in viability (48 to 96 h following exposure to a lethal dose of CDT), HeLa cells undergo death-related phenotypes, including abortive mitosis, a pseudomitotic state where cells containing highly condensed and fragmented nuclei round up and detach from the plate yet fail to undergo mitosis; endoreduplication, where cells continue to undergo completion of new rounds of DNA synthesis but without cell division; and micronucleation, characterized by nuclei displaying an aberrant morphology, progressive lobulation, and finally fragmentation. During this period of declining viability, the accumulation of free 3'-OH groups, indicative of DNA strand breakage, was not observed (3, 4). This may suggest that the DNA-altering effect of CDT (CdtB) is independent of strand breakage, that the damage that occurs is continually being repaired, or that early minimal damage is sufficient to seal the fate of the cell once the checkpoint cascade is enforced.

CELLULAR ENTRY OF CDT

The emerging model for CDT action (Fig. 5) predicts that CDT enters the cell and, at least CdtB, traffics to the nucleus, where it inflicts damage on chromosomal DNA to initiate the mitotic checkpoint or apoptosis. Using a pharmacological blockage, the internalization of Hd-CDT by Hep-2 and HeLa cells was examined by Cortes-Bratti et al. (2). While monitoring the accumulation of phosphorylated Cdc2 as evidence of CDT intoxication, various drug and cellular conditions were tested to characterize the apparent uptake of CDT. By this approach, CDT intoxication was blocked by chlorpromazine and by the expression of a dominant negative dynamin gene, indicating that CDT is probably taken up in clathrin-coated pits (2). Inhibitors of both early and late endosome formation and brefeldin A, a disrupter of Golgi function, all blocked the action of CDT. Together, these data suggest that following binding to a cell surface receptor, CDT enters the cells within endosomes that mature to the late endosomal stage followed by trafficking to the Golgi. From here, it is presumed that CdtB enters the nucleus to impart a disruptive action on the chromosome. Electroporation of CdtB into cells is sufficient to mediate intoxication to the full extent observed with CDT

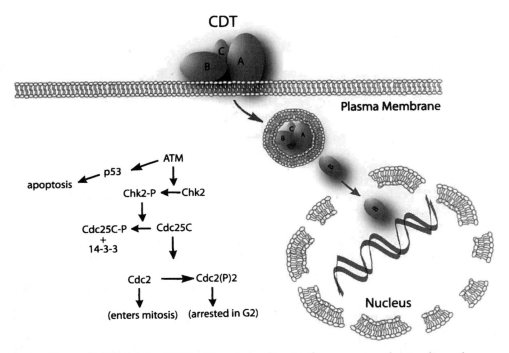

Figure 5 Model for CDT action. According to the current understanding of CDT action, CDT appears to bind to a receptor on the surface of sensitive cells, enters the cell by receptor-mediated endocytosis, and traffics through the cell through early and late endosome, and following involvement of the Golgi, CdtB enters the nucleus. Nuclear CdtB is then responsible for imparting a damaging effect on chromosomal DNA, resulting in growth arrest either by the mitotic checkpoint cascade or DNA damage-mediated apoptosis.

holotoxin treatment (5). This finding suggests that CdtB needs only to be delivered to the cytoplasm to proceed to its final target. Thus, nuclear transport occurs without endosomal involvement. The size of CdtB (approximately 30 kDa) is such that it may simply diffuse from the Golgi into the nucleus. Alternatively, CdtB may contain a nuclear localization sequence or sequences that facilitate nuclear entry. Alternatively, CdtB may bind in the cytoplasm to a nuclear-bound cellular protein and enter the nucleus via the bound carrier. Resolution of this will take direct assessment of the cellular and nuclear trafficking of CdtB.

PERSPECTIVES ON CDT

The identification of CDT production by a growing list of widely divergent gram-negative bacterial pathogens is unparalleled in the investigation of bacterial toxin. Despite direct evidence implicating a role for CDT in pathogenesis, the extreme sensitivity of human T and B cells and dividing cells in general suggests that CDT plays at least an important accessory role during the initiation of pathogenesis, that is, overcoming the host's cellular defenses. In addition, that biologically functional CDT appears in so many pathogenic species suggests that there may be a selective advantage for its maintenance and production.

Recent findings on the role of CdtB in the mechanism of CDT action, the similarity of CDT action to that of IR, the apparent pathway of CDT internalization, and the reconstitution of biological activity from pure CDT subunits constitute major advances in the understanding of CDT. Despite these advances, much remains unclear or unknown. For instance, despite genetic and biochemical evidence suggesting that all three CDT proteins are required for biological action, Aa-CdtB apparently is sufficient to induce apoptosis in HPBMCs. This discrepancy may be due to the differences in Aa-CdtB relative to other CdtBs; or all CdtBs have the capacity to induce apoptosis in HPBMCs, a model that has not been used with other CDTs. Clearly, native and recombinant CDT holotoxin from various CDT-producing species should be purified and analyzed in side-by-side comparisons. Hybrid analysis of CDT subunits from various sources should also be performed. Ultimately, the role of each CDT subunit in cell cycle arrest and apoptosis needs to be clarified. Recent insights into how CDT enters and traffics through the cell give some clues to the mechanism of the intracellular transport of CDT. In addition, these studies provided indirect evidence for the presence of a cell surface receptor for CDT. Studies to identify the putative CDT receptor and a direct examination of CDT uptake and intracellular trafficking are clearly needed. The proposed nuclear import of CDT also needs direct examination.

Several reports indicate that the cellular response to CDT is similar, if not identical, to that following IR or other forms of genotoxic stress. The fact that CdtB possesses DNase activity and the finding that DNase active-site residues are required for CDT biological activity suggest that the capacity of CdtB to damage the host cell chromosome triggers the mitotic checkpoint cascade. The inability to observe significant DNA strand breakage in CDT-treated cells, however, represents a potential discrepancy that needs resolution. The apparent cell-type specificity for the induction of apoptosis versus cell cycle arrest and the activation of p53 following CDT treatment represent additional parallels between toxin action and radiation treatment. The resolution of this observation will yield not only important information regarding CDT but perhaps also the cellular response to DNA damage in general.

Understanding the molecular mechanism of bacterial toxins not only provides clues to the role these molecules play in pathogenesis but also yields insights into the cell biology and biochemistry of eukaryotic cells. In this regard, the study of CDT is no exception. Studies thus far have proved CDT to be a unique bacterial product with a novel mechanism of action. Like other bacterial toxins that may have originated as eukaryotic gene products, CDT turns the table on the host cell by apparently delivering a type I DNase to the nucleus of the cell. The resulting chromosomal damage mediates cell death by triggering either cell cycle arrest or apoptosis. Defining the molecular functions of CdtA and CdtC, identifying the CDT receptor, and characterizing the trafficking of the toxin through the host cell to its target should prove to be exciting and fruitful studies. Finally, studies designed to define the role of CDT in pathogenesis will hopefully provide some insight into the unprecedented production of this novel bacterial toxin by such a wide and varied group of bacterial pathogens.

REFERENCES

1. **Bates, S., and K. H. Vousden.** 1999. Mechanisms of p53-mediated apoptosis. *Cell. Mol. Life Sci.* **55:**28–37.
2. **Cortes-Bratti, X., E. Chaves-Olarte, T. Lagergard, and M. Thelestam.** 2000. Cellular internalization of cytolethal distending toxin from *Haemophilus ducreyi. Infect. Immun.* **68:**6903–6911.
3. **Cortes-Bratti, X., T. Frisan, and M. Thelestam.** 2001. The cytolethal distending toxins induce DNA damage and cell cycle arrest. *Toxicon* **39:**1729–1736.
4. **De Rycke, J., and E. Oswald.** 2001. Cytolethal distending toxin (CDT): a bacterial weapon to control host cell proliferation? *FEMS Microbiol. Lett.* **203:**141–148.
5. **Elwell, C. A., K. Chao, K. Patel, and L. A. Dreyfus.** 2001. *Escherichia coli* CdtB mediates the cytolethal distending toxin cell cycle arrest. *Infect. Immun.* **69:**3418–3422.
6. **Elwell, C. A., and L. A. Dreyfus.** 2000. DNase I homologous residues in CdtB are critical for cytolethal distending toxin-mediated cell cycle arrest. *Mol. Microbiol.* **37:**952–963.
7. **Hassane, D. C., R. B. Lee, M. D. Mendenhall, and C. L. Pickett.** 2001. Cytolethal distending toxin demonstrates genotoxic activity in a yeast model. *Infect. Immun.* **69:**5752–5759.
8. **Lara-Tejero, M., and J. E. Galan.** 2000. A bacterial toxin that controls cell cycle progression as a deoxyribonuclease I-like protein. *Science* **290:**354–357.
9. **Lara-Tejero, M., and J. E. Galán.** 2001. CdtA, CdtB, and CdtC form a tripartite complex that is required for cytolethal distending toxin activity. *Infect. Immun.* **69:**4358–4365.
10. **Pickett, C. L., and C. A. Whitehouse.** 1999. The cytolethal distending toxin family. *Trends Microbiol.* **7:**292–297.
11. **Rotman, G., and Y. Shiloh.** 1999. ATM: a mediator of multiple responses to genotoxic stress. *Oncogene* **18:**6135–6144.
12. **Shenker, B. J., R. H. Hoffmaster, A. Zekavat, N. Yamaguchi, E. T. Lally, and D. R. Demuth.** 2001. Induction of apoptosis in human T cells by *Actinobacillus actinomycetemcomitans* cytolethal distending toxin is a consequence of G2 arrest of the cell cycle. *J. Immunol.* **167:**435–441.
13. **Smits, V. A., and R. H. Medema.** 2001. Checking out the G(2)/M transition. *Biochim. Biophys. Acta* **1519:**1–12.
14. **Wilson, J. F.** 2002. Elucidating the DNA damage pathway. *Scientist* **16:**30–31.

SUGGESTED READING

1. **Alberts, B., A. Johnson, J. Lewis, M. Raff, K. Roberts, and P. Walter.** 2002. The cell cycle and programmed cell death, p. 983–1026. *In* B. Alberts, A. Johnson, J. Lewis, M. Raff, K. Roberts, and P. Walter (ed.), *Molecular Biology of the Cell*, 4th ed. Garland Science, New York, N.Y.

Bacterial Protein Toxins
D. Burns et al., Editors
©2003 ASM Press, Washington, DC 20036

Chapter 19

Proteases

Ornella Rossetto, Fiorella Tonello, and Cesare Montecucco

Proteases are enzymes that catalyze the hydrolysis of peptide bonds in proteins and peptides. Many extracellular bacterial proteases from pathogenic organisms have been demonstrated to play important roles in virulence. They can be classified into four groups based on the essential catalytic residue at their active site. They include serine proteases, cysteine proteases, aspartate proteases, and the metalloproteases. This chapter will focus on bacterial protein toxins that target eukaryotic cells acting as zinc-metalloproteases.

Metalloproteases are hydrolytic enzymes characterized by an active site containing a metal atom, usually zinc. They include aminopeptidases, carboxypeptidases, and endopeptidases, depending on whether they remove an N- or a C-terminal residue or whether they cleave internal peptide bonds of the protein substrate. Zinc-dependent endopeptidases are characterized by the presence of a zinc-binding motif consisting of His-Glu-X-X-His. The determination of their primary sequence has led to the discovery of the metalloproteolytic activity of the bacterial toxins responsible for tetanus, botulism, and anthrax. The protease domain of these toxins enters into the cytosol, where it displays a zinc-dependent endopeptidase activity of remarkable specificity. Tetanus and botulinum toxins cleave three protein components of the neuroexocytosis machinery, leading to the blockade of neurotransmitter release and consequent muscle paralysis. The lethal factor of *Bacillus anthracis* is specific for the mitogen-activated protein kinase (MAPK) kinases that are cleaved within their aminoterminus.

Strains of *Bacteroides fragilis* associated with diarrheal disease in animals and young children have been recognized to secrete an ~20-kDa zinc-dependent metalloprotease toxin termed fragilysin (BFT), which alters the cytoskeletal structure of epithelial cells by cleaving the cell surface protein E-cadherin.

Additional understanding of the mode of action of proteases that contribute to pathogenicity could lead to the development of inhibitors (chelators, surrogate substrates, antibodies, etc.) that could prevent or interrupt the disease process.

Ornella Rossetto, Fiorella Tonello, and Cesare Montecucco • Dipartimento di Scienze Biomediche, Università di Padova, Viale G. Colombo, 3, 35121 Padova, Italy.

TETANUS AND BOTULINUM NEUROTOXINS

The seven botulinum neurotoxins (BoNTs, types A to G) and the single tetanus neurotoxin (TeNT) are responsible for the clinical manifestations of botulism and tetanus, respectively, and constitute a group of metalloproteases endowed with unique properties (Box 8). BoNTs bind to and enter inside peripheral cholinergic terminals and cause a sustained block of acetylcholine release, with ensuing flaccid paralysis and loss of function of some glands. On the contrary, TeNT acts on the central nervous system and blocks neurotransmitter release at the inhibitory interneurons of the spinal cord, resulting in a spastic paralysis. These bacterial

---

BOX 8

Landmarks in tetanus and botulinum research

Tetanus
1884 Tetanus is a transmissible disease (Carle and Rattone)
1884 Identification of a bacterium as the cause of tetanus (Nicolaier)
1889 Isolation and pure culture of *Clostridium tetani* (Kitasato)
1890 Identification of tetanus neurotoxin as the single cause of tetanus disease (Faber, Tizzoni, and Cattani; Kitasato)
1892 Neuroselective binding and spinal cord activity (Bruschettini)
1903 Retrograde transport to the spinal cord (Meyer)
1925 Toxoid and vaccination (Ramon and Descombey)
1955 Inhibitory interneurons of the spinal cord identified as the site of prolonged blockade of neurotransmitter release (Eccles and colleagues)
1961 Binding to ganglioside (van Heyningen)
1976 Retroaxonal transport and transynaptic transfer (Schwab and Thoenen)
1986 DNA sequence of the tetanus neurotoxin gene (Eisel et al.; Fairweather et al.)
1992 Enzymatic action (Schiavo et al.)

1992 Target identification (Schiavo et al.)
1997 Crystal structure of H_C (Umland et al.)

Botulism
1817 Clinical description of botulism (Kerner)
1897 Isolation of *Clostridium botulinum* (van Ermengem)
1923 Specific action on cholinergic synapses (Edmunds)
1949 Neurotransmission blockade revealed (Burgen et al.)
1951 Wound botulism (Davis et al.)
1971 Binding to gangliosides (Simpson)
1973 First clinical use of BoNT/A (Scott et al.)
1976 Infant botulism (Midura and Arnon; Pickett et al.)
1990/93 Complete DNA sequence (various laboratories)
1992/93 Enzymatic action (Schiavo et al.)
1992/93 Targets identified (Blasi et al.; Schiavo et al.)
1998 Crystal structure of BoNT/A (Lacy et al.)
2000 Crystal structure of BoNT/B (Swaminathan and Eswaramoorthy; Hanson and Stevens)

metalloproteases do not act on the cell surface but exert their enzymatic action in the cell cytosol on selected proteins that form the core of the neuroexocytosis machinery. As a net result of such double specificity of binding (presynaptic membrane and substrate) and of their catalytic activity, these neurotoxins are the most potent toxins known, with a mouse 50% lethal dose of few picograms.

The various BoNTs are produced by distinct strains of *Clostridium botulinum* that display heterogeneous bacteriological characteristics. Moreover, several BoNT serotypes can be secreted by other clostridia. In contrast, TeNT is produced by a uniform group of *Clostridium tetani* bacteria. All clostridial neurotoxins (CNTs) are made in the bacterial cytosol and are released into the culture medium after bacterial lysis as a single, inactive, polypeptide chain of 150 kDa that is subsequently cleaved within a surface-exposed loop subtended by a highly conserved disulfide bridge. This generates the fully active neurotoxin, composed of a heavy chain (H, 100 kDa) and a light chain (L, 50 kDa) linked by both non-covalent protein-protein interactions and a conserved interchain S-S bond, whose integrity is essential for neurotoxicity (Fig. 1A). During the intoxication process, the interchain bridge is reduced, and this is a necessary prerequisite for the intracellular action of the toxins.

The crystallographic structures of BoNT/A and BoNT/B have been recently determined and reveal three distinct 50-kDa domains (4, 9): an N-terminal domain endowed with zinc-endopeptidase activity (L chain), a membrane translocation domain (H_N) characterized by the presence of two 10-nm-long α-helices, which are reminiscent of similar elements present in colicins and in the influenza virus hemagglutinin, and a binding domain (H_C), composed of two unique subdomains (H_{CN} and H_{CC}) similar to the legume lectins and Kunitz inhibitor (Fig. 1B).

Their structural architecture, of three domains (L, H_N, and H_C) endowed with different functions, is strictly linked to their mechanism of cell intoxication, which consists of four steps. Nerve cell binding mediated by H_C (step 1) is followed by internalization inside vesicles (step 2), which are endowed with an ATPase proton pump. The acidification of the vesicular lumen causes a rearrangement of the toxin structure, which inserts H_N in the membrane and translocates L in the cytosol (step 3), where it displays its proteolytic activity (step 4). This chapter will deal only with the latter step (for recent discussions of the other steps, see reference 7).

The catalytic activity of these neurotoxins was discovered following the sequencing of the corresponding genes, which began with TeNT and, within a few years, was extended to all CNTs. Sequence comparison revealed a highly conserved 20-residue-long segment, located in the middle of the L chain, containing the His-Glu-Xaa-Xaa-His zinc-binding motif of zinc-endopeptidases. At the center of the long cleft-shaped active site there is a zinc atom coordinated via the two histidines and the glutamic residues of the zinc-binding motif, and by Glu-262 in BoNT/A and Glu-268 in BoNT/B, a residue conserved among all CNTs (the amino acid numbering is that of CNTs with methionine as the first residue). The Glu residue of the motif is particularly important because it coordinates the water molecule, which actually performs the hydrolytic reaction of proteolysis. Its mutation leads to complete inactivation of these neurotoxins. The critical role of Glu-271 of TeNT and Glu-262 in BoNT/A has been shown to be that of pro-

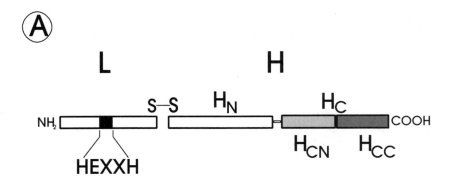

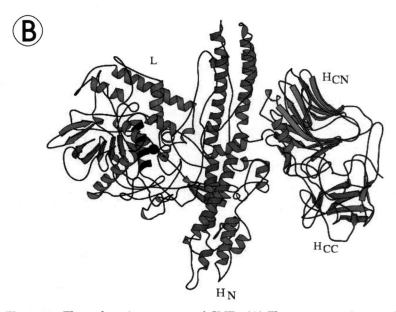

Figure 1 Three-domain structure of CNTs. (A) These neurotoxins consist of three domains of similar size (50 kDa). The N-terminal domain (L) is a zinc-endopeptidase containing the His-Glu-Xaa-Xaa-His zinc-binding motif of zinc-endopeptidases. Its intracellular activity is expressed after reduction of the interchain disulfide bond. H_N, the central domain, is responsible for the membrane translocation of the L chain into the neuronal cytosol. The C-terminal H_C domain consists of two equally sized subdomains. The N-terminal subdomain (H_{CN}) has a structure similar to that of sugar-binding proteins. The C-terminal subdomain (H_{CC}) folds similarly to proteins known to be involved in protein-protein binding functions such as the K^+ channel-specific dendro-toxin. Such structure is consistent with the toxin binding to the presynaptic membrane via a double interaction, most likely with two different molecules of the nerve terminal. (B) Crystallographic structure of BoNT/A oriented as in the top panel: the zinc atom and the α-helix containing the HEXXH motif are shown in black. Notice the unusual 10-nm-long pair of α-helices present in the H_N domain (from reference 4 with permission).

viding a negatively charged carboxylate moiety. This active-site architecture is similar to that of thermolysin and identifies a primary sphere of residues essential to the catalytic function, which coincides with the zinc coordinating residues. From the analysis of the available structures, it appears that a secondary layer of residues is present at the active site of BoNT/A and BoNT/B. Among these residues, Arg-363 and Tyr-366 in BoNT/A play a role in the catalytic activity of this family of metalloproteases. The mutation of this Tyr with an alanine inactivates the TeNT L chain and dramatically reduces the proteolytic activity of the BoNT/A L chain, clearly indicating that this residue plays a critical role in the hydrolysis of the substrate. For the BoNT/B L chain it has been proposed that this Tyr assists the hydrolysis reaction by donating a proton to the amide nitrogen of VAMP Phe-77, stabilizing, together with bound water molecules, the leaving group (Fig. 2). In addition, other fully conserved residues contained inside the active-site pocket, such as Phe-266 of BoNT/A, which appears to shield partially the opposite opening of the mouth of the active-site cavity, play a role in the enzymatic activity of CNTs. The replacement of Phe-266 by an alanine leads to a reduction of the rate of hydrolysis, and it is possible that this mutation alters the distances between the bound substrate, the Arg-363 (postulated to stabilize the transition state complex), and the Tyr-366, which appears to be critical for catalysis.

The light chains of the eight CNTs are remarkably specific proteases: among the many proteins and synthetic substrates assayed so far, only three targets, the

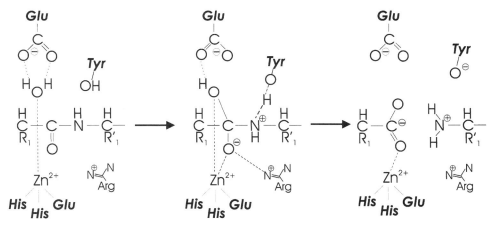

Figure 2 A mechanism for substrate proteolysis by CNTs and by anthrax LF. The active site of the CNTs and of LF is centered around a structural zinc cation coordinated by a strictly conserved HEXXH+E motif. In both cases zinc is directly coordinated by the two histidines of the motif, by a glutamic acid (Glu-268 in BoNT/B and Glu-735 in LF) and by a water molecule that forms a strong hydrogen-bonding interaction with the Glu of the motif. A conserved tyrosine of the active site (Tyr-373 in BoNT/B and Tyr-728 in LF) likely functions as a general acid to protonate the amine leaving group in the transition state, assisting the hydrolysis reaction to yield the cleaved products. A conserved Arg present in the active site of CNTs but not in the one of LF probably has a role in stabilizing the transition state in combination with the zinc ion.

so-called SNARE proteins, have been identified. TeNT and BoNT/B, /D, /F, and /G cleave VAMP, each at a single site; BoNT/A and /E cleave SNAP25, each at a single site; and BoNT/C cleaves both syntaxin and SNAP25. Numerous isoforms of the three SNARE proteins have been identified, and different isoforms coexist within the same cell. Only some of them are susceptible to proteolysis by the CNTs. In general, a SNARE protein is resistant to a neurotoxin because of mutations at the cleavage site and/or in the regions involved in neurotoxin binding (3, 7).

The three SNARE proteins form a heterotrimeric coiled-coil SNARE complex, which induces the juxtaposition of vesicle to the target membrane and is involved in their fusion. The proteolysis of one SNARE protein prevents the formation of the complex and consequently the release of the neurotransmitter (Fig. 3). The SNARE complex is insensitive to CNT proteolysis, as expected on the basis that proteases are known to attack predominantly unstructured exposed loops.

All toxins cleave their targets at distinct sites, except for BoNT/B and TeNT, which attack the same peptide bond in VAMP. The molecular basis of the CNTs' specificity for the three SNAREs is only partially known. Sequence comparison of the SNARE proteins involved in neuroexocytosis revealed the presence of a nine-residue-long motif, characterized by three carboxylate residues alternated with hydrophobic and hydrophilic residues, termed thereafter SNARE motif (6).

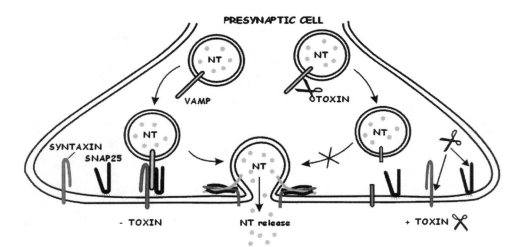

Figure 3 Scheme of a presynaptic cell. The figure shows the normal process of the neurotransmitter (NT) release (left) and the mechanism of action of TeNT and BoNTs (right). Vesicles containing the NT have a transmembrane protein called VAMP that binds specifically two proteins of the cell membrane (syntaxin and SNAP25). The three proteins make a complex that forces the vesicle and cellular membranes to come in close contact and finally to fuse, thus releasing the NT into the intersynaptic space. TeNT and BoNTs are zinc-endopeptidases that cleave VAMP, syntaxin, and SNAP25, thus preventing the NT release.

The motif is present in two copies (V1 and V2) in VAMP and syntaxin and four copies in SNAP25. These segments are contained within regions adopting a helical conformation, and when the motif is plotted as an α-helix, the three negatively charged residues cluster on one face, adjacent to a hydrophobic face of the helix (6).

The various CNTs differ with respect to the specific interaction with the SNARE motif. The three carboxylate residues of the V2 copy of the motif are very important for the proteolysis by BoNT/B and /G of VAMP, whereas those of the V1 copy are implicated in the recognition of BoNT/F and TeNT. BoNT/D shows a particular requirement for the Met-46 present in V1. In view of the fact that V1 is more amino-terminal with respect to V2, these results explain why the minimal length of VAMP segments cleaved by TeNT is longer than that required by BoNT/B. Assuming that the L chains of TeNT and BoNT/B are similar, these results indicate that the V1 and V2 motifs of VAMP are paired and similarly oriented with respect to the Gln-76–Phe-77 bond. In addition, a basic region located after the cleavage site of TeNT and BoNT/B is important for their binding and optimal cleavage of VAMP.

The biochemical and structural properties of CNTs define them as a distinct group of metalloproteases, whose origin cannot, at the present time, be traced in any of the known families of metalloproteases.

THE ANTHRAX LETHAL FACTOR

Anthrax lethal factor (LF) is produced by toxigenic strains of *Bacillus anthracis*, a gram-positive spore-forming bacterium responsible for the anthrax disease. *Bacillus anthracis* secretes an exotoxin complex consisting of three distinct proteins and acting in binary combinations: the protective antigen (PA), which elicits a protective immune response against anthrax; LF; and the edema factor (EF). By itself PA is nontoxic. It binds to a cellular receptor recently identified as a type I membrane protein with an extracellular von Willebrand factor A domain (1). PA mediates the endocytosis and the translocation of EF or LF from the endosomal lumen into the cytosol.

EF is a calmodulin-dependent adenylate cyclase that causes a dramatic elevation of cyclic AMP (cAMP) concentration within the host cell. Intradermal injection of PA plus EF (EdTx, edema toxin) produces the skin edema characteristic of anthrax but is not toxic when injected intravenously. At variance, LF plus PA (LeTx, lethal toxin) injected intravenously causes rapid death of laboratory animals and in vitro lyses some cell lines and primary cultures of murine macrophages. LF is a zinc-binding protein that includes the HEXXH motif and acts in the cytosol via a metalloproteolytic activity. The mature protein contains 776 residues with a molecular size of 85 kDa. The crystallographic structure of LF (5) reveals four distinct domains (Fig. 4). The amino-terminal domain (1 to 254) binds PA and is perched atop the other three domains. The second domain (residues 263 to 297 and 385 to 550) resembles the ADP-ribosylating toxin VIP2 from *Bacillus cereus*, but the active site has been mutated in such a way that it is not expected to have ADP-ribosylating activity. The third domain (303 to 382) is a small α-helical bundle with a hydrophobic core inserted between the second and third helices of the second domain. It is required for LF activity, since insertion

Figure 4 Stereo ribbon representation of LF, which reveals the presence of four distinct domains: the amino-terminal PA-binding domain (1, in black), the second domain resembling the ADP-ribosylating toxin VIP2 from *Bacillus cereus* (2, in light gray), the third domain inserted between the second and third helices of the second domain (3, in gray), and the fourth C-terminal catalytic domain (4, in dark gray). Zinc ion in the active site of the catalytic domain is shown as a black sphere. Domains 2, 3, and 4 together create a long deep groove that holds the N-terminal tail of the MAPKK prior to cleavage (adapted from reference 5).

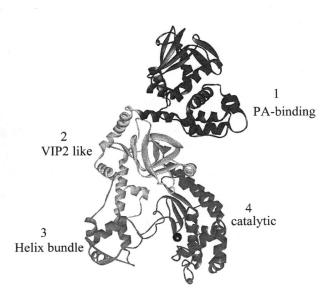

mutagenesis and point mutations of buried residues in this domain abrogate function, and it contributes to substrate specificity. The catalytic domain resides in the C-terminal fourth domain (552 to 776), which consists of a nine-helix bundle packed against a four-stranded β-sheet. No homologies were detected by sequence comparison with other known proteins except for the HEXXH motif. The zinc ion is coordinated tetrahedrally by the two histidines of the motif (His-686 and His-690), by Glu-735, and by a water molecule bound to the glutamic acid of the motif (Glu-687). Substitution of Ala for H686, E687, or H690 in the sequence $_{686}$HEFGH$_{690}$ abolishes the binding of zinc to LF and its toxicity both on macrophage cell lines and on Fisher-344 rat. The hydroxyl group of a tyrosine residue (Tyr-728) forms a strong hydrogen bond with a water molecule and probably functions as a general acid to protonate the amine leaving group (Fig. 2).

The recognition that LF contains the consensus sequence of zinc-metalloproteases started a process that eventually led to the identification of its catalytic activity. LF is a metalloprotease that cleaves members of the MAPK kinases (MAPKKs) family. These proteins play pivotal roles in a variety of signal transduction pathways, aspects of which are critical for cell cycle progression and differentiation. The MAPK pathway relays environmental signals to the transcriptional machinery in the nucleus of the cell modulating gene expression via a burst of protein phosphorylation. Seven different MAPKKs are known, composing three distinct MAPK cascades. LF cleaves all the MAPKKs except MAPKK5, and cleavage invariably occurs within the N-terminal proline-rich region preceding the kinase domain. LF appears to recognize the following consensus motif of the MAPKKs: a cluster of positively charged residues within the

P7 to P4 positions, P2 is generally an aliphatic lateral chain, and the P1' is always a hydrophobic residue (10). MAPKK cleavage disrupts a sequence involved in directing specific protein-protein interactions necessary for the assembly of signaling complexes. Elevated MAPK expression has been detected in a variety of human tumors, and LeTx was shown to be effective in inhibiting MAPK activation in V12 H-ras-transformed NIH 3T3 cells, potently inhibiting tumor growth (9). Although the signaling pathways involving MAPKKs have a crucial role in the activation of macrophages and are directly involved in the production of cytokines such as tumor necrosis factor (TNF), interleukin (IL) -1, and IL-6, the link between the cleavage of MAPKKs by LF, macrophage lysis, and pathogenesis remains unclear. Macrophage cell lines resistant to the lytic effect of LeTx and peritoneal macrophages isolated from mouse strains insensitive to LeTx are still sensitive to the protease activity of LF on their MAPKKs. Mutations in Kif1C, a kinesin-like motor protein of the UNC104 subfamily, was suggested to be responsible for differences in mouse macrophage susceptibility to LeTx, but the mechanism by which Kif1C would function downstream in the intoxication pathway and protect the macrophages is still unclear. The release of proinflammatory cytokines potentially elaborated by macrophages upon treatment with LeTx could account for shock, and mice depleted of macrophages have been reported to become insensitive to LeTx challenge. However, there are conflicting data concerning the modulation of TNF and IL-1 by LeTx. Whereas one laboratory reported that sublytic doses of LeTx induced production of these cytokines, a lack of effect and a LeTx inhibition of the production of NO and TNF and of the lipopolysaccharide-induced production of mRNA coding for TNF-α, IL-1, and IL-6 were reported by others. These findings suggest that LeTx may act to impair host response in reducing the inflammatory response and thus enhance bacterial growth and virulence in the initial phase of infection.

FRAGILYSINS

Bacteroides fragilis is a member of the normal microflora of most mammals and is the most commonly isolated anaerobe from human clinical specimens. Some enterotoxigenic strains (ETBF) produce a toxin (fragilysin, BFT) implicated as a cause of diarrheal disease in farm animals and humans. The three highly homologous genes coding for BFTs known so far are termed *bft-1* (from strain VPI 13784), *bft-2* (from strain 86-5443-2-2), and *bft-3* (from strain Korea 419). These genes are included in a 6-kb chromosomal region unique to ETBF strains termed the *Bacteroides fragilis* pathogenicity island. The BFT genes consist of one open reading frame coding for proteins of nearly 400 residues. BFT is synthesized as a preproprotein peptide, and it belongs to the intramolecular chaperone protease family (8). As depicted in Fig. 5, BFT is composed of an 18-residues signal sequence and a 193-residues-long propeptide, which is essential for the correct folding of the enzymatic domain (186 residues). To account for a large variability in the net amount of secreted toxin, differences must exist in the intracellular synthesis, processing, and secretion of BFTs among the various enteropathogenic *Bacteroides fragilis* strains, but the mechanism accounting for this variation and its relationship with pathogenesis remain to be investigated in detail.

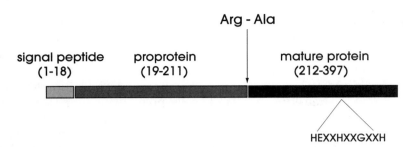

Figure 5 Schematic structure of fragilysin. The predicted amino acid sequences of the *bft-1*, *bft-2*, and *bft-3* genes suggest that the encoded proteins are preproprotein toxins. Cleavage between Arg and Ala residues appears to release the mature BFT protein. Each protein contains a HEXXHXXGXXH motif (and a conserved methionine near the C terminus), suggesting that they belong to the metzincins group of metalloproteases.

The analysis of the BFT genes indicates that, at variance from BoNT and LF, they belong to the metzincins group of metalloproteases, characterized by the common presence of the motif HEXXHXXGXXH and by the presence of a conserved methionine near the C terminus. The catalytic zinc atom is coordinated by three conserved histidines and the water-binding glutamate of the motif. BFTs have both autoproteolytic activity and in vitro proteolytic activity for substrates such as actin, gelatin, casein, and azocoll. The *bft-2* gene additionally includes a 20-residues-long COOH-terminal amphipathic segment, which has been suggested to mediate the oligomerization and the membrane insertion of the protein toxin with creation of an ion channel. This hypothesis was put forward to account for the fact that BFT-2 is more toxic than BFT-1 or BFT-3 when tested on HT29/C1 cells (8). The activity of BFT has been demonstrated in ligated intestinal loops isolated from different animal species and has been shown to stimulate dose-dependent secretion of fluids containing sodium, chloride, and proteins. At higher doses, BFT causes a hemorrhagic inflammation. Accordingly, BFT stimulates T84 cells to secrete the proinflammatory cytokine IL-8 and transforming growth factor-β, which promote the repair of "wounded" epithelium in a dose-dependent manner. However, there are no data, as yet, to indicate whether BFT stimulates proinflammatory cytokine secretion in vivo.

The most appropriate in vitro system to assay for BFT activity consists of polarized epithelial cell monolayers grown on filters (HT29, HT29/C1, T84, MDCK, Caco-2) that form tight junctions and develop a transepithelial resistance, which is a sensitive measure of the tight junctional sealing. In fact, the main activity of BFT in these cells is the decrease of the transepithelial resistance in a dose- and time-dependent manner. Moreover, BFT induces time- and concentration-dependent changes in the structure of filamentous actin, without altering the total F-actin content of the cells. In parallel, BFT stimulates a rapid and sustained increase in the volume of HT29/C1 cells, and the electron microscopy of BFT-treated monolayers of T84 cells reveals that cell swelling occurs with a loss of the microvilli of the apical membrane and that the tight junctions and the zonula adherens develop concentration-dependent changes. Interestingly, BFT is more active when placed on the basolateral membrane of polarized (i.e., tight junction-forming) epithelial cells. T84 cells treated for 48 h or more, even with low concentrations of BFT, undergo cell death with features of apoptosis. However, BFT-induced morphological changes could be detected within approximately 15 min after addition of the toxin. Unlike CNTs and LF, BFT does not appear to enter cells but acts on the cell surface of selected targets, which include the tight junctions. Of the tight junctional proteins tested, only E-cadherin was cleaved by BFT, and cleavage could be detected within 1 min from toxin addition, in close agreement with the rapidity of BFT biologic activity. BFT cleavage of E-cadherin is a two-step process in which the extracellular domain of E-cadherin is first degraded in an ATP-independent manner (directly mediated by BFT), followed by the degradation of its intracellular domain in an ATP-dependent manner, most likely mediated by one or more intracellular proteases.

These data suggested the following model for the pathogenesis of *Bacteroides fragilis*-induced intestinal secretion (8). The toxigenic bacterial strains are presumed to attach to the apical membrane of intestinal epithelial cells and secrete BFT, which may diffuse through the zonula occludens to reach and cleave E-cadherin. Cleavage of the extracellular domain of E-cadherin is followed by loss of the intracellular domain of this molecule. Because the intracellular domain of E-cadherin is tethered to the apical network of F-actin in the epithelial cell, loss of these protein-protein interactions may precipitate focal morphological changes in the apical cellular cytoskeleton with loss of microvilli and decreased barrier function, which allows the delivery of BFT to the basolateral membranes of the intestinal epithelial cells. Here, further cleavage of E-cadherin may increase the apical morphological changes initially stimulated by BFT. These dramatic changes in the apical membrane of epithelial cells caused by BFT are hypothesized to alter the function of one or more ion transporters, resulting in net intestinal secretion. Concomitantly, protein synthesis is stimulated in the intestinal epithelial cells, resulting in the secretion of IL-8, which recruits polymorphonuclear leukocytes to the intestinal submucosa. This resulting inflammatory response is predicted to contribute to the intestinal secretory response. However, it is not yet known if ETBF stimulates intestinal inflammation in humans, and several aspects of this model still require direct experimental testing.

BFT is the first bacterial toxin known to remodel the intestinal epithelial cytoskeleton and F-actin architecture via cleavage of a cell surface molecule and represents the prototype of a novel class of bacterial toxins that act without cell internalization and covalent modification of intracellular substrates. Hence, CNTs are well-defined substrate-specific toxins that act intracellularly, whereas BFT may be the first recognized substrate-specific cell surface protease toxin. In analogy with tetanus, botulism, and anthrax, one is tempted to suggest that BFT is an essential virulence factor. However, wild-type ETBF strains and isogenic mutants of the same strain differing only in the in-frame deletion of the *bft* gene have not yet been compared for their virulence. Moreover, current knowledge of the pathogenicity of other gastrointestinal pathogens suggests a word of caution in attributing a predominant importance to single virulence factors. Much remains to be learned about the biological and physiological activities, mechanism of action, and role of these proteins in animal and human disease. Further studies will seek to characterize the signal transduction events and the cytoskeletal changes induced by BFT that promise to be a valuable addition for cell biologists and physiologists studying epithelia.

REFERENCES

1. **Bradley, K. A, J. Mogridge, M. Mourez, R. J. Collier, and J. A. Young.** 2001. Identification of the cellular receptor for anthrax toxin. *Nature* **414:**225–229.
2. **Duesbery, N. S., J. Resau, C. P. Webb, S. Koochekpour, H. M. Koo, S. H. Leppla, and G. F. Vande Woude.** 2001. Suppression of ras-mediated transformation and inhibition of tumor growth and angiogenesis by anthrax lethal factor, a proteolytic inhibitor of multiple MEK pathways. *Proc. Natl. Acad. Sci. USA* **98:**4089–4094.
3. **Humeau, Y., F. Doussau, N. J. Grant, and B. Poulain.** 2000. How botulinum and tetanus neurotoxins block neurotransmitter release. *Biochimie* **82:**427–446.
4. **Lacy, D. B., and R. C. Stevens.** 1999. Sequence homology and structural analysis of the clostridial neurotoxins. *J. Mol. Biol.* **291:**1091–1104.
5. **Pannifer, A. D., T. Y. Wong, R. Schwarzenbacher, M. Renatus, C. Petosa, J. Bienkowska, D. B. Lacy, R. J. Collier, S. Park, S. H. Leppla, P. Hanna, and R. C. Liddington.** 2001. Crystal structure of the anthrax lethal factor. *Nature* **414:**229–233.
6. **Rossetto, O., G. Schiavo, C. Montecucco, B. Poulain, F. Deloy, L. Lozzi, and C. C. Shone.** 1994. SNARE motif and neurotoxins. *Nature* **372:**415–416.
7. **Schiavo, G., M. Matteoli, and C. Montecucco.** 2000. Neurotoxins affecting neuroexocytosis. *Physiol. Rev.* **80:**717–766.
8. **Sears, C. L.** 2001. The toxins of *Bacteroides fragilis*. *Toxicon* **39:**1737–1746.
9. **Swaminathan, S., and S. Eswaramoorthy.** 2000. Structural analysis of the catalytic and binding sites of *Clostridium botulinum* neurotoxin B. *Nat. Struct. Biol.* **7:**693–699.
10. **Vitale, G., L. Bernardi, G. Napolitani, M. Mock, and C. Montecucco.** 2000. Susceptibility of mitogen-activated protein kinase kinase family members to proteolysis by anthrax lethal factor. *Biochem. J.* **15:**739–745.

SUGGESTED READING

1. **Hanson, M. A., and R. C. Stevens.** 2000. Cocrystal structure of synaptobrevin-II bound to botulinum neurotoxin type B at 2.0 Å resolution. *Nat. Struct. Biol.* **7:**687–692.
2. **Rawlings, N. D., and A. J. Barrett.** 1995. Evolutionary families of metallopeptidases. *Methods Enzymol.* **248:**183–228.

Bacterial Protein Toxins
D. Burns et al., Editors
©2003 ASM Press, Washington, DC 20036

Chapter 20

Bacterial Toxins that Modulate Rho GTPase Activity

James B. Bliska and Gloria I. Viboud

A large number of bacterial toxins target small GTPases of the Rho subclass. This fact underscores the importance of Rho GTPases as key regulators of diverse cellular processes. Bacterial toxins employ several mechanisms to activate or inactivate Rho GTPases. Some toxins are enzymes that covalently modify GTPases (see chapters 15 and 16, this volume). This chapter will describe a distinct class of toxins that mimic endogenous regulatory factors of Rho GTPases. The actions of these toxins are noncovalent and are therefore reversible. These "modulating toxins" encode domains that function as guanine nucleotide exchange factors (GEFs) or as GTPase-activating proteins (GAPs). Modulating toxins are injected into host cells by bacterial type III secretion systems (see chapter 7, this volume). They allow bacterial pathogens to control host cell actin dynamics during infection. The study of bacterial GEFs and GAPs is shedding new light on mechanisms of bacterial pathogenesis. In addition, these toxins represent powerful tools for the study of actin dynamics in eukaryotic cells.

REGULATION AND FUNCTION OF Rho GTPases

Proteins that bind to and hydrolyze the nucleotide GTP (also called GTPases) are key regulators of many cellular processes. Rho GTPases comprise a subgroup of the Ras superfamily of small GTP-binding proteins (2). Rho GTPases are key regulators of actin dynamics and transcriptional responses in eukaryotic cells. In essence, they act as on/off switches by flipping back and forth between two distinct conformations. When bound to GTP, they adopt an active (or on) conformation. GTP hydrolysis switches them to an inactive (or off) conformation (Fig. 1). Two conserved domains of Rho GTPases, known as switch I and switch II, undergo important conformational changes during this cycle. The switch I domain makes contact with "downstream" signaling proteins (also known as

James B. Bliska and Gloria I. Viboud • Department of Molecular Genetics and Microbiology, Center for Infectious Diseases, State University of New York at Stony Brook, Stony Brook, NY 11794-5222.

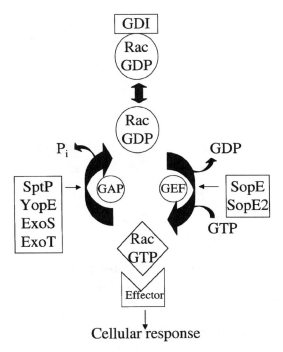

Figure 1 Factors that modulate Rho GTPase activity. As shown for Rac1, Rho GTPases cycle between an active GTP-bound form (diamond) and an inactive GDP-bound form (oval). GEFs facilitate the release of GDP and the binding of GTP by Rac1. The active form of Rac1 interacts with effector proteins to stimulate cellular responses. GAPs accelerate hydrolysis of GTP. Rac1 is maintained in the inactive form by interaction with GDIs. The *Salmonella* sp. proteins SopE and SopE2 function as GEFs for Rac1 and Cdc42. *Salmonella* sp. SptP functions as a GAP for Rac1 and Cdc42. *Pseudomonas* sp. ExoS and ExoT and *Yersinia* sp. YopE function as GAPs for Rac1, Cdc42, and RhoA.

effectors, not to be confused with type III toxins that are often referred to as effectors) that are regulated by the GTPase. Switch I also contributes to the formation of a magnesium (Mg^{2+})-binding pocket. Insertion of Mg^{2+} into this pocket is critical for high-affinity nucleotide binding. Switch II plays a key role in the GTP hydrolysis reaction (see below).

Binding of GTP to the nucleotide-free form of the GTPase is favored by a high cytoplasmic concentration of GTP relative to GDP. Three types of endogenous eukaryotic proteins regulate GTPase cycling between GTP-bound and GDP-bound forms (Fig. 1). The rate of GTP hydrolysis is accelerated by interaction with GAPs. GAPs act in two ways to increase GTPase activity (9). The first involves the positioning of a nucleophilic water molecule that attacks the γ-phosphate of GTP. A glutamine residue in switch II plays a central role in the positioning of this water molecule. The intrinsic rate of GTP hydrolysis is slow due to flexibility of this glutamine. By stabilizing the switch II glutamine and water molecule, GAPs help the GTPase adopt a transition state that is required for efficient GTP hydrolysis. Second, GAPs donate a catalytic arginine residue that is needed to complete the active site of the GTPase. This key arginine residue is positioned within an exposed loop and is inserted "finger-like" into the GTPase active site. The arginine and surrounding sequences comprise a conserved "arginine finger" motif that is found in all Rho GAPs (see below).

Rho GTPases can be stabilized in the inactive GDP-bound form by guanine nucleotide dissociation inhibitors (GDIs) (Fig. 1). In addition, GDIs prevent the GTPase from binding membranes, which is necessary to activate signaling. This involves sequestration of a geranylgeranyl lipid that is added posttranslationally to the carboxy terminus of Rho GTPases (10).

GEFs facilitate the removal of GDP from the GTPase (Fig. 1). Rho GEFs of eukaryotic origin contain a conserved domain known as Dbl-homology (DH) domain that encodes the exchange activity. Eukaryotic GEFs also contain a second type of domain known as a pleckstrin-homology (PH) domain. PH domains appear to be involved in subcellular localization or intramolecular regulation of exchange activity. The structure of the GTPase Rac1 bound to the GEF Tiam1 shows the critical features of the reaction (14). Contacts between the Tiam1 DH domain and Rac1 alter the conformations of switch I and switch II such that binding of Mg^{2+} and nucleotide is precluded.

The Rho GTPase subfamily is composed of 10 members; the best characterized are RhoA, Rac1, and Cdc42 (2). Studies carried out with RhoA, Rac1, and Cdc42 under defined experimental conditions in Swiss 3T3 fibroblast cells have shown that each GTPase can be linked to a specific signaling pathway that triggers a distinct mode of actin polymerization. RhoA activation triggers formation of actin stress fibers. Rac1 activation stimulates the formation of actin-rich lamellipodia, which are thin veil-like extensions of the membrane (also known as membrane ruffles). Cdc42 activation stimulates the formation of peripheral actin-rich filopodia, which are thin spike-like extensions of the membrane. An additional complexity is that Rho GTPases may activate each other in a hierarchical fashion. For example, in Swiss 3T3 cells, Cdc42 can activate Rac1, and in turn Rac1 can activate RhoA. Studies carried out in macrophages have uncovered a division of labor among Rho GTPases for different phagocytic pathways (4). RhoA is specifically required for complement receptor 3-dependent phagocytosis, whereas Rac1 and Cdc42 are required for Fc receptor-mediated phagocytosis. Rho GTPases interact with a multitude of different downstream effectors to trigger different modes of actin polymerization. As a discussion of these effectors is beyond the scope of this chapter, the reader is referred to a recent review on the subject (2).

BACTERIAL GEFs

SopE and SopE2 (Fig. 2) are two highly related proteins (69% identical) that are delivered into host cells by the SPI type III system of *Salmonella enterica* (11, 13). Although SopE and SopE2 function as GEFs (Fig. 1), they lack sequence relatedness to the DH or PH domains that are invariably found in eukaryotic GEFs. SopE, SopE2, and a third protein, SopB, are responsible for triggering the actin polymerization and membrane ruffling that drives *Salmonella* sp. invasion into host cells. Mutants of *Salmonella* sp. that lack SopE, SopE2, and SopB invade cells with very low efficiency (5). SopB has inositol phosphatase activity and stimulates actin polymerization by a mechanism distinct from that of SopE and SopE2.

SopE is carried by a bacteriophage, and this phage is not present in all *S. enterica* strains. SopE-expressing strains are associated with severe epidemics of disease, suggesting that acquisition of SopE "improves" the virulence of *Salmonella* sp. How could expression of two seemingly redundant GEFs increase *Salmonella* sp. virulence? Recent biochemical studies reveal that SopE and SopE2 have different specificities for Rho GTPases in vitro and in vivo (5). SopE acts

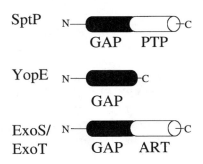

SopE/ SopE2   GEF

SptP   GAP   PTP

YopE   GAP

ExoS/ ExoT   GAP   ART

Figure 2 Physical organization of toxins with GEF or GAP domains. The *Salmonella* sp. SopE and SopE2 proteins contain a single GEF domain. *Salmonella* sp. SptP is a bifunctional toxin with an amino-terminal GAP domain and a carboxy-terminal PTP domain. *Yersinia* sp. YopE contains a single GAP domain. ExoS and ExoT of *Pseudomonas* sp. are bifunctional, with amino-terminal GAP and carboxy-terminal ADP-ribosyltransferease (ART) domains. The amino-terminal sequences of these bacterial toxins (thin black lines) contain signals for recognition and transport by type III secretion systems.

equally well on Rac and Cdc42, while SopE2 acts preferentially on Cdc42. Studies using dominant negative forms of Rho GTPases show that blocking Cdc42 reduces *Salmonella* sp. uptake fivefold, while blocking Rac1 function reduces uptake twofold. Although Cdc42 activation appears to be of prime importance for uptake, activation of Rac1 increases uptake efficiency. In support of this idea, SopE is twofold more efficient than SopE2 in triggering bacterial invasion. Thus, stimulation of Rac by SopE appears to increase the efficiency of uptake in a subtle but important manner.

The mechanism of GTP exchange mediated by SopE or SopE2 has not been elucidated. However, the kinetic parameters describing the SopE-Cdc42 interaction and the SopE-mediated exchange reaction on Cdc42 are similar to those previously observed for members of the Ras GTPase superfamily with their cognate GEFs (5). In addition, like DH domains, SopE appears to be largely composed of α-helical domains. The data suggest that SopE and SopE2 closely mimic host cell GEFs.

BACTERIAL GAPs

Several toxins secreted by type III systems have GAP activity for Rho GTPases (Fig. 1). These are SptP from *S. enterica*, ExoS and ExoT from *Pseudomonas aeruginosa*, and YopE from *Yersinia pestis*, *Yersinia pseudotuberculosis*, and *Yersinia enterocolitica* (Fig. 2) (11). The GAP domains of these toxins share sequence similarity, a GAP-like arginine finger motif, and a common mechanism of action (see below). In addition, all three GAP domains function to block actin polymerization in host cells (Table 1). However, as we will see, their precise functions during pathogenesis may differ.

Table 1 Functional comparison of Rho GAP domains in bacterial toxins

Toxin	YopE	SptP	ExoS	ExoT
Pathogen	*Yersinia* sp.	*Salmonella* sp.	*Pseudomonas* sp.	*Pseudomonas* sp.
Specificity	RhoA, Rac1, Cdc42	Rac1, Cdc42	RhoA, Rac1, Cdc42	RhoA, Rac1, Cdc42
Catalytic arginine	R144	R209	R146	R149
Role in pathogenesis	Antiphagocytosis, inhibition of pore formation	Down-regulation of membrane ruffling after invasion	Antiphagocytosis	Antiphagocytosis, inhibition of wound repair

SptP of *S. enterica*

It has long been known that the dramatic membrane ruffling induced during *Salmonella* sp. invasion of epithelial cells is a reversible phenomenon. The mechanism of this reversion, which involves down-regulation of membrane ruffling, was recently elucidated and shown to require the GAP activity of SptP (6). SptP is a dual-function enzyme. The N-terminal region contains a GAP domain, and the C-terminal region contains a protein tyrosine phosphatase (PTP) domain (Fig. 2). An *sptP* null mutant of *Salmonella* sp. is able to stimulate membrane ruffling and invade epithelial cells but is defective for down-regulation of membrane ruffling. The activity of the SptP GAP domain is required for this down-regulation. Thus, the SptP GAP domain functions to counteract the effects of SopE and SopE2 in vivo. In vitro studies confirm that the N-terminal domain of SptP is a potent GAP for Rac1 and Cdc42, but not for RhoA or Ras. The data suggest that continued activation of Cdc42 and Rac1 following successful invasion may be detrimental to *Salmonella* sp. pathogenesis. For example, excessive membrane ruffling may harm the host cell, which serves as a protective niche for the intracellular organism (11). How does *Salmonella* sp. orchestrate this precise control over Cdc42 and Rac1 activity? One possibility is that *Salmonella* sp. effectors are delivered in a temporal manner such that SopE or SopE2 enters the cell before SptP. Another important question is the role of the PTP domain in SptP function. The linkage of both domains into a single protein suggests a functional association between GAP and PTP activities. However, PTP activity is clearly dispensable for down-regulation of membrane ruffling. Recent experiments indicate that PTP activity of SptP acts to shut down mitogen-activated protein kinase pathways that are activated in response to *Salmonella* sp. invasion (8).

YopE of *Y. pestis*, *Y. pseudotuberculosis*, and *Y. enterocolitica*

The YopE toxin of *Yersinia* spp. is composed of a single GAP domain (Fig. 2). Straley and coworkers have shown that YopE is a critical virulence determinant, as inactivation of this factor increases the 50% lethal dose of *Y. pestis* by 4 orders of magnitude. The action of YopE in cultured mammalian cells triggers a dramatic rounding and detachment from the extracellular matrix. This process, originally described as "cytotoxicity," does not result from cell necrosis but rather from complete disruption of the actin cytoskeleton (3). YopE also has potent antiphagocytic activity, which likely plays a key role in pathogenesis during the extracellular growth phase of the organism (4). In vitro, YopE exhibits GAP activity for Cdc42, Rac1, and RhoA, but not Ras (11). The cell-rounding activity of YopE results from down-regulation of RhoA, as cultured cells expressing a constitutively activated form of RhoA do not round up after infection with *Yersinia* sp. YopE also inhibits Rac1-dependent phagocytosis and membrane ruffling in cultured cells. In addition, by blocking actin polymerization at the site of bacterial-host cell interaction, YopE appears to protect the host cell membrane from damage due to insertion of the type III translocation machinery (12). In this context, YopE may have a "host-protective" role like that proposed for the GAP domain of SptP.

ExoS and ExoT of *P. aeruginosa*

P. aeruginosa secretes two type III toxins with GAP activity, ExoS and ExoT (Fig. 2). These proteins share 75% amino acid identity (4). The carboxy-terminal regions of these toxins encode an ADP-ribosyltransferase activity (see chapter 15). The GAP domains are localized within the amino-terminal regions of ExoS and ExoT (1). Like YopE, the GAP domains of ExoS and ExoT are active against Rac1, Cdc42, and RhoA, but not Ras. The GAP domains of ExoS and ExoT elicit a cell-rounding activity similar to that observed for YopE. ExoS and ExoT are considered to function as anti-internalization factors during *P. aeruginosa* pathogenesis (4). Interestingly, the GAP activity of ExoT has been shown to antagonize wound repair, a healing response of cell monolayers that requires Rho GTPases and actin polymerization (7). Inhibition of wound repair could contribute to the injury of epithelial barriers by this opportunistic pathogen. It appears that most *P. aeruginosa* strains code for ExoT. The majority of isolates code for ExoS as well, and in strains such as these, ExoT and ExoS appear to be coordinately expressed. The coexpression of ExoS and ExoT by many clinical isolates of *P. aeruginosa* leads one to conclude that these proteins are not functionally redundant. A major difference between them is that ExoT possesses only about 0.2% of the catalytic ADP-ribosyltransferase activity of ExoS. It would be interesting to know if the GAP domains of these toxins have subtle differences in selectivity for Rho GTPases. Another important issue is whether the GAP and ADP-ribosyltransferase domains of these toxins are functionally related.

Structure and Mechanism of Bacterial GAPs

As noted by Galán and coworkers, the bacterial GAP toxins contain sequences related to the arginine finger motif of Rho GAPs (Fig. 3) (11). The conserved arginine residues within these motifs were shown to be essential for GAP activity in these toxins. This indicated that bacterial GAPs were likely to use a mechanism similar to that of eukaryotic GAPs to accelerate GTP hydrolysis by Rho proteins.

The three-dimensional structures of ExoS bound to Rac1 and SptP bound to Rac1 in the presence of GDP have been determined (11, 15). The complexes reveal that bacterial GAPs use unique structural features to achieve the same outcome as eukaryotic GAPs. The GAP domains of ExoS and SptP are similarly composed of a four-helix bundle, packed into a cylindrical shape. The key features of the interaction with Rac1 are also similar for both toxins. As shown for ExoS in Fig.

Figure 3 Alignment of the arginine finger motif of Rho GAPs with sequences of ExoS, SptP, and YopE. The bottom line shows a consensus sequence obtained by alignment of 15 Rho GAPs of eukaryotic origin (9). Sequences from SptP, ExoS, and YopE are shown aligned with this consensus sequence. Invariant residues are shown in bold; a, aromatic; h, hydrophobic.

SptP	$_{206}$**GPLR**S**LM**$_{212}$
ExoS	$_{143}$**GALR**S**LA**$_{149}$
YopE	$_{141}$**GPLR**G**SI**$_{147}$
RhoGAP consensus	**G**ha**R**h**SG**

4a (see Color Plates following p. 256), helices 1, 3, 4, and 7 are arranged in an antiparallel configuration to form the main bundle. Three smaller helices (H2, H5, and H6) cap the top of the cylinder. Loops connecting H2 to H3 and H5 to H6 protrude or "bulge" from the cylinder toward the GTPase. Both bulges, as well as several of the helices, make important contacts with the switch domains of the GTPase or the GDP. The residues in Rac that are contacted by ExoS are conserved in both RhoA and Cdc42. This conservation of key contact residues is consistent with the known specificity of ExoS for Rac1, Cdc42, and RhoA. Of particular interest is the positioning of the catalytic arginine of ExoS (and SptP) that becomes inserted into the Rac1 active site. Instead of being positioned at the end of an exposed loop, which is the arrangement in eukaryotic GAPs, the arginine extends from the side of helix 3 (Fig. 4a and b) (see Color Plates following p. 256). Thus, the bacterial GAPs use a novel structural motif to precisely mimic the activity of an endogenous GAP.

Comparison of the ExoS-Rac1 complex with the SptP-Rac1 complex reveals a subtle divergence in key contacts with switch I and II (Fig. 4b). For example, key residues in ExoS helix 1 that contact switch II (Met-112 and Lys-119) do not appear to be conserved in SptP (or YopE) (Fig. 4b). This divergence may reflect the different specificities of these GAPs for Rho GTPases. In the future, it may

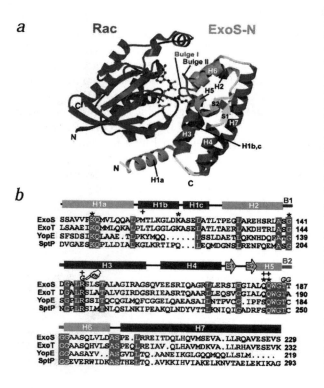

Figure 4 Secondary and tertiary structure of the GAP domain of ExoS. (a) Ribbon diagram of the GAP domain of ExoS in a complex with Rac1, GDP, and aluminum fluoride (AlF3). AlF3 is required to trap the ExoS-Rac1 complex in a transition state. Rac1 is colored blue. Bulge I is colored red, and bulge II is colored green in ExoS. The catalytic arginine in ExoS (Arg-146), the GDP, and the AlF3 are shown as ball-and-stick models. (b) Secondary structure assignments for ExoS GAP domain and sequence alignments of ExoS, ExoT, YopE, and SptP. Invariant residues are shown as white on green or red (catalytic Arg). Asterisks and plus signs indicate residues involved in hydrophilic or hydrophobic interactions, respectively. G indicates residues that contact GDP. Reprinted from reference 15 with permission. (See Color Plates following p. 256.)

be possible to explore the molecular basis for selectivity of these GTPases using the divergence in key contacts with switch I and II as a guideline for mutagenesis.

Analysis of GAP activity

GAP activity can be determined with a filter-binding assay as follows: Rho GTPases (4 μM) are loaded with GTP for 5 min at 30°C in 50 mM Tris-HCl, pH 7.6; 2 mM EDTA; 100 mM NH_4Cl; 0.5 mg/ml of bovine serum albumin; 1 mM dithiothreitol; and 10 μM [γ-^{32}P]GTP. Then 3 mM $MgCl_2$ and 1 mM GTP are added to initiate the intrinsic GTPase activity of the Rho GTPase or with 5, 10, and 20 nM GAP protein to measure GAP stimulation of intrinsic GTPase activity. After 5 min, 50 μl of the reaction is spotted on a nitrocellulose filter, and the filter is washed with 10 ml of 50 M Tris-HCl, pH 7.6; 100 mM NH_4Cl; 1 mM $MgCl_2$; and 7 mM β-mercaptoethanol. The amount of [γ-^{32}P]GTP remaining bound to the Rho GTPase is determined by scintillation counting. Measurement of the stimulation of GTP hydrolysis of Rho GTP is determined after 5-min incubations, using a concentration of GAP protein that stimulates the hydrolysis of less than 25% of the available Rho GTP.

Analysis of GEF activity: Ras nucleotide exchange

Ras is loaded with ^{35}S-GDPβ and assayed for nucleotide dissociation as follows: 2.5 μM ^{35}S-GDPβ-Ras is incubated in 40 mM Tris, pH 7.6, containing 2 mM dithiothreitol and 10 mM $MgCl_2$ alone (intrinsic) or in the presence of the GEF protein (GEF stimulated). GDP dissociation is determined as the rate of release of ^{35}S from Ras with a filter-binding assay essentially as described above for the GAP assay.

REFERENCES

1. **Barbieri, J. T.** 2000. *Pseudomonas aeruginosa* exoenzyme S, a bifunctional type-III secreted cytotoxin. *Int. J. Med. Microbiol.* **290**:381–387.
2. **Bishop, A. L., and A. Hall.** 2000. Rho GTPases and their effector proteins. *Biochem. J.* **348**(Pt 2):241–255.
3. **Bliska, J. B.** 2000. Yop effectors of *Yersinia* spp. and actin rearrangements. *Trends Microbiol.* **8**:205–208.
4. **Ernst, J. D.** 2000. Bacterial inhibition of phagocytosis. *Cell. Microbiol.* **2**:379–386.
5. **Friebel, A., H. Ilchmann, M. Aepfelbacher, K. Ehrbar, W. Machleidt, and W. D. Hardt.** 2001. SopE and SopE2 from *Salmonella typhimurium* activate different sets of RhoGTPases of the host cell. *J. Biol. Chem.* **276**:34035–34040.
6. **Fu, Y., and J. E. Galan.** 1999. A *Salmonella* protein antagonizes Rac-1 and Cdc42 to mediate host-cell recovery after bacterial invasion. *Nature* **401**:293–297.
7. **Geiser, T. K., B. I. Kazmierczak, L. K. Garrity-Ryan, M. A. Matthay, and J. N. Engel.** 2001. *Pseudomonas aeruginosa* ExoT inhibits in vitro lung epithelial wound repair. *Cell. Microbiol.* **3**:223–236.
8. **Murli, S., R. O. Watson, and J. E. Galán.** 2001. Role of tyrosine kinases and the tyrosine phosphatase SptP in the interaction of *Salmonella* with host cells. *Cell. Microbiol.* **3**:795–810.
9. **Scheffzek, K., M. R. Ahmadian, and A. Wittinghofer.** 1998. GTPase-activating proteins: helping hands to complement an active site. *Trends Biochem. Sci.* **23**:257–262.
10. **Seabra, M. C.** 1998. Membrane association and targeting of prenylated Ras-like GTPases. *Cell Signal.* **10**:167–172.

11. **Stebbins, C. E., and J. E. Galán.** 2001. Structural mimicry in bacterial virulence. *Nature* **412:**701–705.

12. **Viboud, G. I., and J. B. Bliska.** 2001. A bacterial type III secretion system inhibits actin polymerization to prevent pore formation in host cell membranes. *EMBO J.* **20:** 5373–5382.

13. **Wallis, T. S., and E. E. Galyov.** 2000. Molecular basis of *Salmonella*-induced enteritis. *Mol. Microbiol.* **36:**997–1005.

14. **Worthylake, D. K., K. L. Rossman, and J. Sondek.** 2000. Crystal structure of Rac1 in complex with the guanine nucleotide exchange region of Tiam1. *Nature* **408:** 682–688.

15. **Wurtele, M., E. Wolf, K. J. Pederson, G. Buchwald, M. R. Ahmadian, J. T. Barbieri, and A. Wittinghofer.** 2001. How the *Pseudomonas aeruginosa* ExoS toxin downregulates Rac. *Nat. Struct. Biol.* **8:**23–26.

Bacterial Protein Toxins
D. Burns et al., Editors
©2003 ASM Press, Washington, DC 20036

Chapter 21

Staphylococcal and Streptococcal Superantigens

Patrick M. Schlievert

Superantigens (SAgs) represent a large class of bacterial and viral proteins that cause a variety of human diseases by overactivation of the immune system. Of these proteins, the best characterized are the bacterial SAgs made by *Staphylococcus aureus, Streptococcus pyogenes,* and *Mycoplasma arthritidis.* However, there are also important viral SAgs, including those from murine mammary tumor viruses. A list of the known and putative SAgs is provided in Table 1. In this chapter the SAgs of *Staphylococcus aureus* and *Streptococcus pyogenes* will be discussed. The reader is referred to excellent reviews for discussions of other SAgs (2).

FAMILY OF STAPHYLOCOCCAL AND STREPTOCOCCAL SAgs

Staphylococcal and streptococcal SAgs include staphylococcal toxic shock syndrome toxin-1 (TSST-1); a biologically inactive variant referred to as TSST-ovine; staphylococcal enterotoxins (SEs) A to Q, excluding F (which is TSST-1); streptococcal pyrogenic exotoxins (SPEs) A, C, and G to J; streptococcal superantigen (SSA); and streptococcal mitogenic exotoxin Z (SMEZ) (1–3, 7). Two other putative streptococcal SAgs, SPEs B and F (mitogenic factor), have been reported to be SAgs, but these proteins also have known enzymatic functions, cysteine protease and DNase, respectively. Since their superantigenic properties are in debate, SPEs B and F will not be discussed, but readers are referred to reviews that discuss their properties (1, 7).

The classic SAgs form a large family of related proteins that lack detectable enzymatic function but rather function to cross-link major histocompatibility complex (MHC) class II molecules on antigen-presenting cells (APCs) with certain T-cell receptors (TCRs), primarily those of the CD4 lineage. The consequence of alteration of immune system function by SAgs is massive cytokine release from both macrophages and T cells that results in a variety of toxic shock syndrome (TSS)-like symptoms, including hypotension, erythematous rash, and fe-

Patrick M. Schlievert • Department of Microbiology, University of Minnesota Medical School, 420 Delaware St. SE, MMC 196, Minneapolis, MN 55455.

Table 1   Known and proposed bacterial and
viral T-cell SAgs

Bacterial
Staphylococcal enterotoxins
Staphylococcal TSST-1 and -ovine
Streptococcal pyrogenic exotoxins
Streptococcal SAg
Streptococcal mitogenic exotoxin Z
Mycoplasma arthritidis mitogen
Yersinia enterocolitica SAg
Yersinia pseudotuberculosis mitogen
Viral
Murine mammary tumor virus SAgs
Rabies virus SAg
Epstein-Barr virus-associated SAg

ver. Thus, these toxins are the major causes of staphylococcal and streptococcal TSSs and related illnesses. They are also likely to be causative agents of a variety of immune dysregulatory illnesses that lack known causes. Finally, the classical SEs (serotypes A to E) have the additional property of being able to cause emesis (vomiting) and diarrhea when given orally.

BIOCHEMISTRY OF SAgs

Staphylococcal and streptococcal SAgs are single-polypeptide-chain, nonglyco-sylated proteins of 22 to 30 kDa (Table 2). With only a few exceptions, they have neutral or basic isoelectric points (pIs). One exception for example, SPE A, occurs as two interconvertible molecular forms with pIs of 4.5 and 5.5. As with SPE A, most, if not all, of the SAgs display similar microheterogeneity, and most also occur with multiple allelic variants among bacterial strains. For example, four allelic variants of SPE A have been described, numerous alleles of *sec* (the gene for SE C) and *smeZ* exist, and two forms of TSST-1 have been described, TSST-1 and its variant TSST-ovine, which differ from each other by seven amino acids. As indicated above, there are a large number of serologically distinct SAgs, and even within some SAg types there is differential antibody reactivity among variants, dependent on allelic variation. For example, there is only one serotype of SPE A, but there are a large number of serovariants of SE C.

The amino acid composition of the SAgs is relatively typical for bacterial proteins, but some of the toxins have cysteines, which in the classical SEs (serotypes A to E) form a cystine loop structure. There is incomplete evidence that this loop structure contributes to the emetic activity of the SEs, but it cannot be the sole determinant of emesis, since SPE A also has the cystine loop but is not emetic. Many of the SAgs bind metal ions, mostly zinc, but in one case iron. In some cases zinc binding is critical for interaction of the toxins with MHC class II, but in other instances the role of zinc binding is unclear. SPE A has the ability to bind iron, but the role of this interaction in activity is unknown. SAgs are tightly folded molecules, resistant to heat denaturation (for example, TSST-1 may

Table 2 Properties of staphylococcal and streptococcal SAgs

Subfamily of SAgs	SAg	Molecular weight	Isoelectric point	Gene location
I	TSST-1	22,000	7.2	SaPI[a]
	TSST-ovine	22,000	8.5	?
II	SE B	28,500	8.5	SaPI
	SE C1[b]	27,500	8.5	SaPI
	SE G	27,000	?	Chromosome
	SPE A	26,000	5.0[c]	Phage
	SSA	27,000	7.6	Phage
III	SE A	27,000	6.8	Phage
	SE D	27,000	7.2	Plasmid
	SE E	26,500	8.5	Phage
	SE H	25,000	6.0	?
	SE J	27,000	?	Plasmid
IV	SPE C	24,500	6.7	Phage
	SPE G	24,500	7.5	Chromosome
	SPE J	24,500	7.7	Phage
	SMEZ	24,000	6.2	Chromosome
V	SE I	27,000	6.0	Chromosome
	SE K	26,000	7.2	SaPI
	SE L	26,000	8.5	SaPI
	SE Q	26,000	7.2	SaPI

[a] *Staphylococcus aureus* pathogenicity island.
[b] Some variant forms have isoelectric points of about 7.5 (SE C2, for example).
[c] When multiple forms are present due to microheterogeneity, the average isoelectric point is given.
Note: Some SAgs have not been assigned to subfamilies, such as SPE H (molecular weight 23,500, isoelectric point 7.5) and SPE I (molecular weight 23,500, isoelectric point 8.0), and SEs M, N, O, and P.

be heated to 100°C for 1 h without loss of activity) and acid (stability in the gut for example). SAgs are also resistant to trypsin digestion, and the SEs and at least SPE A are also highly resistant to pepsin treatment.

The primary amino acid sequences of SAgs share obvious similarities, such as SEs A, D, E, and J forming a group of related sequences and SEs B, C, and G and SPE A forming another, but other SAgs share very little primary amino acid sequence similarity with members of the family (TSST-1, for example). This divergence of primary sequence combined with shared activities leads to the hypothesis that the toxins may fold into similar three-dimensional structures. Crystal structure determination of numerous SAgs has verified this hypothesis.

The three-dimensional structure of the prototype SAg, TSST-1, was determined independently by two research groups to approximately 1.7 Å resolution (Fig. 1). TSST-1 has a short N-terminal α-helix that leads into a barrel structure (domain B), referred to as an OB fold. An OB fold is one of the major protein folds in nature and derives its name from oligosaccharide binding. Many bacterial toxins, including cholera toxin, staphylococcal nuclease, and verotoxins, share this domain in their host cell receptor-binding regions. After the OB fold TSST-1 has a central, long, diagonal α-helix that is closed on the front but open

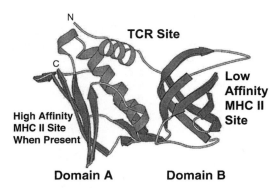

Figure 1 Ribbon diagram of TSST-1. The α-carbon backbone of the toxin is shown, highlighting the A and B domains of the toxin in the standard view. The TCR-binding domain on TSST-1 is on the top back of the toxin. The site for binding MHC class II is the low-affinity site to the right of the toxin in domain B. Some SAgs, but not TSST-1, have a zinc-dependent, high-affinity MHC class II binding site in the position shown to the left of TSST-1.

on the back (Fig. 1). A wall of β-strands completes domain A. Domain A of TSST-1 comprises another major protein fold referred to as β grasp. Other bacterial proteins, notably Fc immunoglobulin-binding proteins, also have β-grasp domains. TSST-1 appears more likely the progenitor SAg, which arose from the joining of two major protein folds with two binding (B) domains but no enzymatic function. Thus, the toxin may be referred to as a BB-type toxin, rather than the classic toxins with AB structural elements. In this way, SAgs are hormone-like in their activities.

The other known staphylococcal and streptococcal SAgs share the basic TSST-1 structure, but the other toxins have additional features. SEs A to G, SPE A, and SSA have the cystine loop structure that is at the top of domain B in the standard front view (Fig. 1). Numerous SAgs—including SEs A, D, E, H to L, and Q; SPEs C, G, H, I, and J; and SMEZ—appear to have a zinc-binding region in domain A (Fig. 1) that provides the high-affinity MHC class II interaction site. Other SAgs also have the ability to bind zinc in other locations, such as SE C and SPE A, but this interaction is not required for biological activity. Based on primary sequence similarities and three-dimensional structural analysis, a table of SAg subfamilies was constructed (Table 2). Group I includes only TSST-1. Group II includes the SE B subfamily, based on highly similar primary sequences, structural relatedness, and the presence of the cystine loop. Group III consists of SAgs related to SE A in primary sequence, structure, and the presence of both the cystine loop and a high-affinity, zinc-dependent MHC class II site. Group IV is considered the SPE C subfamily, based on structural relatedness combined with lacking the cystine loop but having the high-affinity, zinc-dependent MHC class II site. Finally, group V SAgs form a new subfamily and are defined by toxins having the zinc-dependent MHC class II site and lacking the cystine loop, but these toxins have a 12-amino-acid extension not present in the other SAgs. In addition, the group V SAgs are recently described. An interesting recent evolutionary relatedness study showed that all SAgs have 16 conserved residues that are in the same position in space, allowing most of the toxins to be modeled. There are exceptions, such as SPE I, that do not model well.

GENETICS OF SAgs

With the exception of SPE G and SMEZ, the genes for SAgs are variable traits. The toxins encoded as variable traits, SE A, possibly SE E, and SPEs A, C, H, and I (H and I are linked), are encoded by lysogenic bacteriophages (Table 2). Many of these bacteriophages appear to be defective and are not induced by agents such as mitomycin C; remnant bacteriophage DNA is located adjacent to each toxin gene. In addition, and like many other bacteriophage-encoded toxins, these SAgs are adjacent to the phage attachment site for insertion into the bacterial chromosome, leading to the hypothesis that the phages acquired the toxin genes by abnormal excision from a progenitor bacterial strain. A strain has not been identified that contains phage-encoded toxins independent of the presence of phage DNA; thus, evidence for this hypothesis is lacking. The phages that encode all of these toxin genes are related and package DNA by a headful mechanism, which is circular and terminally redundant. The genes for two SEs, D and J, are encoded on 27.6-kb plasmids that also encode penicillin and cadmium resistance (Table 2).

Most of the staphylococcal SAgs are encoded on pathogenicity islands (SaPIs) (Table 2). TSST-1 is encoded on SaPIs 1, 2, and bovine. SaPIs are approximately 15.2 kb in size and are present in a single, but not necessarily the same, site in the chromosome. At least SaPI1 has been mobilized and packaged into phage heads by helper bacteriophages for transfer to other bacterial strains. SE B is encoded on SaPI 3 and SE C on SaPI 4. Other SE genes such as *sek*, *sel*, and *seq* are distributed among the four SaPIs. The SE genes present in PIs localize to either end of the DNA elements. The genetic elements of SaPI 3 are shown in Fig. 2.

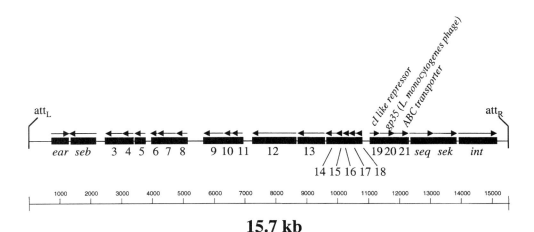

15.7 kb

Figure 2 Genetic structure of SaPI 3. The PI is 15.7 kb. The attachment sites left and right (att$_L$ and att$_R$) for integration into the bacterial chromosome are indicated. Putative genes (identified only as open reading frames greater than 5,000 base pairs) are numbered, including 19 (which has similarity to cI repressor from phage lambda), 20 (which has similarity to a phage protein from *Listeria monocytogenes*), and 21 (which has similarity to an ABC transporter). Direction of transcription of genes is indicated by arrows. *ear* encodes a protein of unknown function that is secreted in very high concentrations. *seb*, *seq*, and *sek* encode staphylococcal enterotoxins; *int* encodes a putative integrase.

HISTORY OF SAgs

Before the description of this family of proteins as SAgs in 1990, about 80 articles were published every 5 years related to their function. After their definition as SAgs, the number of publications rose to more than 800 every 5 years. This increase clearly resulted from intense interest in their novel mechanism of interaction with the immune system.

The classic SEs (serotypes A to E) were originally defined by their abilities to cause emesis and diarrhea in monkeys when administered orally. These SEs cause staphylococcal food poisoning in humans, an illness characterized by vomiting and diarrhea, without fever, 2 to 8 h after eating food contaminated with the preformed toxins. Acid and protease resistance likely contributes to the ability of SEs to cause food poisoning. Although the mechanism of emesis induction and diarrhea is unclear, the vagus and sympathetic nerves appear to be important targets. In addition, local elevation of prostaglandins and leukotrienes also correlates with activity. These SEs are potent pyrogens when administered intravenously, and it is hypothesized that food poisoning is characterized by a lack of fever because small amounts of toxin are sufficient to cause vomiting and diarrhea, but insufficient to induce fever, or because the toxins cannot easily cross mucosal surfaces to gain access to the circulation. Recently described SE-like toxins, such as SE I, L, and Q, lack or have severely reduced emetic ability.

Mutagenesis studies provided evidence that the cystine loop of SE C1 is required for emesis. When the two cysteine residues were mutated to alanines, the emetic activity was lost, demonstrating the importance of this structure. It was also shown that if the cysteines were altered to serines, emetic activity was retained. Thus, the cystine loop itself does not appear to be necessary, but rather, an unknown structure is required that can be maintained by the presence of cysteines or serines. Interestingly, SPE A, which also has a cystine loop, lacks emetic activity, again emphasizing the loop itself is insufficient to account for the activity. The recently described SEs like SAgs (I, L, and Q) and TSST-1 lack the cystine loop structure, consistent with their lack of emetic activity. Future comparison of the structures of classic SEs with related but nonemetic SAgs may identify structural elements associated with the cystine loop that control this activity.

The SPEs were originally defined in the 1920s as erythrogenic toxins (scarlet fever toxins) based on their ability to induce a macular (flat) erythematous sunburn-like rash (the scarlet fever rash) in volunteers and in scarlet fever patients. For many years this was the primary toxicity associated with erythrogenic toxins, and thus, these proteins were not considered medically important. Dennis Watson and his colleagues in the 1940s and 1950s described a family of toxins defined as SPEs based on their capacities to induce fever in rabbits and enhance susceptibility of the animals to the lethal effects of endotoxin (up to 1 million-fold). SPEs were unable to induce an erythematous rash when injected intradermally in rabbits, and thus, the investigators did not believe SPEs were the erythrogenic toxins. In later studies the toxins were demonstrated to be the same molecules when investigators showed their serological relatedness and defined the mechanism by which rash production was generated (discussed later). SPEs have also been described as lymphocyte mitogens and blastogen A in the literature, but these are now recognized as the same proteins.

TSST-1 was first described independently by Schlievert and colleagues as staphylococcal pyrogenic exotoxin C (because of relatedness in properties to SPE C from group A streptococci) and by Bergdoll et al. as SE F (because of relatedness in activities to SEs). In 1984 at an international TSS conference, it was decided to call the toxin TSST based on its ability to cause TSS, its inability to cause emesis when given orally to monkeys, and its function as a toxin. The "1" was added to the name in case other serotypes of the toxin were identified. To date, TSST-ovine has been identified as the only variant of TSST-1. TSST-1 is now recognized as the cause of most, if not all, menstruation-related TSS and half of nonmenstrual illness. Association of TSST-1 only with menstrual TSS may reside in the ability of this toxin to cross vaginal mucosal surfaces more effectively than other SAgs. The mechanism underlying this enhanced mucosal transport ability of TSST-1 compared to other SAgs has not been defined. The SEs cause the remainder of staphylococcal TSS cases, and the SPEs cause streptococcal TSS.

STRUCTURAL IMMUNOBIOLOGY OF SAgs

Before their identification as SAgs, these toxins had been described as nonspecific T-lymphocyte mitogens that depended on APCs for activation. In 1990, Marrack and Kappler (6) identified the novel mechanism of SAg activation of T cells. It was shown that SAgs activate subsets of T cells dependent on the composition of the variable part of the β-chain of the typical α/β T-cell receptor (TCRVβ). Thus, TSST-1 activates any T lymphocyte that has TCRVβ 2, regardless of the antigenic specificity of the T cell. The toxin does not, however, stimulate T cells bearing the other 24 or more TCRVβs of individual T cells. Thus, in a patient with TSS, this stimulation of T cells bearing TCRVβ 2 leads to those cells proliferating to expand from 10 to 15% of T cells to 60 to 70%, which is referred to as TCRVβ skewing. To put this in perspective, a typical antigenic peptide in the MHC class II groove will activate only 1 of 10,000 T cells. Other SAgs recognize different TCRVβs, and indeed, all SAgs have unique patterns of stimulation of T cells. The result, however, is the same, extensive proliferation of T cells, primarily of the CD4 lineage, and massive release of cytokines from T cells and macrophages. Marrack and Kappler also showed that activation of T cells by SAgs depended on interaction with MHC class II.

SAgs interact with TCR outside the typical TCR region for contact with antigenic peptide and MHC class II. Structural studies of the interaction of TSST-1, SE B (Fig. 3A), SE H, and SPE C (Fig. 3B) with MHC class II yielded information on the heterogeneity of these interactions. TSST-1 and SE B interact with MHC class II molecules on APCs with micromolar affinities. There is some specificity in the interaction in that the toxins exhibit preference for types of MHC (for example, TSST-1 binds HLA DR>DP>DQ). TSST-1 and SE B interact primarily with the α-chain of the MHC class II molecule outside the typical antigenic peptide site, though TSST-1 also interacts with the antigenic peptide and to some extent with the β-chain. Domain B (OB fold) of the toxins mediates the majority of this binding, referred to as the low-affinity site. SE B binds to the α1-domain of the MHC molecule, contacting residues in the first and third turns of the β-sheet and the N terminus of the α-helix. Although the TSST-1 and SE B binding sites overlap, SE B interacts away from the peptide-binding groove,

300

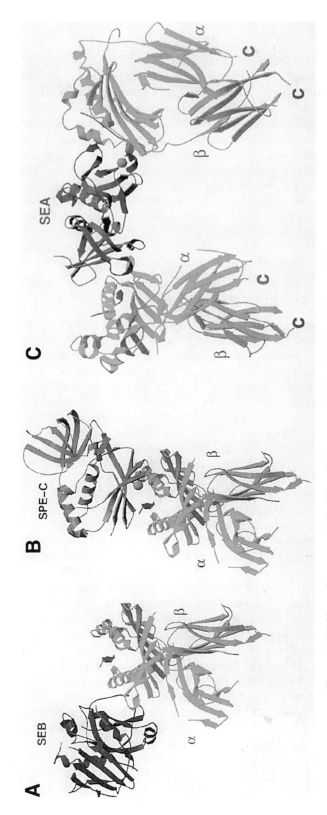

Figure 3 Ribbon diagrams of the SAgs (darker gray) SE B (A), SPE C (B), and SE A (C) complexed with MHC class II molecules (lighter gray). The α and β indicate the chains of the MHC class II molecules. Small ribbon structures in the groove of the MHC class II molecules are the antigenic peptides. Sphere in SPE C and SE A structures is the zinc atom in the high-affinity MHC class II interaction site.

whereas the TSST-1 interaction site (with MHC class II) extends over nearly half of the binding groove, contacting the α-helix of the α1-domain, the bound peptide, and a part of the α-helix of the β1-domain of the MHC.

SE C and SPE A bind MHC class II molecules primarily in domain B. Co-crystallization of SPE C with HLA-DR2a and SE H with HLA-DR1 showed that these two toxins have high-affinity MHC class II binding sites in domain A (β-grasp), which depended on coordination of a zinc cation for stable interaction. His-167, His-201, and Asp-203 form three parts of the coordination site of SPE C (3, 5). The fourth site is provided by HLA-DR2a β1-chain His-81. Unlike TSST-1 and SE B, SPE C and SE H interact with the β-chain of MHC class II (for this high-affinity site). For SPE C, as much as 35% of the interaction with the MHC class II β-chain also includes the antigenic peptide in the groove.

There is also evidence that the SE A (Fig. 3C) subfamily of SAgs also interacts with MHC class II through a similar high-affinity site. The SE A subfamily uses the low-affinity site for MHC class II interaction, through the MHC class II α-chain. Although SPE C interacts with MHC class II through the high-affinity site, SPE C also uses the low-affinity site. For example, Asn-38 of SPE C lies within its putative low-affinity site, and mutation of Asn-38 to aspartic acid results in the loss of superantigenicity. SAgs with the zinc-dependent MHC class II sites may have both binding domains. Superantigenicity of SAgs increases 100-fold if the putative high-affinity MHC class II sites are present.

Molecular interactions between SAgs and TCRVβs have been studied through cocrystallization studies (3, 4). SEs B (Fig. 4) and C and SPE A interact with TCRVβ in a groove between the two domains of the toxins, on the top front in the standard view. For the TCRVβ chain (murine Vβ8.2Jβ2.1Cβ1), residues in the complementarity-determining regions CDR2 and to a lesser extent CDR1 and hypervariable region (HV) 4 interact with SE C3. SE B also binds with residues in CDR2, HV4, and framework regions 2 and 3. SPE A interacts with TCRVβ similarly with only minor differences. Asn-25 in SEs B and C and the corresponding residue in SPE A (Asn-20) are critical interaction residues. It should be noted that the interaction of SAgs with TCRs with affinities of 10^{-4} to 10^{-6} M is similar to TCR interaction with antigenic peptides (10^{-4} to 10^{-7} M), but weaker than antigen-antibody affinities, which range from 10^{-8} to 10^{-11} M.

Although SPE C interacts with MHC class II likely through low- and high-affinity MHC class II sites, the interaction with TCRVβ (in this case human Vβ2.1Jβ2.3Dβ2.1Cβ2) overlaps with the SE B subfamily but is more complicated. SPE C interacts with each of the three TCRβ-chain CDR loops, as well as HV4. Cocrystallization studies of TSST-1 with TCRVβ 2 have been unsuccessful thus far, but mutagenesis studies indicate that this toxin interacts with TCR differently than those described above. TSST-1 interacts with the top back of the TCR, rather than the top front as for the other toxins. The interaction of TSST-1 with TCR and MHC II would prevent the two immune molecules from contacting each other, unlike the other SAgs. Thus, whereas most SAgs form a wedge along the side of the MHC class II:TCRVβ interaction site, TSST-1 fits completely between the two immune molecules (the molecules can be viewed as beads on a string, e.g., MHCII:TSST-1:TCRVβ).

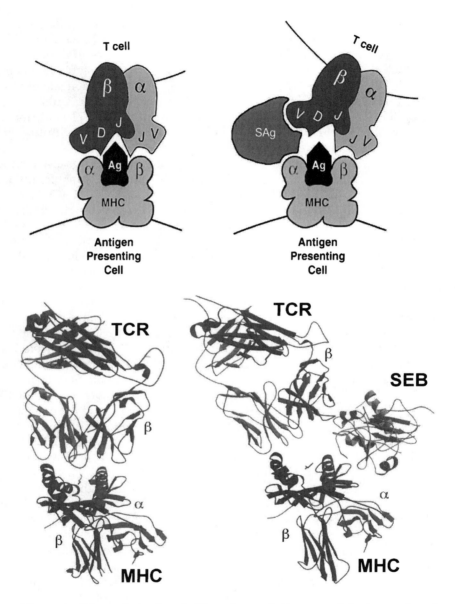

Figure 4 Schematic (top) and ribbon (bottom) diagrams of the interaction of antigenic peptide (Ag, small ribbon in bottom structures) and SAg, (SE B) with the TCR α- and β-chains and the α- and β-chains of MHC class II molecules.

FUNCTIONAL IMMUNOBIOLOGY OF SAgs

With the exception of emesis, all biological activities of SAgs depend on inter-actions with T cells, primarily CD4, and APCs and macrophages. Their disease potentials result from cytokine release, including interleukin 1 (IL-1) and tumor necrosis factor α (TNF-α) from macrophages and IL-2, TNF-β, and interferon-γ from T cells. The massive release of cytokines requires simultaneous interaction of SAg with both cell types. The activation of these immune cells depends on the same intracellular signaling steps required for typical antigenic peptide ac-tivation, with activation apparently much greater in magnitude.

The major biological properties of SAgs are listed in Table 3. Pyrogenicity depends on release of IL-1 (endogenous pyrogen), and possibly TNF-α, from macrophages. Pyrogenicity is controlled by the hypothalamus, where cytokines signal the release of prostaglandin E_2, which alters levels of serotonin and nor-epinephrine, with subsequent effects on α-adrenergic nerve receptors. SAgs in-duce high levels of IL-1 and TNF-α peripherally, which also have the capacity to induce fever. However, SAgs administered centrally in cerebrospinal fluid also induce fever, raising the possibility that they exert their pyrogenic effect by direct stimulation of the hypothalamus.

The ability of SAgs to enhance lethal endotoxin shock appears to depend on the effects of the toxins on liver cells combined with the massive release of TNF-α from macrophages as a result of the synergistic effects of interferon-γ (from CD4 T cells) and SAg. This is observed after first treating the animals (usually rabbits) with SAg and then 4 h later treating the same animals with a nonlethal dose of endotoxin. Within 2 days, depending on the doses of toxins, the animals will succumb to a TSS-like illness. Other agents also enhance the susceptibility of animals to lethal endotoxin shock by altering liver function. These include the mushroom poison α-amanitin, lead acetate, CCl_4, ethanol, and hepatitis viruses. In addition, fatty replacement of the liver is common in TSS patients, and like endotoxin shock, liver triaditis is also common. Although the precise effect of

Table 3 Biological activities of staphylococcal and streptococcal SAgs

Superantigenicity
  Pyrogenicity
  Enhancement of endotoxin shock[a]
  Scarlet fever rash
  Immunoglobulin suppression
  Enhancement of anaphylaxis
  Reactivation of septic arthritis (TSST-1)[b]
Skin desquamation
Mucosal transport (TSST-1)
Lethality in rabbits in subcutaneous mini-osmotic pumps[a]
Emesis (SEs A to E)
Cardiotoxicity (SPEs A and C)

[a] Both of these effects depend on cytokine release and may include superantigenic effects.
[b] (Toxins) refer to unique effects of those toxins.

SAgs on the liver remains unknown, endotoxin enhancement effect results from synergistic release of TNF-α from macrophages, making it likely that the liver's role is Kupffer cell release of the cytokine.

SAgs induce erythematous rashes depending on amplification of preexisting delayed-type hypersensitivity (DTH) to the toxins or other agents. If naive rabbits are challenged with TSST-1 or SPE A alone, the animals do not develop erythroderma (Table 4). This is consistent with the earlier work of Dennis Watson, who identified the SPEs on the basis of their lack of erythema induction. However, if the rabbits are first immunized one time with SAg to induce DTH and then challenged with the homologous toxin, the animals develop typical erythematous scarlet fever rashes. Thus, the SAg induces CD4 T cells to release massive amounts of DTH cytokines in the presence of preexisting DTH to homologous or heterologous antigens. This is the basis for the Dick test, which has been used to determine susceptibility to scarlet fever. Important also, this skin reactivity can be neutralized by the presence of antibodies that inhibit specific SAgs. The data also suggest that the first time individuals encounter SAgs, they may develop TSS-like illnesses without a rash. Finally, these studies suggest that it is possible to develop DTH to SAgs but difficult to develop neutralizing antibodies. This appears to be the case, where 85% of staphylococcal TSS patients remain susceptible to TSS after recovery, a significant recurrence rate. The lack of development of neutralizing antibodies against the toxins has been attributed to a suppressive effect mediated by interferon-γ.

Although desquamation of skin is a common feature of SAg-associated illnesses, the cause of this effect is uncertain. It is clearly intraepidermal, and SAgs can bind to epithelial cells. There are multiple hypotheses for the skin peeling that occurs. The first of these suggests that desquamation is secondary to edema as a result of TNF effects on the vascular endothelium. It is also possible that SAg interaction with epidermis or immune cells in the skin causes peeling indirectly through activation of proteases or through immediate cytokine effects.

Lethality due to SAgs administered in subcutaneous mini-osmotic pumps was first demonstrated by Parsonnet et al. (8). These pumps are designed to release constant amounts of SAgs over a 7-day interval, during which the animals progressively develop lethal TSS symptoms. SAgs (up to 5 mg) are not lethal to rabbits (the most appropriate model animal for the study of TSS) when given as bolus injections. In contrast, administration of small doses of SAgs over a few

Table 4 Skin test reactivity of staphylococcal and streptococcal SAgs

Sensitizing agent	Challenge agent	Reaction diameter (mm)[a]
None	SPE A	0
SPE A	SPE A	39
None	TSST-1	0
PPD[b]	PPD	3
PPD	TSST-1 + PPD	26

[a] Erythema and edema after intradermal injection of agent.
[b] PPD, purified protein derivative tuberculin.

days is highly lethal, and treatment with multiple SAg types at the same time is more lethal than with individual SAg. In the Parsonnet model, approximately 75 μg of TSST-1 is 100% lethal by 7 days. Although mice are susceptible to the superantigenicity of SAgs, they are resistant to the toxins' lethal effects. However, mice can be made susceptible by prior treatment with D-galactosamine to cause liver necrosis. However, liver necrosis is not a clinical feature of TSS. Despite this, mice are attractive animals for use, at least in studying the superantigenicity of SAgs because of the availability of inbred strains.

Lethality associated with SAgs depends on release of TNF by both T cells and macrophages, whether through direct SAg-induced immune cell activation or indirectly through endotoxin enhancement. The massive release of these cytokines can cause hypotension through capillary leak mediated by the nitric oxide pathway, with consequent effects on the endothelium. Thus, TSS and related illnesses are considered capillary leak syndromes, and administration of fluids and electrolytes to patients is an obligatory therapy. If blood pressure can be maintained and further toxin synthesis prevented by treatment with antibiotics, the toxins will be cleared by the kidneys.

ILLNESSES CAUSED BY OR ASSOCIATED WITH SAgs

SAgs are most often associated with TSS illnesses. As indicated above, staphylococcal TSS is induced by TSST-1 and SEs, primarily B and C. Staphylococcal TSS is defined by the criteria listed in Table 5 as acute onset of high fever, hypotension, erythematous rash, desquamation of the skin upon recovery, and any three of a list of variable multiorgan components. Probable TSS is defined as the same illness except that one of the defining criteria is absent. It has recently been proposed that toxin-mediated illness be used in those cases that do not meet either of the above definitions, yet in which toxin-producing *Staphylococcus aureus* is present in the absence of protective antibody. The effects of SAgs on the immune system can account for the clinical features of TSS, including fever, rash, hypotension, and many of the variable multiorgan components.

Streptococcal TSS is a related illness, caused primarily by group A streptococci that make SAgs, but there is evidence that other groups of β-hemolytic streptococci (B, C, and G particularly) make SAgs and are also associated with the same illness (9). The criteria that define streptococcal TSS are listed in Table 5 and include hypotension and any two of a variety of other features combined with the presence of group A streptococci. Importantly, streptococcal TSS is most often associated with invasion of the host by the causative bacterium, whereas staphylococcal TSS is most often associated with nonsystemic focal and mucosal staphylococcal infections. Thus, necrotizing fasciitis or myositis is often associated with group A streptococcal TSS but is rarely seen with staphylococcal TSS. Risk factors for streptococcal TSS include open wounds, chickenpox lesions, and the use of nonsteroidal anti-inflammatory agents. In patients who develop the illness, entry is usually through an open wound. Nonsteroidal anti-inflammatory agents are associated with streptococcal TSS because they mask pain associated with the developing infection.

There are two major subsets of staphylococcal TSS, menstrual and nonmenstrual. Menstrual TSS is defined as occurring a day or two before onset of or

Table 5 Diagnostic criteria for staphylococcal and streptococcal TSS

Staphylococcal TSS[a]	Streptococcal TSS
Fever	Isolation of group A streptococci from:
Hypotension	Sterile site for a definite case
Diffuse macular rash with subsequent	Nonsterile site for a probable case
desquamation	Hypotension
Three of the following organ systems in-	Two of the following symptoms:
volved:	Renal dysfunction
Liver	Liver involvement
Blood	Erythematous macular rash
Renal	Coagulopathy
Mucous membranes	Soft tissue necrosis
Gastrointestinal	Adult respiratory distress syndrome
Muscular	
Central nervous system	
Negative serologies for measles,	
leptospirosis, and Rocky Mountain	
spotted fever as well as negative blood	
or cerebral spinal fluid cultures for	
organisms other than *Staphylococcus*	
aureus	

[a] Proposed diagnostic criteria for staphylococcal toxin-mediated illness include (i) isolation of *Staphylococcus aureus* from a mucosal or normally sterile site, (ii) production of TSS-associated SAg by isolate, (iii) lack of antibody to the implicated toxin at the time of acute illness, and (iv) development of antibody to the toxin during convalescence.

during menses, with the peak onset being days 3 to 4, and coincides with peak vaginal levels of *Staphylococcus aureus*. Approximately 10 to 15% of women of menstrual age have *Staphylococcus aureus* vaginally during menstruation, and one-third of the strains make TSST-1. Thus, 3 to 5% of women have TSST-1-positive organisms vaginally. Despite this, TSS is relatively rare, with only a few thousand cases occurring each year in the United States. There are at least two reasons for this discordance. First, TSS occurs primarily in menstruating women in the 15 to 25 age group. There is an age-dependent development of neutralizing antibodies against TSST-1 such that 75% of women in this age group are protected from the toxin. (It is unclear under what circumstances individuals have this age-related development of neutralizing antibodies, but it is likely due to unapparent infections by toxin-producing organisms.) Second, menstrual TSS is associated primarily with use of tampons, with risk increasing with increased absorbency, and thus, additional factors must occur for TSS. Tampon use provides environmental conditions that favor TSST-1 production. TSST-1 is made well at 37°C and at pHs between 6.5 and 8.0, consistent with pH of 7.0 found vaginally during menstruation. TSST-1 production requires media containing protein and is repressed by glucose. Finally, TSST-1 requires oxygen and CO_2 for production. It has been reported that a two-component regulatory system in *Staphylococcus aureus,* designated SrrA/B, functions directly or indirectly to sense oxygen levels, and operates as a repressor of TSST-1 production when oxygen is below 2 to 3%. Since the vagina is anaerobic in the absence of tampon use and becomes oxygenated with tampon introduction, it is hypothesized that the role of tampons in TSS is introduction of oxygen. It was also proposed that the higher-absorbency tampons (marketed from 1976 to 1985) introduce more oxygen than those of lower absorbency, sufficient to cause significant TSST-1 production and consequent illness in the susceptible host. In other studies, it has also been shown that surfactants (wetting agents) used in some high-absorbency tampons also induced TSST-1 production.

There are a large number of subsets of nonmenstrual TSS, as would be expected, since *Staphylococcus aureus* is associated with many types of infection. For example, influenza TSS occurs after superinfection of the upper and/or lower respiratory tract with SAg-producing *Staphylococcus aureus* in the susceptible host as a consequence of having a respiratory viral infection. This illness may be 90% fatal in children.

Staphylococcus aureus is a major cause of furuncles and postsurgical infections (the latter estimated to be in excess of 800,000 per year in the United States), and TSS may occur in association with either infection. Again, many of the environmental conditions that favor SAg production are present in these infections. However, there are two surprising findings that bear mentioning. First, it is generally assumed that abscesses are anaerobic, which would not favor SAg production. However, this is not the case with staphylococcal abscesses, where oxygen levels are 60% of atmospheric air. Second, abscesses due to TSS organisms are not inflammatory, which contrasts perceptions of *Staphyloccus aureus* infections. *Staphylococcus aureus* is generally considered a pyogenic coccus, and SAgs induce proinflammatory cytokines. However, it is likely that the massive over-stimulation of the immune system by SAgs actually interferes with polymorpho-nuclear leukocyte (PMN) infiltration into abscess sites. This lack of inflammation is also often seen in tissue areas of the advancing spread of group A streptococci in streptococcal TSS with necrotizing fasciitis, which is due in large part to TNF-α.

There are numerous other subsets of staphylococcal TSS, and more are defined each year. *Staphylococcus aureus* can cause a wide variety of infections that can be associated with TSS. Furthermore, the amount and types of SAgs made, combined with host susceptibility, will likely determine the extent of illness that occurs, from clinically recognizable to many that remain undefined.

It is hypothesized that these SAgs contribute to a variety of autoimmune diseases (for example, rheumatic fever, multiple sclerosis, rheumatoid and other forms of arthritis, and Crohn's disease). There is also evidence to suggest that these toxins cause guttate (tear drop-shaped lesions) psoriasis, which typically follows group A streptococcal infections in susceptible individuals, atopic dermatitis, and many cases of Kawasaki syndrome (a TSS-like illness lacking hypotension that occurs in children usually less than 4 years of age and having significant associated coronary artery aneurysms). Clearly, the full association of SAgs with such illnesses is incomplete.

WHY ARE SAgs MADE BY STAPHYLOCOCCI AND STREPTOCOCCI?

Staphylococci and streptococci are common pathogens of humans, and as such have myriad cell surface virulence factors that contribute to colonization and immune avoidance. The SAgs may be made in an attempt to interfere with normal immune function, and particularly to prevent PMN influx into the area of infection, exerting their effects at a distance. Because the organisms cause so many types of infections, it may be that numerous SAgs are needed to ensure that the organisms can cause infections, even if the host has antibodies to one or a few toxins. Most of the SAgs are made during the postexponential phase of growth (excluding SE A, K, and Q, which are exponential phase regulated), when cell numbers are highest and the organisms are most likely to spread to new hosts. Thus, when the organisms begin their colonization of a new host, they are most likely making SAgs as a consequence of their having been in postexponential phase. *Staphylococcus aureus* strains that make TSST-1, for example, emerged in 1972 and peaked in numbers by 1976 as dominant strains, suggesting that high production of SAgs gives the organisms a significant survival advantage.

SUMMARY

We have gained significant insight into the actions and disease associations of SAgs. These are potent modulators of immune system function. We know that the organisms that produce SAgs also have myriad undefined SAg-like molecules that have been detected by genome sequencing studies, referred to as SET proteins, that will likely give rise to new SAg variants. Thus, these proteins are likely to be the subjects of intense study well into the future.

REFERENCES
1. **Dinges, M. M., P. M. Orwin, and P. M. Schlievert.** 2000. Exotoxins of *Staphylococcus aureus. Clin. Microbiol. Rev.* **13**:16–34.
2. **Kotzin, B. L., D. Y. Leung, J. Kappler, and P. Marrack.** 1993. Superantigens and their potential role in human disease. *Adv. Immunol.* **54**:99–166.
3. **Li, H., A. Llera, E. L. Malchiodi, and R. A. Mariuzza.** 1999. The structural basis of T cell activation by superantigens. *Annu. Rev. Immunol.* **17**:435–466.
4. **Li, H., A. Llera, D. Tsuchiya, L. Leder, X. Ysern, P. M. Schlievert, K. Karjalainen, and R. A. Mariuzza.** 1998. Three-dimensional structure of the complex between a T cell receptor beta chain and the superantigen staphylococcal enterotoxin B. *Immunity* **9**:807–816.
5. **Li, Y., H. Li, N. Dimasi, J. K. McCormick, R. Martin, P. Schuck, P. M. Schlievert, and R. A. Mariuzza.** 2001. Crystal structure of a superantigen bound to the high-affinity, zinc-dependent site on MHC class II. *Immunity* **14**:93–104.
6. **Marrack, P., and J. Kappler.** 1990. The staphylococcal enterotoxins and their relatives. *Science* **248**:705–711.
7. **McCormick, J. K., J. M. Yarwood, and P. M. Schlievert.** 2001. Toxic shock syndrome and bacterial superantigens: an update. *Annu. Rev. Microbiol.* **55**:77–104.
8. **Parsonnet, J., Z. A. Gillis, A. G. Richter, and G. B. Pier.** 1987. A rabbit model of toxic shock syndrome that uses a constant, subcutaneous infusion of toxic shock syndrome toxin 1. *Infect. Immun.* **55**:1070–1076.
9. **Schlievert, P. M., M. Y. Kotb, and D. L. Stevens.** 2000. Streptococcal superantigens: streptococcal toxic shock syndrome. *In* M. W. Cunningham and R. S. Fujinami (ed.), *Effects of Microbes on the Immune System.* Lippincott Williams & Wilkins, Philadelphia, Pa.

SUGGESTED READING
1. **Leung, D. Y. M., B. T. Huber, and P. M. Schlievert.** 1997. *Superantigens: Relevence to Human Disease and Basic Biology.* Marcel Dekker, Inc., New York, N.Y.

SECTION V
BACTERIAL TOXINS: FRIENDS AND FOES

At the dawn of the 20th century, little was known about the molecular basis of bacterial toxins. Specific knowledge of mechanisms of toxin action was still decades away. During the century, our understanding of pathogenic mechanisms and the role of bacterial toxins in disease expanded dramatically. We now know the three-dimensional structure of a number of bacterial toxins, we have information concerning the molecular events that occur as toxins cross membrane barriers, and we understand the enzymatic mechanisms by which toxins disrupt critical cellular pathways.

This knowledge has provided us with the ability to use toxins for purposes that have saved the lives of many individuals and improved the quality of life for countless others. Our ability to turn toxins into vaccines has had dramatic effects on the morbidity of a number of diseases. For example, in the United States in the last century, vaccines based on toxins decreased the annual morbidity due to pertussis by 95%, due to tetanus by 98%, and due to diphtheria by 99.99%. Future advances in the use of bacterial toxins for medicinal purposes will no doubt lead to other exciting applications of these powerful molecules.

Unfortunately, the increased knowledge of the mechanisms of bacterial pathogenesis and of the role of bacterial toxins in disease has led to the implementation of improved strategies to use bacteria as agents of warfare and terrorism. Today, humans continue to exploit toxins for destructive purposes, applying the same tools that were initially developed to use toxins for the betterment of humankind.

In this section, we review how bacterial toxins have been used to cure or prevent devastating diseases and how toxins might be used to manipulate the immune system to make it stronger and more effective. We also review the history of the use of bacterial toxins as biological weapons and as tools of terror.

Bacterial Protein Toxins
D. Burns et al., Editors
©2003 ASM Press, Washington, DC 20036

Chapter 22

Toxins as Vaccines and Adjuvants

Mariagrazia Pizza, Vega Masignani, and Rino Rappuoli

Several bacteria produce potent toxins that are entirely, or in part, responsible for the severe diseases caused by the microorganisms. For instance, the clinical manifestations of diphtheria, tetanus, botulism, and anthrax can be reproduced by the systemic administration of the toxins, whereas cholera can be reproduced by oral administration of cholera toxin. Most of these diseases can be prevented by the administration of toxin-neutralizing antibodies, and some of them (diphtheria and tetanus) are routinely prevented by immunization with partially purified and detoxified toxins (toxoids). In other cases, such as in pertussis, the toxin is the main but not the only molecule contributing to toxicity. Administration of pertussis toxin by the systemic route reproduces some of the systemic symptoms of the disease, but not the local toxicity and cough. However, even in this case, administration of toxin-neutralizing antibodies reduces the duration and the severity of the disease. Immunization with detoxified toxin either alone or combined with other purified antigens from *Bordetella pertussis* prevents disease (11).

This chapter will describe those toxins classically used to prevent disease and the approaches for their future use. In 1890, practical application of bacterial toxins was pioneered by Behring and Kitasato, who succeeded in immunizing animals with diphtheria toxin that had been detoxified by treatment with iodine trichloride and allowed the development of antitoxic horse serum for the passive immunization against diphtheria. Active immunization was also pioneered by E. von Behring, who in 1913 tested the use of toxin-antitoxin mixtures, a procedure adopted by H. Park to immunize children in New York schools in 1922. Large-scale, safe vaccination was introduced after the discovery of Ramon in 1924 that diphtheria toxin could be detoxified by a simple chemical treatment: exposure to formaldehyde. This process was rapidly applied also for the detoxification of tetanus toxin, allowing the development of diphtheria and tetanus vaccines, which were introduced for universal immunization in the subsequent decades. Today, 80 years after their initial discovery, diphtheria and tetanus vaccines are

Mariagrazia Pizza, Vega Masignani, and Rino Rappuoli • IRIS, Chiron SpA, Via Fiorentina 1, 53100 Siena, Italy.

still produced by the same process described by Ramon in 1924, and even most of the recently developed acellular pertussis vaccines contain a pertussis toxin detoxified by formaldehyde treatment, as described by Ramon. Following the initial discovery, formaldehyde was frequently used to inactivate other toxins and to kill bacteria and viruses used in vaccines containing inactivated microorganisms. Despite successes in using formalin or other chemical treatments to inactivate bacterial toxins and whole microorganisms, this process is not ideal. In fact, the chemical treatment modifies the antigenic properties of these molecules and, therefore, large amounts of immunogen are required for immunization and the immunity induced is of low affinity. In addition, formaldehyde-induced chemical reaction may be reversible, and the inactivated microorganisms and toxins may revert to toxicity. This is no longer a problem today because rigorous quality control of the products guarantees absence of reversion, but it has been a serious problem in the past. In fact, several examples of reversion are available, with severe intoxication and death of vaccinated healthy people.

Today, it is possible to inactivate toxins and microorganisms by using genetic tools. Therefore, in addition to traditional diphtheria and tetanus vaccines, the acellular pertussis vaccine will be described. This represents the first vaccine produced by genetic inactivation of a bacterial toxin. The effectiveness of this novel method of toxin inactivation has been validated in animal models and in phase I, phase II, and phase III clinical studies, proving that genetic detoxification of bacterial toxins produces vaccines that are safe and more immunogenic than those obtained by conventional chemical treatment. Finally, the genetic inactivation of cholera and *Escherichia coli* heat-labile enterotoxins and their potential use as vaccines against diarrhea and as mucosal adjuvants will be described.

DIPHTHERIA AND TETANUS TOXOIDS

Diphtheria toxin is a 535-amino-acid single-chain protein secreted into the growth medium by toxinogenic strains of *Corynebacterium diphtheriae* when grown in the absence of iron. Following proteolytic cleavage, diphtheria toxin is divided into two fragments, A and B, of 21 and 37 kDa, respectively (Fig. 1a) (5). Functionally, the crystallographic structure of diphtheria toxin has revealed the presence of three separate domains, of which the catalytic domain with ADP-ribosyltransferase activity (C) corresponds to the A subunit, whereas the translocation domain (T) and the carboxy-terminal, receptor-binding domain (R) are contained in fragment B (1, 7).

Tetanus neurotoxin is produced by *Clostridium tetani* as a single-chain protein of 150 kDa that, following proteolytic cleavage, is divided into fragments H (heavy) and L (light), held together by a disulfide bridge (Fig. 1b). The H chain is composed of fragments B and C, which mediate to an undefined, cellular receptor and the transmembrane translocation of the L chain to the cytosol. Its overall structure is similar to that of A/B toxins, where the toxic subunit A is represented by the L chain, whereas the B domain is constituted by B and C fragments. The tetanus neurotoxin L chain is a 50-kDa fragment containing the −HExxH−motif typical of metalloproteases.

To produce conventional diphtheria vaccine, the hypertoxinogenic *Corynebacterium diphtheriae* strain PW8 is grown in fermentors for 36 to 48 h until the

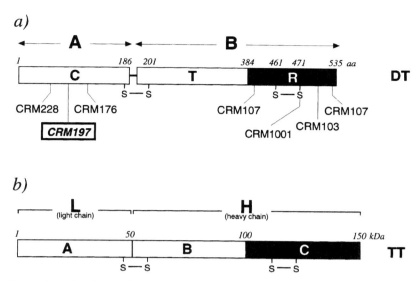

Figure 1 (a) Schematic representation of diphtheria toxin (DT) and of the CRM derivatives. (b) Schematic representation of tetanus toxin (TT).

concentration of the toxin in the supernatant reaches 150 to 250 Lf/ml (1 Lf = 2.5 μg of diphtheria toxin). Although several semisynthetic media for the industrial growth of diphtheria have been described, many manufacturers still grow the PW8 strain in a medium containing enzymatic digests of beef. When the fermentation is complete, the bacteria are removed by centrifugation or filtration, and formaldehyde is added to the supernatant to a final concentration of 0.75%. The supernatant is then stored for 4 to 6 weeks at 37°C to allow complete detoxification of diphtheria toxin. The toxin may also be partially purified before formalin detoxification.

Similarly, to produce the conventional tetanus vaccine, the highly toxinogenic Harvard strain of *Clostridium tetani* is grown in a semisynthetic medium (usually modified from Mueller and Miller) in a fermentor for about 1 week, until bacteria lyse and release tetanus toxin into the supernatant. The average yield obtained under these conditions is approximately 60 to 80 Lf/ml (1 Lf = 2–2.5 μg of tetanus toxin). The culture is then filtered, and the filtrate containing the toxin is detoxified by adding formaldehyde to a final concentration of 0.5%. The pH is adjusted to 7.6, and the supernatant is then stored at 37°C for 4 weeks to allow the complete detoxification of tetanus toxin.

Formaldehyde treatment of diphtheria and tetanus toxins involves a cross-linkage between an ε-amino group of lysine and a second amino group, a histidine, a tyrosine, or a tryptophan. These reactions can occur between amino acids on the same toxin molecule, resulting in internal cross-linking of the protein; between two toxin molecules, resulting in dimerization; or between the toxin and small peptides present in the medium. When the formaldehyde is added to the culture supernatant, the primary reaction that occurs is the cross-linkage with peptides in the medium. In the case of diphtheria, these peptides are unnecessary antigenic determinants of bovine origin, which might be responsible for some of the side effects associated with diphtheria and tetanus vaccination, especially in the adult population that has been repeatedly exposed to bovine antigens. To avoid this potential problem, toxins can be purified before formaldehyde treatment. Several methods have been described for the purification of diphtheria and tetanus toxins and are generally based on diafiltration of culture supernatant, precipitation by ammonium sulfate, and, if necessary, purification by gel filtration or ion-exchange chromatography. With these methods, diphtheria and tetanus toxins can be purified to 85 to 95% purity, representing approximately 2,300 and 1,800–2,000 Lf/mg of protein nitrogen for tetanus and diphtheria, respectively, by ammonium sulfate precipitation, whereas the conventional vaccines have a purity of approximately 60%. The partially purified or highly purified toxin preparations can then be detoxified by the conventional formaldehyde treatment.

Although production of diphtheria and tetanus toxoids starting from purified proteins would be preferable, thus far, these procedures have not found large-scale application. This process has been discouraged because of the necessity of handling large quantities of concentrated, highly toxic materials, the low yield in purified tetanus toxin, and the requirements of the regulatory agencies.

CRM197 and Toxoids as Carriers for Conjugate Vaccines

Modern genetics of bacterial toxins started with a pivotal experiment of T. Uchida in the laboratory of A. M. Pappenheimer, Jr., in 1972. Uchida and Pappenheimer used nitrosoguanidine to mutagenize the phage containing the gene encoding diphtheria toxin and isolated a number of phages encoding mutated forms of diphtheria toxin termed cross-reacting materials, or CRMs (Fig. 1a). The CRMs proved essential for the characterization of diphtheria toxin, and one, CRM197, is used today for vaccination. Among these CRMs, CRM176 contained a Gly-128 to aspartic acid mutation that reduced the enzymatic activity of fragment A by a factor of 10. CRM228 contained two mutations in fragment A and three in fragment B, which resulted in inactive A and B fragments. CRM1001 had an enzymatically active fragment A and a fragment B that bound the toxin receptors but was unable to translocate fragment A across the eukaryotic cell membrane due to a Cys-471 replacement by tyrosine. CRM103 and CRM107 were unable to bind surface receptors on eukaryotic cells due to a Ser-508 replacement by phenylalanine in CRM103 and to a Leu-390 and Ser-525 replacement by phenylalanine in CRM107. CRM45 was a C-terminal deletion peptide caused by a nonsense mutation that introduced a stop codon at position 387. Possibly the most important CRM identified was CRM197, which contains a single Gly-52 replacement by glutamic acid in fragment A. This eliminated NAD$^+$ binding and made CRM197 enzymatically inactive. Upon discovery of this mutant, it was immediately obvious that CRM197 was an ideal candidate for novel vaccines against diphtheria. Being nontoxic, CRM197 could be safely purified from the culture supernatant, avoiding the cross-linking to peptones during the formaldehyde treatment, did not need chemical detoxification, and did not have any risk of reversion to toxicity. However, CRM197 did not replace the classical diphtheria toxoid, since it was more sensitive to proteases and therefore less immunogenic than diphtheria toxoid for primary immunization. Furthermore, it had production yields that were low compared to those of diphtheria toxin. Production yields were enhanced by isolation of C7 strains of *Corynebacterium diphtheriae* that contained two phages encoding CRM197 stably integrated into the chromosome and by optimization of fermentation conditions. Stability and immunogenicity were significantly increased by a very mild treatment of purified CRM197 with formaldehyde in amounts unsuitable for detoxification but that have been shown to potentiate stability and immunogenicity of a number of toxins present in vaccine formulations. CRM197 has proven to be safe and effective in boosting diphtheria immunity in a phase I trial in adults.

Although the use of CRM197 as diphtheria vaccine has not yet found its application, CRM197 was developed as a carrier for conjugate vaccines against encapsulated bacteria. In these vaccines a bacterial capsular polysaccharide or oligosaccharides deriving from it are covalently linked to a carrier protein, which provides T-cell epitopes and enables the polysaccharide to elicit a T-cell-dependent immunity, and therefore, to elicit immunologic memory and to be immunogenic in infants. CRM197 is suitable for fine conjugation chemistry, as this chemistry is more difficult with proteins that have been previously treated with formaldehyde, such as diphtheria and tetanus toxoids. The conjugate vaccines available today and their carrier molecules are listed in Table 1. As shown, CRM197 is used today in at least three widely used vaccines and is delivered annually to millions of infants and adolescents.

Recombinant Fragment C of Tetanus Toxin

Although the absence of efficient genetic mechanisms to manipulate *Clostridium tetani* prevented the isolation of tetanus toxin CRM-like proteins, expression of specific domains of tetanus toxin in *E. coli* has been successful. In fact, the receptor-binding domain, fragment C, initially purified following papain treatment and subsequently expressed as a soluble form by *E. coli* and *Pichia pastoris*, retains the biological and immunological properties of native fragment C, including internal disulfide bridge structure, ganglioside-binding activity, and the ability to elicit protective antibodies in mice. Fragment C has been also successfully expressed in *Salmonella enterica* servovar Typhimurium, where recombinant strains stimulate mice to produce antitetanus antibody that protects from challenge with tetanus toxin. Fragment C has been also expressed in attenuated strains of *Salmonella typhi*, the human pathogen responsible for typhoid fever, suggesting the possibility of developing oral typhoid-tetanus vaccines. Recently, recombinant tetanus toxin CRM has been generated by site-directed mutagenesis. Noteworthy is the Glu-234 replacement by alanine, which renders the L chain nontoxic and, in combination with native H chain, can generate protective immunity in mice.

PERTUSSIS TOXIN AND ACELLULAR PERTUSSIS VACCINES

Pertussis, or whooping cough, is a disease caused by *B. pertussis*, a gram-negative bacterium, which adheres to the cilia of the upper respiratory tract of humans, colonizes this tissue, and releases a number of virulence factors responsible for the local and systemic damages associated with the disease (paroxysmal cough, accompanied by whoops, vomiting, cyanosis, and apnea) (12). Vaccination is the

Table 1 Conjugate vaccines currently available

Vaccine (trademark)	Polysaccharide/ oligosaccharide	Carrier protein	Spacer and conjugation technology
Haemophilus influenzae vaccine			
PRP-D (Prohibit-Connaught)	Polysaccharide	Diphtheria toxoid	Spacer of 6 carbon atoms
PRP-T (ActHIB–Aventis-Pasteur)	Polysaccharide	Tetanus toxoid	Spacer of 6 carbon atoms
PRP-T (Hiberix-GSK)	Polysaccharide	Tetanus toxoid	Spacer of 6 carbon atoms
PRP-OMP (PedVax–Merck Sharp & Dohme)	Polysaccharide	Vesicles of the meningococcus B outer membrane	Thioether
HbOC (HibTITER–WLV)	Oligosaccharide	CRM197	–
Vaxem Hib (Chiron)	Oligosaccharide	CRM197	Spacer of 6 carbon atoms
Meningococcus C vaccine			
Menjugate (Chiron)	Oligosaccharide	CRM197	Spacer of 6 carbon atoms
Meningitech (WLV)	Oligosaccharide	CRM197	–
Neivac-C (Baxter)	Polysaccharide	Tetanus toxoid	Reductive amination
Pneumococcus vaccine			
Prevnar (WLV)	Oligosaccharide/ Polysaccharide	CRM197	Reductive amination

only way to control pertussis. Mass vaccination using killed bacteria (cellular vaccine) was introduced in the 1950s and reduced by 99% the incidence of the disease in infants. In the 1980s, the use of this vaccine in developed countries decreased dramatically because of concerns of potential side effects associated with vaccination. This stimulated the search for an acellular vaccine that retained efficacy but was less reactive than the whole-cell vaccine. Several molecules produced by *B. pertussis* were identified as candidates for inclusion in acellular vaccine against whooping cough. This included molecules involved in the adhesion of the bacteria to the eukaryotic cells and to the cilia of the upper respiratory tract (e.g., filamentous hemagglutinin and pertactin) and molecules that cause local and systemic damage of the host (pertussis toxin, or PT).

PT is a 105-kDa multisubunit protein that is secreted into the extracellular medium by *B. pertussis*. PT is an AB class, ADP-ribosylating toxin and is composed of six subunits, named S1 through S5, where S4 is present in two copies (Fig. 2) (7). The genes encoding PT are organized into an operon and are con-

Figure 2 Schematic representation of the genes encoding the five subunits of PT. The mutations introduced by site-directed mutagenesis into the S1 subunit are also reported.

tained within the *B. pertussis* chromosome. The five subunits are independently secreted into the periplasmic space, where the toxin is assembled and then released in the culture medium by a specialized type IV secretion apparatus. The S1 subunit ADP-ribosylates GTP-binding proteins G_i, G_o, and transducin, uncoupling the G-proteins from their receptors, whereas the pentamer $S2$-$S3$-$(S4)_2$-$S5$ constitutes the receptor-binding domain. PT alters the response of eukaryotic cells to exogenous stimuli with a variety of in vivo phenotypes, such as leukocytosis, histamine sensitization, and increased insulin production (9). One unusual activity of PT in vitro is the clustering of CHO cells (4). The pentamer binds the receptors on the surface of eukaryotic cells and allows the toxic subunit S1 to reach its intracellular target proteins through a mechanism of receptor-mediated endocytosis. In contrast to the other ADP-ribosyltransferases, where all the biologic activities are mediated by the enzymatically active domain A, in the case of PT, this toxin possesses other nontoxic activities, such as the mitogenic activity on T cells, which are mediated exclusively by the receptor-binding domain B.

PT plays a central role in the pathogenesis of whooping cough and induces protective immunity against infection. As for the other toxins, to be included in vaccines, PT needs to be detoxified. In the early 1980s, during the development of acellular pertussis vaccines, a number of chemical methods (formaldehyde, hydrogen peroxide, tetranitromethane, and glutaraldehyde) were used to detoxify PT. Although many of the vaccines in use today contain a formaldehyde-detoxified PT, genetic engineering was used to detoxify PT. Following cloning of the genes coding for PT, the S1 subunit was expressed in *E. coli* and residues were found to be essential for ADP-ribosylation activity, including Arg-9, Asp-11, Arg-13, Trp-26, His-35, Phe-50, Glu-129, and Tyr-130 (Fig. 2). By photoaffinity labeling with NAD^+, Glu-129 was shown to be equivalent to the catalytic residues Glu-148 of DT and Glu-553 of exotoxin A of *Pseudomonas aeruginosa*. When the above amino acid changes were introduced into the PT operon in the *B. pertussis* chromosome, a number of correctly assembled mutant PT proteins were obtained that had toxicities ranging from 0.1 to 10% of wild-type PT. These reductions in activity were useful for biochemical analysis, but were too small to be used in a vaccine. Subsequently, combinations of the above mutations were engineered and several double-mutated forms of PT were identified that did not possess detectable ADP-ribosyltransferase activity. One of these (PT-9K/129G), containing the mutations Arg-9 → Lys and Glu-129 → Gly (Fig. 3) (see Color

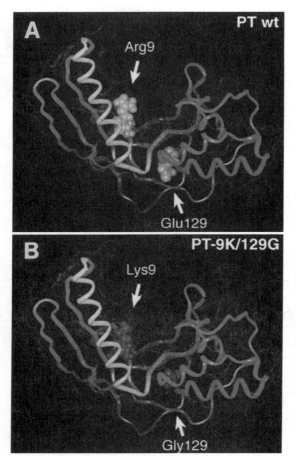

Figure 3 Computer modeling of the S1 subunit of (A) wild-type PT and of (B) PT-9K/129G mutant represented as an α-carbon trace. The NAD-binding site is highlighted in yellow; the catalytic residues in position 9 and 129 are shown in yellow for the wild-type PT and in red for the PT-9K/129G mutant, respectively. (See Color Plates following p. 256.)

Plates following p. 256), had lost the toxic properties of PT, but maintained immunological activity (Table 2) (6).

In 1990, the National Institute of Allergy and Infectious Diseases performed a large-scale phase II trial in the United States to compare the safety and immunogenicity of acellular pertussis vaccines. A control whole-cell vaccine and 13 acellular vaccines were tested; 11 of the acellular vaccines contained chemically detoxified PT and 2 vaccines contained the genetically detoxified PT. The results showed that the acellular vaccines were less reactive and more immunogenic than the traditional whole-cell vaccines and that genetically detoxified PT induced titers of anti-PT neutralizing antibodies 5- to 20-fold higher than those induced by equivalent doses of chemically detoxified forms of PT present in the other vaccines (Fig. 4). These data were confirmed by efficacy trial studies, in which the vaccine containing the genetically detoxified PT induced an overall protection from disease of 84% and protection started from the first vaccination dose.

The influence of chemical treatment of PT on immunogenicity was clarified in a comprehensive study using several anti-PT monoclonal antibodies. Most of

Table 2 Toxic and nontoxic properties of PT and PT-9K/129G mutant (7)

Property	Native PT	PT-9K/129G mutant
Toxic properties of PT		
CHO cell-clustered growth (ng/ml)	0.005	>5,000[a]
Histamine sensitization (μg/mouse)	0.1–0.5	>50
Leukocytosis stimulation (μg/mouse)	0.02	>50
Anaphylaxis potentiation (μg/mouse)	0.04	>7.5
Enhanced insulin secretion (μg/mouse)	<1	>25
IgE induction (in vitro) (ng/ml)	0.8	>100
IgE induction (in vivo) (ng/rat)	10	>200
Long-lasting enhancement of nerve-mediated intestinal permeabilization of antigen uptake (ng/rat)	1	>200
Inhibition of interleukin 1-induced interleukin 2 release in EL4 6.1 cells (μg/ml)	0.1	>100
Inhibition of neutrophil migration (μg/rat)	0.2	>1.2
Lethal dose (μg/kg)	15	>1,500
ADP-ribosylation (ng)	1	>20,000
Nontoxic properties of PT		
T-cell mitogenicity (μg/ml)	0.1–0.3	0.1–0.3
Hemagglutination (μg/well)	0.1–0.5	0.1–0.5
Mitogenicity for PT-specific T cells (μg/ml)	3	3
Platelet activation (μg/ml)	5	5
Mucosal adjuvanticity (μg/mouse)	3	3
Affinity constant (monoclonal 1B7, anti-S1)	2.4×10^8	6.1×10^8
Affinity constant (polyclonal anti-PT)	2.0×10^{10}	9.8×10^9

[a] Means that no effect was observed at the highest dose reported that was used in the assay.

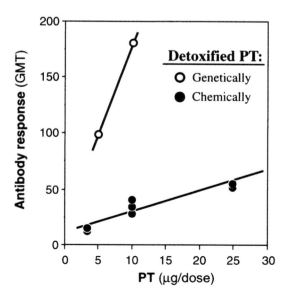

Figure 4 Antibody response induced by the chemically and genetically detoxified PT.

the antibodies failed to recognize chemically detoxified PT molecules, whereas the binding to the genetically detoxified PT was fully conserved and in some cases even enhanced. This indicated that chemical detoxification modifies the native epitopes of PT, which are conserved in genetically detoxified PT.

MUTANTS OF LT AND CHOLERA TOXIN FOR MUCOSAL DELIVERY OF VACCINES

Mucosal vaccination is the only method known to induce immunity at mucosal sites to block entry of pathogens. However, in spite of the obvious advantages of mucosal vaccination, most vaccines are still administered by the subcutaneous or intramuscular route because mucosal vaccination is not efficient in eliciting an immune response. Mucosal surfaces provide a physical barrier between the external environment and the body being constantly exposed to substances, which are substantially ignored by the healthy immune system that induces an immunological tolerance. Several approaches to deliver vaccines at mucosal surfaces have been based on the use of live-attenuated bacteria or plants as the delivery system for viral and bacterial antigens. The most powerful mucosal immunogens and adjuvants recognized to date are cholera toxin (CT) and *E. coli* heat-labile enterotoxin (LT). Although LT and CT have been extensively investigated as mucosal immunogens and adjuvants in animal models, their use in humans has been hampered by their toxicity. Knowledge of the molecular structure of LT and CT has allowed a rational design of LT and CT nontoxic derivatives that are still active as mucosal immunogens and adjuvants.

CT and *E. coli* LT share an identical mechanism of action and homologous primary and three-dimensional structures (2, 7, 10) (Fig. 5a and b). CT is produced by *Vibrio cholerae,* the etiological agent of cholera, whereas LT is produced by enterotoxigenic strains of *E. coli.* The corresponding genes of CT and LT are organized in a bicistronic operon and are located on a filamentous bacteriophage and on a plasmid, respectively. The two toxins are organized in an AB_5 architecture, where the B domain is a pentamer that binds the receptor on the surface of eukaryotic cells and domain A bears the enzymatic activity and is thus responsible for toxicity. For activity, the A subunit is cleaved and reduced at the disulfide bridge between cysteines 187 and 199 to produce two fragments, the enzymatic subunit A_1 and the linker fragment A_2 (Fig. 5a and b). Whereas in CT, the proteolytic process is performed during biosynthesis by an endoprotease, in the case of LT, it occurs by extracellular processes; in both cases, the reduction is thought to take place at the surface of the target cell.

The enzymatically active domain A ADP-ribosylates GTP-binding proteins such as G_s, G_t, and G_{olf}. Upon ADP-ribosylation of G_s, in particular, the adenylate cyclase is permanently activated, causing an intracellular accumulation of cyclic AMP that results in intestinal fluid accumulation and watery diarrhea. A peculiar feature of CT and LT is that the basal ADP-ribosyltransferase activity is enhanced by interaction with 20-kDa guanine-nucleotide-binding proteins, known as ADP-ribosylation factors (ARFs). After receptor binding, the holotoxins are internalized and undergo retrograde transport through the Golgi to the endoplasmic reticulum (ER), where they are likely retained by the carboxy-terminal "KDEL" ER retention sequence; the A subunits cross the membrane and are released in

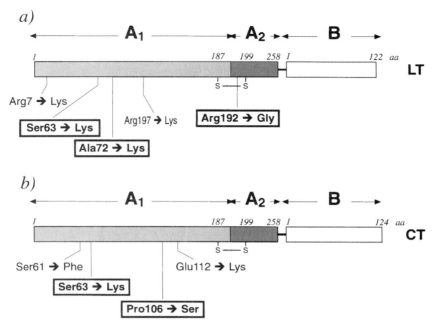

Figure 5 Schematic representation of (a) LT and (b) CT and of their mutant derivatives.

the cytoplasm where ARF is bound. The exact localization of the ARF-binding site is still unknown, but it has emerged from recent studies that the two domains (the NAD-binding and ARF-binding) are independent and located in different regions of the A domain. The B-subunit domain remains in the lumen of the Golgi and is subsequently degraded. The B domain is composed of five identical subunits with a molecular size of 11.5 kDa that are arranged in a symmetric shape around a central pore inside which the C-terminal portion of the catalytic domain, A_2, is inserted. Although still well conserved in terms of quaternary structure, CT and LT B domains have a lower degree of primary sequence homology than the corresponding A domains (3).

To study the structure-function of CT and LT and to find molecules that are nontoxic but still active as mucosal adjuvants and immunogens, more than 50 site-directed mutations have been generated within these toxins (Fig. 5) (8). Mutations that inactivate the toxin include LTK63, LTK97, and LTK7, which are located near the catalytic domain. Mutations that yield a toxin with residual toxic activity include LTK72 (20- to 100-fold reduction in toxicity) and LTG192 (10- to 1,000-fold reduction in toxicity). LTK63, LTR72, and LTG192 have been studied extensively in preclinical studies, and LTG192 and LTK63 have also been used in some clinical studies. LTK63 contains a Ser-63 to lysine substitution in the A subunit, has no detectable enzymatic activity and no toxicity in vitro or in vivo, and is considered a complete knockout of enzymatic activity, but indistinguishable from wild-type LT including cell surface receptor and ARF binding. The crystal structure of LTK63 is identical to that of wild-type LT except within the active site, where the bulky side chain of Lys-63 fills the catalytic cavity, making it unsuitable for the enzymatic activity (Fig. 6) (see Color Plates following p. 256). LTR72 contains an A-72 to arginine substitution in the A subunit (Fig. 6). This mutation renders the protein partially inactive, retaining 1% of the wild-type ADP-ribosylating activity, with a 10^4- to 10^5-fold reduction in toxicity for Y1 cells and a 25- to 100-fold reduction in the rabbit ileal loop. LTK63 and LTR72 are immunogenic following systemic and mucosal immunization and induce toxin-neutralizing antibodies, supporting their potential as components of vaccines against enterotoxigenic *E. coli*. LTK63 and LTR72 have also been extensively characterized for their ability to act as mucosal adjuvants. As shown in Table 3,

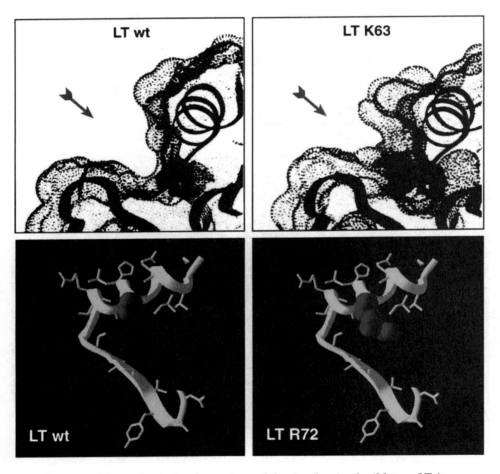

Figure 6 Three-dimensional structure of the A subunit of wild-type LT (top left panel) and LTK 63 mutant (top right panel) represented as α-carbon trace and as solvent-exposed surfaces. The residues in position 63 (serine in the wild-type LT and lysine in the LTK-63 mutant) are indicated by the arrow. Computer modeling of the NAD-binding site of wild-type LT (top left panel) and LT R72 (bottom right panel). The residues in position 72 (alanine in the wild-type LT and arginine in LT R72) are shown in green and red, respectively.

Table 3 Adjuvant activity of LTK63 and LTR72 with a variety of antigens by different immunization routes and in different animal models

Adjuvant		Antigen[a]	Route of immunization	Animal model
LTK63	LTR72			
+	+	Ovalbumin	Intranasal	
+		Fragment C of tetanus toxin	Intranasal	Mice
+		CTL epitope of measles virus	Intranasal	Mice
+		Diphtheria, tetanus, apertussis	Intranasal	Mice
	+	A pertussis	Intranasal	Mice
+	+	FHA, 69K, and PT-9K/129G of *B. pertussis*	Intranasal	Mice
+		Subunit influenza vaccine	Intranasal	Mice
+	+	Subunit influenza vaccine	Intranasal	Mice
+		gD2 of herpes simplex virus	Intranasal	Mice and guinea pigs
+		p24 and gp120 of HIV	Intranasal	Mice
+		Meningococcus C conjugate	Intranasal	Mice and pigs
+	+	Pneumococcus polysaccharide	Intranasal	Mice
+		Ovalbumin	Intravaginal	Mice
+		Subunit influenza vaccine	Intranasal	Rabbits and minipigs
+		CagA, VacA, and urease of *Helicobacter pylori*	Oral	Mice and beagle dogs
+		KLH	Oral	Mice
+	+	Gag p55 of HIV	Intranasal, oral	Mice
+		CTL peptide of respiratory syncytial virus	Intranasal	Mice
	+	HPV-6b virus-like particles	Intranasal	Mice
+	+	Influenza hemagglutinin	Oral	Mice

[a]CTL, cytotoxic T lymphocyte; FHA, filamentous hemagglutinin; HIV, human immunodeficiency virus; HPV-6b, human papillomavirus type 6b.

the adjuvant activity has been tested with a wide range of coadministered antigens (bacterial and viral antigens and synthetic peptides), using different routes of immunization and different animal models. LTR72, the mutant retaining a residual enzymatic activity, is a stronger adjuvant as compared to the fully nontoxic LTK63 mutant, inducing antigen-specific antibody titers comparable to those induced by wild-type LT. On the other hand, the LTK63 is a stronger adjuvant as compared to the B subunit, suggesting that ADP-ribosylation activity is important but not necessary for adjuvanticity (8). These mutant molecules represent not only promising adjuvants for human use, but also an interesting tool to dissect the mechanism of mucosal adjuvanticity and to evaluate the role and the relative contribution of the B subunit, the AB complex, and the enzymatic activity to adjuvanticity. Clinical trials have been planned to evaluate the safety, immunogenicity, and adjuvanticity profiles of LTK63 and LTR72.

An alternative approach to detoxify LT was based on the generation of mutations within the protease-sensitive loop, between A_1 and A_2, rendering the toxin protease resistant (Fig. 5a). LTG192, where Arg-192 is replaced by glycine, is the best-characterized member of this family of mutations. Although LTG192 is resistant to trypsin cleavage in vitro, in vivo proteases other than trypsin cleave the loop and activate the toxin, since toxicity is still detectable. LTG192 is indistinguishable from wild-type LT in terms of immunogenicity and adjuvanticity. In humans, LTR192 was not reactive at low doses but induced diarrhea at higher doses.

With a similar approach, nontoxic or partially detoxified mutations within CT have also been obtained. The best characterized are the nontoxic CTK63 (Ser-63 replaced with lysine, the same mutation in the LTK63) and the CTS106 (Pro-106 replaced with serine) (Fig. 5b). Whereas CTK63 exhibited a very weak adjuvant activity following intranasal immunization in mice compared to the adjuvanticity of the homologous LTK63, CTS106 possessed better adjuvant activity, comparable to that of the LTR72 mutant. These data suggest that enzymatic activity is necessary for the adjuvanticity of CT. The qualitatively different immunological properties exhibited by LT and CT may be due to different receptor-binding activities of the toxins. However, in a different study, nontoxic mutants of CT (CTF61 and CTK112) maintained the adjuvant properties of the wild-type toxin. These contrasting results may reflect the different doses of adjuvant and antigen used and the route of delivery.

Continued expansion of our understanding of the molecular and cellular properties of bacterial toxins will provide insight into new approaches for the use of bacterial toxins to explore the function of eukaryotic physiology and strategies for prevention and treatment of bacterial infections.

REFERENCES

1. **Bennett, M. J., and D. Eisenberg.** 1994. Refined structure of monomeric diphtheria toxin at 2.3 A resolution. *Protein Sci.* **3:**1464–1475.
2. **Dallas, W. S., and S. Falkow.** 1980. Amino acid sequence homology between cholera toxin and *Escherichia coli* heat-labile toxin. *Nature* **288:**499–501.
3. **Domenighini, M., M. Pizza, M. G. Jobling, R. K. Holmes, and R. Rappuoli.** 1995. Identification of errors among database sequence entries and comparison of correct amino acid sequences for the heat-labile enterotoxins of *Escherichia coli* and *Vibrio cholerae*. *Mol. Microbiol.* **15:**1165–1167.
4. **Hewlett, E. L., K. T. Sauer, G. A. Myers, J. L. Cowell, and R. L. Guerrant.** 1983. Induction of a novel morphological response in Chinese hamster ovary cells by pertussis toxin. *Infect. Immun.* **40:**1198–1203.
5. **Pappenheimer, A. M., Jr.** 1977. Diphtheria toxin. *Annu. Rev. Biochem.* **46:**69–94.
6. **Pizza, M., A. Covacci, A. Bartoloni, M. Perugini, L. Nencioni, M. T. De Magistris, L. Villa, D. Nucci, R. Manetti, M. Bugnoli, F. Giovannoni, R. Olivieri, J. T. Barbieri, H. Sato, and R. Rappuoli.** 1989. Mutants of pertussis toxin suitable for vaccine development. *Science* **246:**497–500.
7. **Pizza, M., V. Masignani, and R. Rappuoli.** 1999. Molecular, functional and evolutionary aspects of ADP-ribosylating toxins, p. 45–72. *In* J. E. Alouf and J. H. Freer (ed.), *The Comprehensive Sourcebook of Bacterial Protein Toxins*, 2nd ed. Academic Press, Ltd., London, United Kingdom.

8. **Rappuoli, R., M. Pizza, G. Douce, and G. Dougan.** 1999. Structure and mucosal adjuvanticity of cholera toxin and *Escherichia coli* heat-labile enterotoxins. *Immunol. Today* **20:**493–500.
9. **Sekura, R. D.** 1985 Pertussis toxin: a tool for studying the regulation of adenylate cyclase. *Methods Enzymol.* **109:**558–566.
10. **Sixma, T. K., S. E. Pronk, K. H. Kalk, E. S. Wartna, B. A. van Zanten, B. Witholt, and W. G. Hol.** 1991. Crystal structure of a cholera toxin-related heat-labile enterotoxin from *E. coli. Nature* **351:**371–377.
11. **Trollfors, B., J. Taranger, T. Lagergard, L. Lind, V. Sundh, G. Zachrisson, C. U. Lowe, W. Blackwelder, and J. B. Robbins.** 1995. A placebo-controlled trial of a pertussis-toxoid vaccine. *N. Engl. J. Med.* **333:**1045–1050.
12. **Weiss, A. A., and E. L. Hewlett.** 1986. Virulence factors of *Bordetella pertussis. Annu. Rev. Microbiol.* **40:**661–686.

Bacterial Protein Toxins
D. Burns et al., Editors
©2003 ASM Press, Washington, DC 20036

Chapter 23

Nefarious Uses of Bacterial Toxins

Drusilla L. Burns and Joseph T. Barbieri

Bacterial toxins represent some of the deadliest molecules known to man. Over the ages, evolution has led to precision engineering of these toxins such that only minute quantities of certain bacterial toxins are capable of eliciting lethal outcomes for humans (Table 1). For example, only 100 ng of botulinum toxin can kill a 200-pound person.

As described in chapter 22 of this volume, bacterial toxins can be used for beneficial purposes. Many vaccines are inactivated bacterial toxins, which stimulate an immune response to protect the recipient from that toxin-producing pathogen. Vaccines based on inactivated toxins (toxoids) have dramatically reduced the morbidity and mortality associated with a number of diseases including diphtheria, tetanus, and pertussis. Nontoxic forms of diphtheria toxin, such as CRM197, have been licensed as carriers for the capsular antigen of *Haemophilus influenzae* type b to generate immunity to this organism in children under the age of 18 months. Bacterial toxins have also been used for therapeutic purposes. The ability of botulinum toxin to cause a flaccid paralysis has been exploited to treat disorders such as strabismus (misalignment of the visual axes) and dystonia (sustained abnormal muscle contraction) and has even been used cosmetically as a treatment for wrinkles.

Regrettably, the power of bacterial toxins has also been harnessed for destructive purposes. Bacterial toxins have been utilized directly, that is, as purified proteins, or indirectly, with the pathogen used as a delivery vehicle. A number of toxins and toxin-producing bacteria have been identified as agents that have a high potential to be used as biological weapons. The United States Centers for Disease Control and Prevention has classified biological agents into categories according to their potential risk to public health and national security. Toxins and toxin-producing bacteria that are thought to pose major risks to the public are *Bacillus anthracis*, which causes anthrax in humans and animals, *Yersinia pestis*, which is the causative agent of plague, and botulinum toxin. These agents can

Drusilla L. Burns • Food and Drug Administration, 8800 Rockville Pike, Bethesda, MD 20892. *Joseph T. Barbieri* • Department of Microbiology, Medical College of Wisconsin, 8701 Watertown Plank Rd., Milwaukee, WI 53226-0509.

Table 1 Lethal amounts of some bacterial toxins

Organism	Toxin	Lethal quantity[a] (per kg body weight)	Species
Aeromonas hydrophila	Aerolysin	7 μg	Mouse
Bacillus anthracis	Lethal toxin (LF + PA)	<114 μg	Rat
Bordetella pertussis	Pertussis toxin	15 μg	Mouse
Clostridium botulinum	Botulinum toxin	1 ng	Human
Clostridium tetani	Tetanus toxin	<2.5 ng	Human
Corynebacterium diphtheriae	Diphtheria toxin	≤100 ng	Human
Pseudomonas aeruginosa	*Pseudomonas* toxin A	3 μg	Mouse
Staphylococcus aureus	Enterotoxin B	20 μg	Monkey
Vibrio cholerae	Cholera toxin	250 μg	Mouse

[a] Values taken from reference 2.

be easily disseminated, they result in high morbidity and mortality rates, and they have the potential to cause significant social disruption. Toxins or toxin-producing agents that are moderately easy to disseminate and that have the potential to result in moderate morbidity and low mortality rates include *Salmonella* spp., *Esherichia coli* O157:H7, *Shigella* spp., *Vibrio cholerae*, epsilon toxin of *Clostridium perfringens*, and *Staphylococcus aureus* enterotoxins. *Salmonella* spp., *E. coli* O157:H7, and *Shigella* spp. are primarily considered threats to food safety, whereas *V. cholerae* is considered a threat to the water supply. Our consideration of toxins and toxin-producing bacteria that are potential terrorist weapons will focus on the high-risk agents *B. anthracis*, *Y. pestis*, and botulinum toxin.

HIGH-RISK AGENTS

B. anthracis

As witnessed from the events of autumn 2001 in the United States, *B. anthracis* is a very effective agent of terrorism. The bacterium causes a lethal disease, and dissemination of the organism can result in significant social disruption and panic. Dissemination is facilitated by the fact that the spore form of the organism is stable to adverse environmental conditions and can be manipulated such that spores are readily dispersed by air currents.

Inhalation anthrax is the most lethal form of the disease. *B. anthracis* spores that have been inhaled by the host are taken up by alveolar macrophages in the lung. The spores then germinate and escape into the bloodstream. The vegetative form of the bacteria produces a protective capsule composed of poly-D-glutamic acid, which helps the bacteria evade the host immune system. Once in the bloodstream, the bacteria rapidly proliferate, attaining levels approximating 10^8 organisms per ml of blood. The lethality of *B. anthracis* is believed to be related to the production of copious quantities of anthrax toxin.

Anthrax toxin has an AB structure-function organization. The A and B components are secreted separately and associate on the eukaryotic cell membrane. Anthrax toxin's tripartite composition comprises a single B (binding) component that can deliver two different A (active) components, known as lethal factor (LF) and edema factor (EF), into eukaryotic cells. The B component is known as protective antigen (PA). Electron microscopy studies predict a heptameric structure for PA, and the crystal structure of monomeric PA has provided the coordinates to align PA into a heptameric ring-like structure (Fig. 1) (see Color Plates following p. 256).

Upon contact with a eukaryotic cell, PA binds to a cellular receptor, identified as a type I membrane protein with an extracellular von Willebrand factor A domain. PA is then nicked by the eukaryotic protease furin, promoting formation of the heptamer, which then binds EF and/or LF. Receptor-bound anthrax toxin is internalized by receptor-mediated endocytosis. Upon acidification of the early endosome, a conformational change in PA triggers its insertion into the endosomal membranes, which then stimulates translocation of the A components from the lumen of the endosome into the cytosol of the host cell (Fig. 2).

Both A components display enzymatic activity. LF is a metalloprotease that was initially observed to proteolyze mitogen-activated protein kinase kinase; therefore, the proteolysis of this kinase was implicated in the pathology elicited by LF. However, the mechanism by which LF contributes to disease has not been fully established. Recently, a role has been implicated for a kinesin-like motor protein in the susceptibility of macrophage to the action of LF. EF is an adenylate cyclase. Its entry into a susceptible cell stimulates the production of supraphysiological concentrations of cyclic AMP (cAMP), which disrupts host cell signaling. In some cell lines, EF can stimulate the elevation of cAMP over 1,000-fold relative to normal cellular concentrations. EF is activated by calmodulin, providing posttranslational regulation of the expression of catalysis. Under physiological conditions, LF plus PA cause the rapid death of certain animals, whereas EF plus PA cause edema. The two activities are believed to act synergistically to cause symptoms of disease and resulting death.

Y. pestis

Y. pestis is one of the three species of the Yersinia family: Y. pestis, the causative agent of plague; Y. pseudotuberculosis, responsible for adenitis and septicemia; and Y. enterocolitica, responsible for limiting gastrointestinal disorders. Y. pestis is the major human pathogen within the Yersinia family and has been responsible for several pandemics. During one pandemic, which occurred in the 14th century, 25 million people died in Europe, representing about one-fourth of the population at that time.

The virulence of Y. pestis is associated with the presence of a large plasmid (70 kb) that regulates the secretion of proteins called Yersinia outer membrane proteins (Yops). Yops are delivered directly by the bacterium into the host cell via a type III system (see chapter 7). The Yops function in two capacities to enhance the virulence of the Yersinia, inhibiting the innate immune system and directly damaging the host. YopH is a phosphotyrosine phosphatase that manifests antiphagocytosis activity by dephosphorylating focal adhesion proteins. YpkA is serine/threonine kinase that is a virulence factor in the mouse infection model and has recently been shown to bind RhoA and Rac1. To date, YpkA has an undefined function for its observed virulence within the eukaryotic cell. YopP is an anti-inflammatory molecule that prevents the activation of the transcription

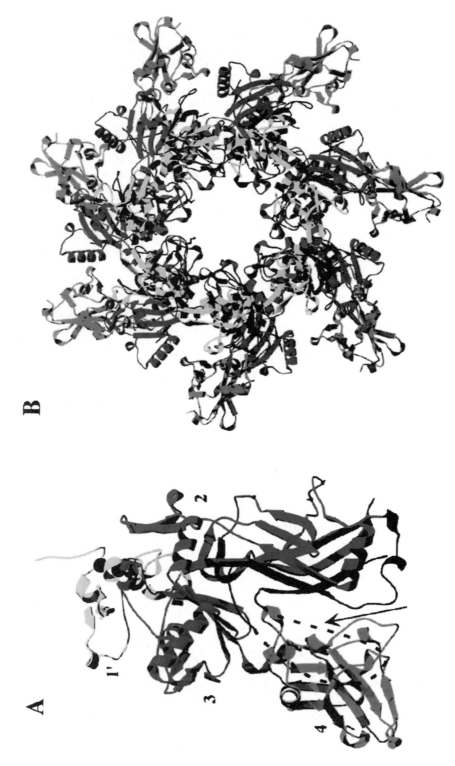

Figure 1   Three-dimensional structure of the PA of anthrax toxin. (A) Three-dimensional structure of monomeric PA. The four domains of the protein are indicated. The dashed line indicated by the arrow represents a loop that is believed to form a segment of a transmembrane β-barrel upon pore formation. (B) Modeled heptamer of PA depicting its predicted ring-like structure. Adapted from reference 8 with permission. (See Color Plates following p. 256).

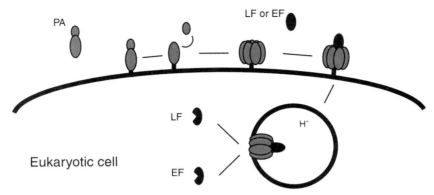

Figure 2 Mechanism of action of anthrax toxin. Anthrax toxin is composed of three proteins, PA, LF, and EF. PA first binds to receptors on the eukaryotic cell. PA is nicked by the eukaryotic protease furin, producing a PA fragment that oligomerizes to form a heptamer and can bind EF and/or LF. The complex is internalized by endocytosis. Exposure of the complex to the acidic environment of the endosome is believed to result in formation of a pore by the PA heptamer, producing a channel through which LF and EF can translocate. Once inside the cytoplasm, LF expresses metalloprotease activity and EF is an adenylate cyclase that synthesizes the second messenger, cAMP.

factor NF-κB. YopE and YopT are antiphagocytosis molecules that inactivate Rho GTPases, the master switch molecules that control cytoskeleton dynamics. In addition, *Yersinia* spp. produce other type III cytotoxins, such as YopM, which have yet-to-be-defined roles in the bacterium's pathogenesis.

Strategies to combat *Y. pestis* include vaccination with components of the type III apparatus. This approach may target not only individual Yops, but also components of the type III apparatus. Identifying agents that neutralize the type III apparatus of *Yersinia* spp. may have an added benefit of having the capacity to neutralize other pathogenic bacteria, since numerous bacteria utilize type III delivery for pathogenesis and some components of the apparatus can be complemented with homologs from other gram-negative bacteria.

Botulinum Toxin

Botulinum toxin is considered a major biowarfare or bioterrorist threat because minute quantities are lethal to humans. Botulinum toxin is a single 150-kDa protein with defined AB structure-function properties. The N-terminal A domain is a metalloprotease that specifically cleaves proteins that are components of vesicle fusion pathways (see chapter 19). Several serotypes of botulinum toxins exist that target different host proteins for proteolysis. The tropism of botulinum toxin for neuronal cells is due to the binding specificity of the B domain.

USE OF BACTERIAL TOXINS IN BIOLOGICAL WARFARE

One of the earliest recorded attempts to use biological organisms in warfare occurred in the 14th century in what is now the Ukraine. Tartar troops who were attacking the city of Kaffa were experiencing an outbreak of plague. The Tartars did not understand the molecular basis of this disease but believed that the corpses of the individuals who succumbed to the disease could transmit the disease. They therefore attempted to initiate a plague epidemic on their enemy by catapulting the cadavers of their deceased comrades over the city walls. In fact, an epidemic occurred within the city and Kaffa was conquered. Whether the source of the disease was the corpses or whether the disease was due to a natural infectious cycle between rodents, fleas, and humans within the city remains speculation. Nonetheless, this incident represents one of the earliest recorded attempts at biological warfare (1).

The use of biological warfare was explored as an option by a number of countries well into the 20th century. However, in response to the chemical warfare of World War I, an international effort was made to limit the use of weapons of mass destruction, including biological agents. Although the 1925 Geneva Protocol for the Protection of the Use in War of Asphyxiation, Poisonous or Other Gases and of Bacteriological Methods of Warfare was ratified by a number of countries, the treaty did not prohibit research for the development or production of biological weapons and, therefore, biological warfare programs continued in a number of countries.

From 1932 through World War II, Japan conducted biological warfare research in Manchuria on prisoners who were infected with a number of toxin-producing pathogens including *B. anthracis*, *Y. pestis*, *Shigella* spp., and *V. cholerae*. At least 10,000 prisoners died during this testing or upon execution following these atrocious acts (1). The Japanese biological warfare program ended after World War II; however, programs in the United States, Canada, the Soviet Union, and the United Kingdom continued until 1972 when the Biological and Toxin Weapons Convention was ratified (3). This treaty, which was signed by 140 countries, required the termination of research on offensive biological weapons and prohibited the possession and stockpiling of pathogens or toxins in quantities that could not be justified for prophylactic or other peaceful purposes. Many, but not all, countries complied with this treaty.

Rather than abolishing its biological warfare program, the Soviet Union expanded its program after 1972 (3). The worst outbreak of inhalation anthrax in recorded history occurred because of this program. The outbreak was not due to the use of biological agents against an enemy, but rather, it occurred because of human error in Sverdlovsk, a Soviet city of approximately 1.2 million people. In 1979, *B. anthracis* was accidentally released from the Sverdlovsk bioweapons production facility (3). Winds spread *B. anthracis* spores in a narrow band south of the city. At least 66 people died in the outbreak and 11 others became ill; most of these individuals lived or worked within 4 km downwind of the facility. As far as 50 km downwind of the facility, livestock died of anthrax (5).

USE OF BIOLOGICAL WEAPONS BY INDIVIDUALS OR GROUPS OF INDIVIDUALS

Although many nations recognize that the use of biological weapons must not be tolerated, individuals and groups within society continue to use these weapons for corrupt purposes.

Covert Use of Toxins

In 1978, a Bulgarian exile, Georgi Markov, was assassinated at a subway stop in London. In this incident, metallic pellets that had been hollowed out, filled with the toxin ricin, and sealed with wax were discharged into Markov from a weapon disguised as an umbrella. Once the pellet entered Markov's body, the wax plug melted and ricin was released. Ricin is a toxin found in castor beans. Although ricin is not a bacterial toxin, its mode of action is almost identical to Shiga toxin (see chapter 17). Ricin is an RNA *N*-glycosidase, which cleaves the *N*-glycosidic bond of an adenosine at position 4324 in the 28S ribosomal RNA of eukaryotic cells, resulting in an inhibition of protein synthesis.

Recent Bioterrorist Attacks

The Aum Shinrikyo cult in Japan attempted to use both chemical and biological agents in the 1990s. This religious cult predicted an Armageddon for Japan and other nations of the civilized world, where followers of Aum Shinrikyo would survive. To fabricate the Armageddon, chemical and biological agents of mass destruction were developed. The cult released Sarin, a chemical toxin, into the Tokyo subway system, killing 12 people and injuring several thousand. Between 1990 and 1995, this group also attempted several unsuccessful acts of bioterrorism that involved the use of botulinum toxin and *B. anthracis* (6).

The bioterrorist attack on the United States in 2001 used *B. anthracis* as an agent of destruction. On October 2, 2001, a 63-year-old photo editor who worked for a Florida newspaper awoke with nausea, vomiting, and confusion. He was diagnosed with inhalation anthrax, and on October 5 he died. This diagnosis perplexed epidemiologists, since the last case of inhalation anthrax in the United States had occurred in 1976 and this individual did not have risk factors for the disease. When a coworker was also diagnosed with inhalation anthrax, it became evident that these were not natural infections, but probably resulted from intentional release of the organism. Subsequently, several cases of cutaneous anthrax occurred in New York, where the source of the *B. anthracis* was believed to be spores contained in letters mailed to news organizations. Two additional letters containing anthrax spores were mailed to United States senators in Washington, D.C. These letters contained a high-quality form of *B. anthracis* spores that were particularly lethal because of their ability to be dispersed and spread by almost imperceptible air currents. One letter was opened by a senatorial staff member, exposing a number of individuals in the immediate area to very large numbers of spores (estimated to be up to several thousand 50% lethal doses). Although many lives were saved by prompt treatment with antibiotics, by the time it was realized that the letters contained a particularly lethal form of spores, the letters had passed through the postal system. The fine particle size of the spores allowed escape during the processing of the envelopes. A number of individuals contracted either cutaneous or inhalation anthrax, most likely from contact with spores released from the envelopes. Five people who contracted inhalation anthrax died.

CONCLUSIONS AND COMMENTS

Controlling naturally acquired bacterial diseases is a difficult problem for scientists and public health officials; the fact that we must also be concerned about limiting morbidity and mortality due to intentional release of pathogenic bacteria or bacterial toxins is almost unthinkable. As so aptly stated by Joshua Lederberg (4), "One sine qua non for the elimination of biological warfare is its utter delegitimation; in the language of the Geneva Protocol of 1925, it must be 'justly condemned by the general opinion of the civilized world.' "

The future holds many challenges to identify new ways to prevent the damage inflicted by these bacterial agents. Continued studies on the basic molecular properties of bacterial toxins will provide insight to develop novel antitoxin therapies, such as the recent recognition that nontoxic forms of bacterial toxins can act in a dominant-negative manner to neutralize toxin action (7). The potential utilization of pathogenic bacteria and their toxins for criminal acts is a formidable problem that requires a concerted effort on the part of the scientific community. Scientific research has the promise of leading to the development of vaccines and therapeutics that will make biological agents useless in the hands of terrorists.

REFERENCES

1. **Christopher, G. W., T. J. Cieslak, J. A. Pavlin, and E. M. Eitzen.** 1997. Biological warfare: a historical perspective. *JAMA* **278:**412–417.
2. **Gill, D. M.** 1982. Bacterial toxins: a table of lethal amounts. *Microbiol. Rev.* **46:**86–94.
3. **Henderson, D. A.** 1999. The looming threat of bioterrorism. *Science* **283:**1279–1282.
4. **Lederberg, J.** 1997. Infectious disease and biological weapons: prophylaxis and mitigation. *JAMA* **5:**435–436.
5. **Meselson, M., J. Guillemin, M. Hugh-Jones, A. Langmuir, I. Popova, A. Shelokov, and O. Yampoloskaya.** 1994. The Sverdlovsk anthrax outbreak of 1979. *Science* **266:**1202–1208.
6. **Olson, K. B.** 1999. Aum Shinrikyo: once and future threat? *Emerging Infect. Dis.* **5:**513–516.
7. **Sellman, B. R., M. Mourex, and R. J. Collier.** 2001. Dominant-negative mutants of a toxin subunit: an approach to therapy of anthrax. *Science* **292:**695–697.
8. **Sellman, B. R., S. Nassi, and R. J. Collier.** 2001. Point mutations in anthrax protective antigen that block translocation. *J. Biol. Chem.* **276:**8371–8376.

INDEX

A

A domain, of ADP-ribosylating toxins, 216, 218, 219

AB toxins, 132, 218–219

AB_n toxins, 132–133

Abrin, pathophysiology of, 252

AC toxin, *see* Adenylate cyclase toxin

Acetylcholine inhibition, botulinum toxin in, 272

N-Acetylglucosaminyltransferases, clostridial toxins as, *see* Clostridial cytotoxins, large

Actin, RhoGPTase action on, 283

Actinobacillus actinomycetemcomitans
 cytolethal distending toxin, *see* Cytolethal distending toxins
 LtxA toxin, 205, 210–211

Actinobacillus pleuropneumoniae, ApxIA, ApxIIA, ApxIIIA, 205

Actions, of toxins, *see also specific toxins*
 ADP-ribosylation, 215–228
 cytolethal distending, 257–270
 deamidation, 237–243
 glucosylating, 229–237
 immune system overactivation by superantigens, 293–308
 membrane-damaging, 203–214
 overview of, 187–188
 pore-forming, 153, 189–202
 proteolytic, 271–282
 Rho GTPase modulation, 283–292
 ribosome-inactivating, 245–255

Acyl-homoserine lactone-based quorum sensing, 55–56

Adaptor proteins, in toxin transport, 74–75

Adenosine diphosphate-ribosylating toxins, *see* ADP-ribosylating toxins

Adenylate cyclase toxin, 149–156
 action of, 151, 153–156, 214
 apoptosis induction by, 212–213

Bordetella pertussis, 19
 calcium-binding domain of, 209
 cell interaction with, 152–153
 discovery of, 150
 domains of, 150–151
 expression of, 204, 206
 gene of, 150
 structure of, 150, 207

Adhesion factors, regulation of, AraC family regulators in, 51–53

ADP-ribosylating factor, in proenzyme activation, 226

ADP-ribosylating toxins, 215–228
 actions of, 215–216
  active sites in, 221–224
  in vivo detection of, 227–228
  mechanisms of, 224–225
 examples of, 217
 NAD affinity of, 225
 stereochemistry of, 224–225
 structures of, 216, 218–219, 221–224
 synthesis of, 225–226
 targets of, 219–221
 as tools for cell physiologic studies, 226–227
 vs. other ADP-ribosylating proteins, 227

Aerolysin, *Aeromonas hydrophila,* 193
 action of, 191
 pore formation by, 195
  advantages of, 189, 191
  consequences of, 200
 receptor for, 198
 secretion of, 82–83

Aeromonas hydrophila
 aerolysin, *see* Aerolysin
 toxin secretion from, 81–83, 89

Agmatine, ADP-ribosylating toxin action on, 220–221

Agrobacterium tumefaciens
 sensor kinases of, 8

Exotoxin A, *Pseudomonas aeruginosa*
  (*continued*)
  receptor for, 135–140
    action of, 135–136
    characteristics of, 136–139
    evidence for, 139–140
    ligands recognized by, 138–139
    mutations of, 137–138
    structure of, 136–137
  recombinant, 134–135
  retrograde transport of, 169
  secretion of, 82–84
  structure of, 135
Exotoxin A ADP-ribosylated elongation
  factor-2, 215, 221
Exs protein family, *Pseudomonas
  aeruginosa*, 46–49
Extracytoplasmic sensor domain, in two-
  component systems, 8–9

F
Factor activating exoenzyme S, in
  proenzyme activation, 226
Fatty acyl groups, of RTX toxins, 207–
  208
FcRn protein, in immunoglobulin
  transcytosis, 177–179
Flagella, structures of, 111–113
Fli proteins, in type III secretion systems,
  109–111
Fluorescence studies, of toxin effects on
  Rho GTPases, 242
Food poisoning, superantigens in, 298
Formaldehyde, for toxin treatment, in
  vaccine manufacture, 313–314
Fragilysins, 280–282
Fur protein, in iron regulation, 32–35

G
G proteins, ADP-ribosylating toxin
  action on, 220, 226–227
Ganglioside GM1, cholera toxin binding
  to, 180–183
Gb3 Shiga toxin receptor, 145–147, 168
Genomic islands, toxin secretion and, 99
Geobacter sulfurreducens, toxin secretion
  from, 82
Glucosylation
  clostridial toxins in, *see* Clostridial
    cytotoxins, large
  of ribosome-inactivating proteins, 248

Glutamic acid, ADP-ribosylating toxin
  action on, 221–222, 224
Glycohydrolase, in ADP-ribosylating
  toxin action, 221
N-Glycosidases, RNA, *see* Ribosome-
  inactivating proteins
Glycosylation
  clostridial toxins in, *see* Clostridial
    cytotoxins, large
  of ribosome-inactivating proteins, 248
Golgi apparatus, in toxin transport, 166–
  168, 182–183
Gram-negative bacteria, toxin transport
  from, channel tunnels in, 76–77
GTPase-activating proteins, 284, 286–291
GTPases, Rho, *see* Rho GTPases
Guanine nucleotide dissociation
  inhibitors, 284–285
Guanine nucleotide exchange factors,
  285–286, 291

H
Haemophilus ducreyi, cytolethal distending
  toxin, *see* Cytolethal distending
  toxins
Heat-labile enterotoxin, *Escherichia coli*
  action of, 222–223, 225, 320–321
  ADP-ribosylating, 217
  as mucosal vaccine adjuvant, 320–324
  structure of, 218–219, 320
Helicobacter hepaticus, cytolethal
  distending toxin, *see* Cytolethal
  distending toxins
Helicobacter pullorum, cytolethal
  distending toxin, *see* Cytolethal
  distending toxins
Helicobacter pylori
  toxin secretion from, 124
  VacA toxin, 193, 200, 201
Hemolysin(s), *see also* Adenylate cyclase
  toxin
  export of, 71–72
  repeat-in-toxin, *see* Repeat-in-toxin
    family
  Serratia marcescens, 200
  Staphylococcus aureus, 193
    action of, 191
    pore formation by, 189, 196–197, 200
  Vibrio cholerae, 200
Hemolytic-uremic syndrome, Shiga
  toxins in, 146, 251–252